D0927971

SCOPE 35

Scales and Global Change

SCOPE 35

Scales and Global Change

Spatial and Temporal Variability in Biospheric and Geospheric Processes

Edited by

Thomas Rosswall

University of Linköping, Linköping, Sweden

Robert G. Woodmansee

*Natural Resource Ecology Laboratory, Colorado State University,
Fort Collins, USA*

and

Paul G. Risser

University of New Mexico, Albuquerque, USA

Published on behalf of the
Scientific Committee on Problems of the Environment (SCOPE)
of the
International Council of Scientific Unions (ICSU)
by

JOHN WILEY & SONS

Chichester · New York · Brisbane · Toronto · Singapore

British Library Cataloguing in Publication Data:

Scales and Global Change: Spatial and temporal variability in
 biospheric and geospheric processes.—(SCOPE; 35).
 1. Man—Influence on nature   2. Climatic changes
 I. Rosswall, T.   II. Woodmansee, Robert G.
 III. Risser, Paul. G.   IV. International Council of Scientific Unions,
 Scientific Committee on problems of the Environment
 V. Series
 333.7   GF75
 ISBN 0 471 91828 8

Library of Congress Cataloging-in-Publication Data:

Scales and Global Change: Spatial and temporal variability in biospheric and
 geospheric processes / edited by Thomas Rosswall, Robert G. Woodmansee, and
 Paul G. Risser.
    p.   cm. — (SCOPE ; 35)
    Proceedings of Workshop on Spatial and Temporal Variability of Biospheric
 and Geospheric Processes: Research Needed to Determine Interactions with
 Global Environmental Change, 10/28–11/1/85, Saint Petersburg, Fla., sponsored
 by Environmental Studies Board of National Research Council, ICSU's Scientific
 Committee on Problems of the Environment, and ICSU's International
 Association for Ecology.
    Includes index.
    ISBN 0 471 91828 8
    1. Biosphere—Congresses.   2. Biological systems—Congresses.   3. Ecology—
 Congresses.   4. Geology—Congresses.   5. Space and time—Congresses.
 I. Rosswall, T. (Thomas)   II. Woodmansee, Robert George.   III. Risser, Paul G.
 IV. Workshop on Spatial and Temporal Variability of Biospheric and Geospheric
 Processes: Research Needed to Determine Interactions with Global Environmental
 Change (1985 : Saint Petersburg, Fla.)   V. National Research Council (U.S.).
 Environmental Studies Board.   VI. International Council of Scientific Unions.
 Scientific Committee on Problems of the Environment.   VII. International
 Association for Ecology.   VIII. Series: SCOPE (Series) ; 35.
 QH313.S69 1988
 574.9—dc19   87-31955
                                                                          CIP

Typeset by Mathematical Composition Setters Ltd
7 Ivy Street, Salisbury, Wiltshire, England, SP1 2AY
Printed by St. Edmundsbury Press,
Bury St. Edmunds, Suffolk

SCOPE 1: Global Environmental Monitoring 1971, 68 pp (out of print)

SCOPE 2: Man-Made Lakes as Modified Ecosystems, 1972, 76 pp (out of print)

SCOPE 3: Global Environmental Monitoring Systems (GEMS): Action Plan for Phase 1, 1973, 132 pp (out of print)

SCOPE 4: Environmental Sciences in Developing Countries, 1974, 72 pp (out of print)

Environment and Development, proceedings of SCOPE/UNEP Symposium on Environmental Sciences in Developing Countries, Nairobi, February 11–23, 1974, 418 pp (out of print)

SCOPE 5: Environmental Impact Assessment: Principles and Procedures, Second Edition, 1979, 208 pp

SCOPE 6: Environmental Pollutants: Selected Analytical Methods, 1975, 277 pp (out of print)

SCOPE 7: Nitrogen, Phosphorus and Sulphur: Global Cycles, 1975, 129 pp (out of print)

SCOPE 8: Risk Assessment of Environmental Hazard, 1978, 132 pp (out of print)

SCOPE 9: Simulation Modelling of Environmental Problems, 1978, 128 pp (out of print)

SCOPE 10: Environmental Issues, 1977, 242 pp (out of print)

SCOPE 11: Shelter Provision in Developing Countries, 1978, 112 pp (out of print)

SCOPE 12: Principles of Ecotoxicology, 1978, 372 pp (out of print)

SCOPE 13: The Global Carbon Cycle, 1979, 491 pp (out of print)

SCOPE 14: Saharan Dust: Mobilization, Transport, Deposition, 1979, 320 pp (out of print)

SCOPE 15: Environmental Risk Assessment, 1980, 176 pp

SCOPE 16: Carbon Cycle Modelling, 1981, 404 pp (out of print)

SCOPE 17: Some Perspectives of the Major Biogeochemical Cycles, 1981, 175 pp (out of print)

SCOPE 18: The Role of Fire in Northern Circumpolar Ecosystems, 1983, 344 pp

SCOPE 19: The Global Biogeochemical Sulphur Cycle, 1983, 495 pp

Funds to meet SCOPE expenses are provided by contributions from SCOPE National Committees, an annual subvention from ICSU (and through ICSU, from UNESCO), an annual subvention from the French Ministère de l'Environnement et du Cadre de Vie, contracts with UN Bodies, particularly UNEP, and grants from Foundations and industrial enterprises.

International Council of Scientific Unions (ICSU)
Scientific Committee on Problems of the Environment (SCOPE)

SCOPE is one of a number of committees established by the nongovernmental group of scientific organizations, the International Council of Scientific Unions (ICSU). The membership of ICSU includes representatives from 68 National Academies of Science, 18 International Unions and 12 other bodies called Scientific Associates. To cover multidisciplinary activities which include the interests of several unions, ICSU has established 10 Scientific Committees, of which SCOPE is one. Currently representatives of 34 member countries and 15 Unions and Scientific Committees participate in the work of SCOPE, which directs particular attention to the needs of developing countries. SCOPE was established in 1969 in response to the environmental concerns emerging at the time: ICSU recognized that many of these concerns required scientific inputs spanning several disciplines and ICSU Unions. SCOPE's first task was to prepare a report on Global Environmental Monitoring (SCOPE 1, 1971) for the UN Stockholm Conference on the Human Environment.

The mandate of SCOPE is to assemble, review, and assess the information available on man-made environmental changes and the effects of these changes on man; to assess and evaluate the methodologies of measurement of environmental parameters; to provide an intelligence service on current research; and by the recruitment of the best available scientific information and constructive thinking to establish itself as a corpus of informed advice for the benefit of centres of fundamental research and of organizations and agencies operationally engaged in studies of the environment.

SCOPE is governed by a General Assembly, which meets every three years. Between such meetings its activities are directed by the Executive Committee.

R. E. Munn
Editor-in-Chief
SCOPE Publications

Executive Secretary: V. Plocq

Secretariat: 51 Bld de Montmorency
               75016 PARIS

Contents

Preface

The impetus for the workshop that resulted in this book was the deliberations in many national and international forums about the research areas to be addressed in a decade-long international programme to study global environmental change. At the request of the Executive Committee of ICSU's Scientific Committee on Problems of the Environment (SCOPE), the US National Committee for SCOPE of the US National Research Council (NRC) began discussions to organize an international workshop intended to identify the contributions to the upcoming International Geosphere–Biosphere Programme: A Study of Global Change (IGBP) that could be made by biological and physical scientists working together. Discussions with a number of scientists led to the consensus that a useful focus for the workshop would be on an issue of extreme concern in conducting the interdisciplinary research required to understand the processes controlling the global environment—how to overcome the disparities in spatial and temporal scales used in different scientific disciplines. The transfer of information between these disciplines is severely constrained by disparities in scale. Thus, a workshop to identify the research needed to deal meaningfully with these scaling problems and with the spatial and temporal variability in biospheric and geospheric processes was organized.

To carry out the organization of the workshop and the identification of participants, an international Steering Committee was formed under the chairmanship of R. G. Woodmansee with T. Rosswall and P. G. Risser serving as co-chairmen. This committee included representatives of SCOPE, ICSU's International Association for Ecology (INTECOL), and members of the US National Committee for SCOPE and its parent body, the Environmental Studies Board of the US NRC. A mix of ecologists, other biological scientists, atmospheric scientists, geomorphologists, and marine scientists from 17 countries were invited to participate.

At the workshop, participants met in a plenary session and in working groups to explore the research needs for understanding interactions between the atmospheric, aquatic, and terrestrial components of the biosphere at different scales. A report authored by P. G. Risser (Risser, 1986) describes the research priorities identified for dealing with the scaling problems and for possible inclusion in the research agenda for a programme on global change. Papers presented at the workshop are published herein.

The workshop was organized with the intention that it be one of several planning efforts toward the elucidation of research priorities for an International Geosphere–Biosphere Programme. We look forward to further discussion and refinement of the topics outlined in Risser (1986) and this volume, as well as other research approaches, in other international forums.

The Steering Committee and the workshop participants wish to acknowledge with gratitude the financial support for the workshop from the US National Aeronautics and Space Administration, the US National Science Foundation, the US Department of Energy, ICSU, and SCOPE. Ruth DeFries, of the US NRC, was especially helpful in all aspects of the workshop, and her invaluable assistance was appreciated by all the participants. Joële Dallançon of SCOPE provided excellent logistical support during the workshop.

We also wish to acknowledge the previous efforts of our colleagues in setting forth the broad perspectives and challenges of the International Geosphere–Biosphere Programme: A Study of Global Change. These efforts stimulated and assisted the deliberations of the workshop. We also thank the many reviewers of papers contained in this volume.

THOMAS ROSSWALL
ROBERT G. WOODMANSEE
PAUL G. RISSER
Co-Chairmen, Workshop
Steering Committee

Risser, P. G. (1986). *Spatial and Temporal Variability of Biospheric and Geospheric Processes: Research Needed to Determine Interactions With Global Environmental Change*. Report of a Workshop. International Council of Scientific Unions Press, Paris: 53 pages.

List of Participants

Steering Committee

ROBERT G. WOODMANSEE

Chairman
Natural Resource Ecology Laboratory
Colorado State University
Fort Collins, CO 80523, USA

PAUL G. RISSER

Co-chairman
Scholes Hall
The University of New Mexico
Albuquerque NM 87131, USA

THOMAS ROSSWALL

Co-chairman
Department of Water in Environment and
    Society
University of Linköping
S-581 83 Linköping, Sweden

PAUL J. CRUTZEN

Max Planck Institute for Chemistry
PO Box 3060
D-6500 Mainz, FRG

JOHN W. B. STEWART

Department of Soil Sciences
University of Saskatchewan
Saskatoon, Sask., Canada S7N 0W0

Participants

DARWIN ANDERSON

Department of Soil Sciences
University of Saskatchewan
Saskatoon, Sask., Canada S7N 0W0

BERT BOLIN

Department of Meteorology
University of Stockholm Arrhenius
    Laboratory
S-106 91 Stockholm, Sweden

WILLIAM C. CLARK   IIASA
                          A-2361, Laxenburg, Austria

DAVID C. COLEMAN   Department of Entomology
                          University of Georgia
                          Athens, GA 30602, USA

ROBERT E. DICKINSON   National Center for Atmospheric Research
                          PO Box 3000
                          Boulder, CO 80307, USA

JOHN A. EDDY   National Center for Atmospheric Research
                          PO Box 3000
                          Boulder, CO 80307, USA

JOHN W. FARRINGTON   Woods Hole Oceanographic Institute
                          Woods Hole, MA 02543, USA

PETER FRITZ   Department of Earth Sciences
                          University of Waterloo
                          Waterloo, Ontario, Canada N2L 3G1

CONGBIN FU   Institute of Atmospheric Physics
                          Academia Sinica
                          Beijing, China

AKIHIKO HATTORI   Ocan Research Institute
                          University of Tokyo
                          Nakano-Ku, Tokyo 164, Japan

JOHN N. R. JEFFERS   Institute of Terrestrial Ecology
                          Merlewood Research Station
                          Grange-over-Sands
                          Cumbria LA11 6JV, England

R. GARY KACHANOSKI   Department of Land Resource Science
                          University of Guelph
                          Guelph, Ontario, Canada N1G 2W1

STEPHAN KEMPE   SCOPE/UNEP International Carbon Unit
                          University of Hamburg
                          Bundesstrasse 55
                          D-2000 Hamburg 13, FRG

VERNON MEENTEMEYER   Department of Geography
                          University of Georgia
                          Athens, GA 30602, USA

IAN NOBLE

Research School of Biological Sciences
Australian National University
Box 4
Canberra, A.C.T., Australia 2601

HANS OESCHGER

Physics Institute
University of Bern
Sidlerstrasse 5
CH-3012 Bern, Switzerland

REUBEN OLEMBO

United Nations Environment Programme
PO Box 30552
Nairobi, Kenya

ROBERT V. O'NEILL

Environmental Sciences Division
Oak Ridge National Laboratory
PO Box X
Oak Ridge, TN 37831, USA

BRYNJULF OTTAR

Norwegian Institute for Air Research
Elvegt. 52
N-2000 Lillestrom, Norway

TREVOR PLATT

Bedford Inst. of Oceanography
PO Box 1006
Dartmouth, Nova Scotia, Canada B2Y 4A

JEFFREY E. RICHEY

College of Ocean and Fisheries Sciences
WH-10
University of Washington
Seattle, WA 98195, USA

LECH RYSZKOWSKI

Dept. of Agrobiology & Forestry
Polish Academy of Sciences
Swierczewskiego 19
60–809 Poznan, Poland

OSWALDO SALA

University of Buenos Aires
Buenos Aires, Argentina

JOSE SARUKHAM

Instituto de Biologia
Apdo. Postal 70–232
Mexico, 04510
D.F., Mexico

DAVID S. SCHIMEL

Natural Resource Ecology Laboratory
Colorado State University
Fort Collins, CO 80523, USA

STANLEY A. SCHUMM   Department of Earth Resources
                          Colorado State University
                          Fort Collins, CO 80523, USA

HERMAN H. SHUGART   Dept. of Environmental Sciences
                          The University of Virginia
                          Charlottesville, VA 22903, USA

STEPHEN SIMKINS   Department of Plant and Soil Sciences
                          University of Massachussetts at Amherst
                          Stockbridge Hall, Amherst, MA 01003, USA

JAI S. SINGH   Department of Biology
                          Banaras Hindu University
                          Varanasi 221005, India

WIM G. SOMBROEK   ISRITC, PO Box 353
                          6700 AJ Wageningen, Holland

MICHAEL J. SWIFT   Department of Biological Sciences
                          University of Zimbabwe
                          PO Box MP167
                          Harare, Zimbabwe

MONICA TURNER   Institute of Ecology
                          University of Georgia
                          Athens, GA 30602, USA

DEAN URBAN   Environmental Sciences Division
                          Oak Ridge National Laboratory
                          PO Box X
                          Oak Ridge, TN 37830, USA

PAUL K. M. VAN DER HEIJDE Holcomb Research Institute Butler University
                          4600 Sunset Avenue
                          Indianapolis, IN 46208, USA

PETER M. VITOUSEK   Dept. of Biological Sciences
                          Stanford University
                          Stanford, CA 94305–2493, USA

JOHN J. WALSH   Department of Marine Science
                          University of South Florida
                          140 7th Avenue, South
                          St. Petersburg, FL 33701, USA

BRIAN H. WALKER   Div. of Wildlife & Rangeland Research
                          C.S.I.R.O.
                          PO Box 84
                          Lyneham, ACT 2602, Australia

PATRICK J. WEBER   Inst. for Alpine & Arctic Research
                                    University of Colorado
                                    Boulder, CO 80309, USA

US Government Participants
ROGER DAHLMAN   Carbon Dioxide Research
                                    Division ER-12
                                    Office of Basic Energy Sciences
                                    Department of Energy
                                    Washington, DC 20545, USA

JAMES GOSZ   Ecosystems Studies
                                    Room 1140
                                    National Science Foundation
                                    1800 G Street, NW
                                    Washington, DC 20550, USA

HELEN MCCAMMON   ER-25
                                    Office of Health and Environmental
                                        Research
                                    Department of Energy
                                    Washington, DC 20545, USA

JARVIS MOYERS   National Science Foundation
                                    1800 G Street, NW
                                    Washington, DC 20550, USA

ROBERT WATSON   NASA Headquarters
                                    Room 13910 B
                                    600 Independence Avenue, SW
                                    Washington, DC 20546, USA

DIANE WICKLAND   NASA Headquarters
                                    600 Independence Avenue, SW
                                    Washington, DC 20546, USA

Staff
RUTH S. DEFRIES   Environmental Studies Board
                                    National Academy of Sciences
                                    2101 Constitution Ave., N.W.
                                    Washington, DC 20418, USA

JOËLE DALLANÇON   SCOPE
                                    51 Blvd de Montmorency
                                    75016 Paris, France

Scales and Global Change
Edited by T. Rosswall, R. G. Woodmansee and P. G. Risser
© 1988 Scientific Committee on Problems of the Environment (SCOPE)
Published by John Wiley & Sons Ltd.

CHAPTER 1

Spatial and Temporal Variability of Biospheric and Geospheric Processes: A Summary

PAUL G. RISSER

Scholes Hall,
The University of New Mexico,
Albuquerque,
NM 87131, USA

THOMAS ROSSWALL

Department of Water in Environment and Society,
University of Linköping,
S–581 83 Linköping,
Sweden

ROBERT G. WOODMANSEE

Natural Resource Ecology Laboratory,
Colorado State University,
Fort Collins,
CO 80523, USA

INTRODUCTION

Many of the most pressing environmental problems facing mankind today are global in nature. An understanding of global environmental processes, a prerequisite for finding scientific solutions to these problems, will be found only by combining many disciplines, including ecology, oceanography, meteorology, geomorphology, and geology. As a result of the widespread recognition of the need for an international research programme on geosphere–biosphere interactions, the concept of the International Geosphere–Biosphere Program was developed (Malone and Roederer, 1985; National Research Council, 1983, 1986). The plan for this ambitious global research programme includes the following steps:

(a) designing specific studies of the most important processes in global environmental change

1

(b)   developing models of interfacial problems incorporating existing and newly derived data

(c)   performing carefully designed tests of these models

(d)   making observations based on these process studies models, and experimental tests.

At each step, integrating data from different disciplines and from different spatial and temporal scales will be important components of the research designs (Risser, 1986). This book is based on a planning workshop designed to assist in the formulation of appropriate research approaches to be used in the coherent investigation of geospheric–biospheric processes.

Except for cumulonimbus thunderstorm clouds which reach heights greater than 10 km above the earth's surface, most of the clouds and humidity are found in the lower few kilometres of the global atmosphere. The most rapid and turbulent fluxes of atmospheric materials are found in the planetary boundary layer at heights of one to several kilometres although the stratosphere, which extends to about 50 km, has significant photochemical connections to the lower atmosphere layers. However, these processes, and the resulting chemical fluxes, demonstrate marked vertical and horizontal variation, and, in addition, this variation occurs on time scales from seconds to decades or longer (Bolin and Cook, 1983).

Vegetation, as an indicator of terrestrial ecosystems on the earth's surface, responds to changes in atmospheric conditions, but it also influences atmospheric processes by affecting fluxes of energy, momentum and water. Similarly, oceans play equally interactive roles by influencing not only atmospheric conditions, but also terrestrial ecosystems. Geological processes, especially those relating to topography and soils, cause certain characteristic responses in climate and vegetation, and, in turn, geologic processes are influenced by vegetation and climate. Thus, the interactions among the biospheric components are complicated, and operate over a broad array of spatial and temporal scales. This initial chapter will amply describe this complexity, largely from the point of view of each discipline. In this introductory chapter, the most important themes and challenges will be presented and distilled in order to focus on common components of an integrated geosphere–biosphere research programme.

ATMOSPHERE

In the 1960s the first mathematical attempts were made at describing the atmospheric patterns of the globe. The general circulation models (GCMs) began with equations of existing weather prediction models, to which were added equations for the dynamics of water vapour, including evaporation, precipitation, and movement, and equations for atmospheric radiation (Dickinson, 1988). These models describe processes in the atmosphere and the oceans on scales larger than a few hundred kilometres in the horizontal directions and one to a few in the vertical. Thus, long term integrations will not permit horizontal spatial resolutions of smaller than about 100 km (Bolin, 1988).

Though there are mesoscale climate models and smaller scale models which focus on a localized set of conditions, it is now clear that some components of this finer scale resolution must be incorporated into general circulation type models. The necessity of this model enhancement can be easily demonstrated by recognizing the need to extend the current numerical weather prediction approaches to periods of several weeks or longer. This improvement in general circulation models must incorporate fluctuations in the hydrological cycle, radiation, and surface fluxes (Dickinson, 1988). Furthermore, as noted below, understanding the behaviour of atmospheric gases involves the expansion of circulation models to include interactions among the physical, chemical, and biological processes found in the lower atmosphere, top layer of soils, uppermost layers of ocean sediments, oceans and terrestrial ecosystems' portions of the biosphere (Bolin, 1988).

Improvement in modelling the atmosphere will involve the need to characterize many atmospheric processes at scales smaller than 100×100 km. Current models will require greater definition in the description of the radiative properties of clouds, and the vertical transfer of heat, moisture, and momentum by moist and dry convection. In linking small scale atmospheric process models to larger circulation models, the following issues are of particular importance (Dickinson, 1988):

(a)   identifying appropriate parameterizations for the energy-transfer processes within vegetation canopies that yield satisfactory descriptions of the vegetation–atmosphere exchanges of heat and moisture.

(b)   providing adequate descriptions of the soil and plant root resistance processes that limit the rate at which water can be transferred from the soil to the leaves and then to the atmosphere (Dickinson, 1984; Bolin, 1988).

The challenge to describing the fluxes does not only concern the development of methods to statistically characterize these terms over heterogeneous surfaces of the globe, but also to capture the importance of long-term changes of the biota and soils. These changes are often obscured by variations occurring on relatively short time scales such as daily cycles, weather variations, and various annual cycles. Thus, the crucial issues for models in the realm of atmospheric–terrestrial and marine ecosystem interactions involve the interplay between abiotic factors that describe the atmosphere and inorganic substrate with the characteristics of the biota including the organic component of the soil (Bolin, 1988).

TRACE GASES

It is now possible to state unequivocally that the atmospheric constituents are changing dramatically, and that, in some cases, these changes involve magnitudes that lead to rates approximately doubling on time scales of decades. Carbon dioxide has increased from concentrations of less than 280 ppm a few hundred years ago to present values of about 345 ppm (Neftel *et al.*, 1985). Notable and perhaps deleterious increases have also been measured for chlorofluorocarbons (CCl_2F_2 and CCl_3F), methane (CH_4), tropospheric ozone (O_3) and nitrous oxide (N_2O). The sources of these increases are not completely understood, but some are generated from human activities such as combustion of fuel, agricultural practices, and land use changes.

Various models have been used to predict the probable warming of the atmosphere caused by increases in CO_2 and other trace gases. Most suitable advanced global circulation models which include reasonable continuity and a seasonal cycle for solar radiation (Manabe and Stouffer, 1980; Hansen *et al.*, 1984; Washington and Meehl, 1984) indicate a steady state global warming of from 2.0 to 4.8 °C average temperature for a doubling of CO_2. However, virtually all model simulations predict greater temperature increases at high latitudes, especially in the winter, because of more stable temperature stratification and ice–snow feedbacks.

Other compounds produced at the earth's surface by biological and anthropogenic processes are not uniformly distributed in the atmosphere. Oxides of nitrogen (NO and NO_2) are primarily produced from specific sources (e.g. power plants, industrial centres) and since the lifetime of NO_x is only a few days, the dispersion of these molecules in the atmosphere is uneven. In addition, however, physical–chemical processes of the interactions between the biospheric and atmospheric processes also lead to a non-homogeneous distribution of trace gases. For example, although only about ten percent of all atmospheric ozone is found in the troposphere, this small amount of ozone is extremely important because with ultraviolet light it produces hydroxyl radicals (OH). In the background troposphere, two-thirds of the OH react with CO, one-third with CH_4, and small fractions with other atmospheric gases. The concentration of NO, which as just noted is short-lived and spatially heterogeneous, plays a significant role in determining the oxidation pathways of methane and carbon dioxide. It is likely that global increases in CH_4 have contributed to increases in O_3 concentration in NO-rich environments (particularly in the highly industrialized regions of the northern hemisphere) and decreases in hydroxyl concentrations in NO-poor environments elsewhere. Thus, lesser quantities of industrial and natural gases may be oxidized in the tropics by reactions with OH which, over time, may lead to an increase in atmospheric trace gases in this region (Crutzen, 1988).

Understanding these atmospheric–biospheric chemical systems demands resolution of the temporal and spatial heterogeneous distribution and sources of short-lived, reactive compounds. Indeed, the discontinuity of both emission sources and atmospheric chemical interactions is a major challenge to the formulation of photochemical–meteorological models describing the transport and photochemistry of these reactive compounds. Given this complexity, it may be necessary to use other approaches, such as the incorporation of stochastic simulation of boundary layer and convective processes which may be derived from statistical analysis of results obtained by large scale meteorological models (Crutzen, 1988).

OCEANS

Oceans play major roles in numerous biospheric processes, but the magnitude and even the direction of these roles is just now being recognized because of the recent availability of adequate instrumentation. Approximately 30% of the global annual plant carbon fixation comes from the seas which also form an integral part of the N flux of the earth (Lewis *et al.*, 1986; Walsh and Dieterle, 1988). The ocean is also the main sink for CO_2 produced by human activities, and because of its size and heat capacity, the ocean dampens changes in the earth's energy balance (Oeschger, 1988). However, the general patterns have not been measured nor understood specifically enough to predict the response of the biosphere to future modifications of the environment. For example, uncertainties in the magnitudes of the annual uptake of both carbon and nitrogen have been caused by the inability to accurately specify temporal and spatial heterogeneities of phytoplankton biomass and associated productivities in the ocean shelf regions. Under some conditions the *Nimbus-7* coastal zone scanner can now measure oceanic chlorophyll concentration to ±30%. Significant improvements in productivity estimates will be achieved when satellite measurements of ocean colour can be combined with *in situ* observations. In fact, the applicability of more sophisticated regional productivity models to the oceans is still restricted by scarcity of information about the coupling of physical dynamics with biological processes on the appropriate time and space scales (Walsh and Dieterle, 1988).

GEOLOGY AND SOILS

Much of the attention on geosphere–biosphere issues has focused on the atmosphere and oceans where the global nature of the process is so obvious because of the broad expanses and the lack of conspicuous boundaries. For the reverse reasons, global processes in groundwater systems have proven particularly challenging. Spatial scales range from less than a nanometre when considering interactions between water molecules and dissolved chemicals to hundreds of kilometres when assessing and managing regional groundwater systems. As in atmospheric and oceanic systems, expanding the spatial and temporal scales of groundwater systems demands consideration of different characteristics. For example, larger spatial scales mean increasing importance of geological heterogeneities and anisotropy, and of the effects of long-term recharge variations on the water balance of a system for longer time periods (van der Heijde, 1988).

The complexity of the groundwater system offers many challenges to the construction of models, especially as groundwater processes relate to climatic conditions and geomorphology. A major problem is how to distinguish among variables that can be considered as constants or as being uniform across discrete intervals of time and space dimensions, and those non-uniform variables that must be included in models. An example involves the modelling of infiltration into the soil and subsequent percolation toward the saturated zone (van der Heijde, 1988). Runoff from precipitation is divided into a single horizontal segment and infiltration into a one- or two-dimensional vertical component. The infiltrated water percolates to the groundwater where a two-dimensional horizontal or three-dimensional model is used. For each of these submodels different timesteps are invoked, from hourly for surface runoff and daily for percolation, to weekly or monthly for the flow in the saturated zone. To address global issues it will eventually be necessary to develop integrated models which account for the relevant variables on medium-scale drainage basins and physiographically defined spatial units. Since at any given moment some portions of the geomorphic system are eroding and some are aggrading, the definition of the spatial scale and the recognition of the importance of episodic events, are critical in the connections between superficial and groundwater processes (Schumm, 1988).

VEGETATION

The vegetation plays a dynamic role in numerous of these processes, e.g. moderating surface flows of water and sediments, producing gases to the atmosphere, modifying soil processes, and responding to various environmental conditions. These interactions occur over a wide range of space and time scales (Delcourt *et al.*, 1983; Emanuel *et al.*, 1985). The successional response of vegetation to disturbance and changing environmental conditions occurs in a number of ways, such as changes in productivity, species composition, and rates of material cycling. These successional processes have been modelled under a variety of conditions, but it is now clear that these successional models must be made more responsive to changes in atmospheric conditions. A significant need appears to be the realistic input of synoptic climatological variables that are associated with vegetation successional processes (Shugart *et al.*, 1988).

Successional models now incorporate the consequences of the natural history characteristics of plants as well as their physiology and demography. Transect models include both the mechanistic formulation of important population processes and the realism of relatively small-scale spatial heterogeneity. Competition among individuals is usually modelled as a function of the proximity and size of neighbouring individuals. Despite the realism of these vegetation models, atmospheric perturbations have traditionally been treated in a simplistic fashion and furthermore, the relevant atmospheric variables have been assumed to exert primarily first-order effects rather than acting through intermediate agents such as defoliators, herbivores, or disease (Shugart *et al.*, 1988). Thus, the next generation of terrestrial ecosystem or vegetation models must recognize the importance of climatic history in explaining the present vegetation patterns, must realistically couple vegetation responses to changing climatic conditions, and must describe how the spatial heterogeneity of vegetation affects biospheric processes.

ANALYTICAL CHALLENGES

Inherent in the studies described and proposed in this volume is the need to relate measurements of geospheric–biospheric processes made in several different scientific disciplines. A major challenge to these data interrelations is that collections of many of these previous measurements have not followed the formulation of multidisciplinary *a priori* hypotheses and clear statements about the relevant inferences (Jeffers, 1988). Furthermore, the necessary measurements to test the most important interdisciplinary ideas range over broad spatial and temporal scales. Thus, the analytical challenge is to adopt valid methods of sampling within appropriate experimental designs in order to measure the relevant spatial and temporal heterogeneity. A portion of the success in these interdisciplinary studies will depend upon the selection of coherent experimental variables which bridge the procedures used and the measurements made by two or more disciplines involved in the analysis (Allen and Starr, 1982; O'Neill, 1988).

In some cases the challenge of linking space and temporal scales will be met by new methodology such as remote sensing techniques capable of refined measurements of the chemical constituents of the soil, vegetation, and atmosphere. Other experimental procedures will involve the careful construction of models for systems in which the models are specifically designed to simulate the dynamics of several common variables crucial to the biospheric system in question. In virtually all approaches, however, invoking detailed mechanistic information about the processes will be essential to designing appropriate experiments and building reasonable models (Woodmansee, 1988).

An example of a successful experimental and analytical approach involves the measurement of fluxes of gaseous nitrogen compounds from terrestrial ecosystems. As noted earlier, these gases influence the climate as greenhouse gases, participate in the formation and destruction of ozone, contribute to atmospheric acidity, and are significant vectors for loss and gain of nitrogen from terrestrial ecosystems (Lacis *et al.*, 1981; Crutzen, 1983; Bolin and Cook, 1983). Nitrogen-gas fluxes typically display a high degree of spatial and temporal variability (Schimel *et al.*, 1986). In evaluating this variability, Schimel *et al.* (1988) noted that in a Swedish meadow, a significant amount of the fine scale variation in N_2O production could be explained by spatial autocorrelation and by correlation with soil water. In the shortgrass prairie, differences in N_2O production seasonally and between sites were closely coupled to soil mineral nitrogen dynamics (Schimel *et al.*, 1985). The data suggested that nitrification was the dominant vector for N_2O production, and that patterns of temperature and water availability affect N_2O production by controlling nitrogen mineralization and nitrification. From an experimental and analytical viewpoint, the common technique employed in these studies was that the difficult-to-measure gas flux rates were related to more readily measured soil and landscape properties using statistical or modelling techniques (Schimel *et al.*, 1987). Moreover, the choice of the appropriate predictor values depends upon a thorough knowledge of the processes involved in the phenomenon under study.

This book describes an astonishingly broad array of scientific issues that must be studied to understand the global geospheric–biospheric processes that will control the habitability of the earth. These issues irreverently cross the boundaries of typical disciplines, and, thus, mandate that experimental and analytical procedures must do the same. This challenge demands new measurements, more insightful mathematical models, and broader collaborative experimental approaches. The chapters in this volume carefully outline many of the required scientific challenges, and both individually and collectively suggest experimental approaches for the solution of these issues which are of such great global significance.

REFERENCES

Allen, T . F. H. and Starr, T. B. (1982). *Hierarchy: Perspectives for Ecological Complexity*. University of Chicago Press, Chicago, Illinois.

Bolin, B. (1988). Linking terrestrial ecosystem process models to climate models. (Chapter 7, this volume.)

Bolin, B., and Cook, R. B. (Eds.) (1983). *The Major Biogeochemical Cycles and their Interactions*. SCOPE 21. John Wiley and Sons, Chichester.

Crutzen, P. J. (1983). Atmospheric interactions—homogeneous gas reactions of C, N, and S containing compounds. In Bolin, B., and Cook, R. B. (Eds.) *The Major Biogeochemical Cycles and Their Interactions*, John Wiley and Sons, New York.

Crutzen, P. J. (1988). Variability in atmospheric-chemical systems (Chapter 6, this volume).

Delcourt, H. R., Delcourt, P. A., and Webb III, T. (1983). Dynamic plant ecology: the spectrum of vegetation change in space and time. *Quaternary Science Review*, 1, 153–175.

Dickinson, R. E. (1984). Modeling evapotranspiration for three-dimensional global climate models. In Hansen, J. E., Takahashi, T. (Eds.), *Climate Processes and Climate Sensitivity*, pp. 58–72. M. Ewing Series 5, American Geophysical Union, Washington, DC.

Dickinson, R. E. (1988). Atmospheric systems and global change. (Chapter 5, this volume.)

Emanuel, W. R., Shugart, H. H., and Stevenson, M. P. (1985). Climatic change and the broadscale distribution of terrestrial ecosystem complexes. *Climate Change*, 7, 29–43.

Hansen, J., Lacis, A., Rind, D., Russell, G., Stone, P., Fung, I., Ruedy, R., and Lerner, J. (1984). Climate sensitivity: analysis of feedback mechanisms. In Hansen, J. E., and Takahashi, T. (Eds.), *Climate Processes and Climate Sensitivity*., M. Ewing Series 5. American Geophysical Union, Washington, DC.

van der Heijde, P. K. M. (1988). Spatial and temporal scales in groundwater modelling. (Chapter 11, this volume.)

Jeffers, J. N. R. (1988). Statistical and mathematical approaches to issues of scales in ecology. (Chapter 4, this volume.)

Lacis, A., Hanson, G., Lee, P., Mitchell, T., and Lebedeff, S. (1981). Greenhouse effect of trace gases, 1970–1980. *Geophys. Res. Lett.*, 8, 1035–1038.

Lewis, M. R., Harrison, W. G., Oakey, N. S., Hebert, D., and Platt, T., (1986). Vertical nitrate fluxes in the oligotrophic ocean. *Science*, 234, 870–873.

Malone, T. F., and Roederer, J. G. (Eds.) (1985). *Global Change: Proceedings of a Symposium Sponsored by ICSU during its 20th General Assembly in Ottawa, Canada*. ICSU Press Symposium Series No. 5. Cambridge University Press.

Manabe, S., and Stouffer, R. J. (1980). Sensitivity of a global climate model to an increase of CO_2 concentration in the atmosphere. *J. Geophys. Res.*, 85, 5529–5554.

National Research Council. (1983). *Toward an International Geosphere–Biosphere Program: a Study of Global Change*. Commission on Physical Sciences, Mathematics, and Resources. National Academy Press, Washington, DC.

National Research Council. (1986). *Global Change in the Geosphere–Biosphere: Initial Priorities for IGBP*. Commission on Physical Sciences, Mathematics, and Resources. National Academy Press, Washington, DC.

Neftel, A., Moor, E., Oeschger, H., and Stauffer, B. (1985). Evidence from polar ice cores for the increase of atmospheric CO_2 in the last two centuries. *Nature*, 315, 45–57.

Oeschger, H. (1988). The ocean system—ocean/climate and ocean/CO_2 interactions. (Chapter 15, this volume.)

O'Neill, R. V. (1988). Hierarchy theory and global change. (Chapter 3, this volume.)

Risser, P. G. (Compiler.) (1986). *Spatial and Temporal Variability of Biospheric and Geospheric Processes: Research Needed to Determine Interactions with Global Environmental Change*. ICSU Press, Paris.

Schimel, D., Stillwell, M. A., and Woodmansee, R. G. (1985). Biogeochemistry of C, N, and P in a soil catena of the shortgrass steppe. *Ecology*, 66, 276–282.

Schimel, D. S., Parton, W. J., Adamsen, F. J. Woodmansee, R. G., Senft, R. L. and
    Stillwell, M. A. (1986). The role of cattle in the volatile loss of nitrogen from a
    shortgrass steppe. *Biogeochemistry*, **2**, 39–52.
Schimel, D. S., Simkins, S., Rosswall, T., Mosier, A. R., and Parton, W. J., (1988).
    Scale and the measurement of nitrogen-gas fluxes from terrestrial ecosystems.
    (Chapter 10, this volume.)
Schumm, S. A. (1988). Variability of the fluvial system in space and time. (Chapter 12,
    this volume.)
Shugart, H. H., Michaels, P. J., Smith, T. M., Weinstein, D. A., and Rastetter, E. B.
    (1988). Simulation models of forest succession (Chapter 8, this volume.)
Walsh, J. J., and Dieterle, D. A. (1988). Use of satellite ocean colour observations to
    refine understanding of global geochemical cycles. (Chapter 14, this volume.)
Washington, W. M., and Meehl, G. A. (1984). Seasonal cycle experiment on the
    climate sensitivity due to a doubling of CO_2 with an atmospheric general circulation
    model coupled to a simple mixed layer ocean model. *J. Geophys. Res.*, **89**,
    9475–9503.
Woodmansee, R. G. (1988). Ecosystem processes and global change. (Chapter 2, this
    volume.)

Scales and Global Change
Edited by T. Rosswall, R. G. Woodmansee and P. G. Risser
© 1988 Scientific Committee on Problems of the Environment (SCOPE)
Published by John Wiley & Sons Ltd.

CHAPTER 2

Ecosystem Processes and Global Change

ROBERT G. WOODMANSEE

Natural Resource Ecology Laboratory,
Colorado State University,
Fort Collins, CO 80523, USA

ABSTRACT

Man's responsibility to evaluate and forecast his changing environment is being recognized as a major scientific and societal need for the coming decades. Not only small-scale or local environments, but the entire global environment requires evaluation. To accomplish the required evaluations, scientists and political decision-makers must acquire knowledge that will allow the integration of information ranging from small geographic scales (metres to hectares) and short temporal scales (minutes to hours) to large spatial scales (regions and the globe) and long temporal scales (decades to centuries). This integration is necessary to explain the role of biogenic processes in the changing global environment.

Discussions in this paper focus on problems and questions that relate ecological processes occurring at various hierarchical levels to the levels above and below that of current attention. The discussions emphasize that resolution of our environmental problems will require transdisciplinary team research. New institutional structures will probably be required to accommodate such research.

INTRODUCTION

Humans, largely through scientific achievements, have developed into a species that has significantly and profoundly influenced their surrounding environment, from local to global levels of resolution. Humans have had a great effect on their fellow species of organisms, both directly, by harvest, and indirectly, by altering habitat. Yet, even though our conceptual, explanatory models of causality are complex and scientifically sophisticated, they are still simplistic and imperfect and, especially on a large spatial scale (regional or global), often are based on little more than dogma and intuition. With our best current scientific knowledge, we can explain only local environmental responses for short periods of time. When events occur that we do not understand, we declare them stochastic or too complex to explain and state that more data are needed to better understand the phenomena of interest. Our typical response is to conduct reductionist-oriented research about narrow but tractable topics. These topics rarely address problems of large-scale ecosystems such as landscapes, regions, and, indeed, the globe.

Because we have and will continue to alter our environment, we must accept the responsibility to explain rationally how we interact with it from local to global scales of resolution. We must learn to manage and direct those interactions so that we can maintain a hospitable environment. A major aim of research must be to gain the knowledge necessary to explain our environmental interactions at many spatial and temporal scales and to develop appropriate models to communicate that knowledge to decision makers and other scientists. Specifically, we must gain knowledge of biogeochemical cycles, primary production, and physical processes such as erosion and sedimentation, for time scales greater than a few years and spatial scales larger than a few square metres.

The primary goal of this chapter is to discuss examples of small-scale ecological phenomena and how they influence and interact with the broader environment, especially at the scale of landscapes and larger. This discussion will focus on the scientific process required to understand biospheric interactions. To accomplish this goal, I will:

1. Define the terminology of the ecological hierarchy used in this paper

2. address specific global-scale problems that require ecological process-level information for solution at the landscape level

3. evaluate environmental factors that control certain important patch and flowpath (defined below) processes and how they interact in time and space

4. relate how the results of path- and flowpath-scale ecological processes interact with the broader environment to contribute to landscape-, regional-, and global-scale phenomena

5. emphasize the necessity for scientists from many disparate disciplines to work together to solve major environmental problems.

Most ecological research has focused on problems that were perceived to be tractable in two- or five-year time frames because of funding constraints, and at spatial scales of a few square metres or, at most, a few hectares, because of our inability to deal effectively with heterogeneity. This focus is inadequate to assess the issues that truly concern large-scale ecosystem processes and dynamics. The *scope* of future research must extend to ecosystems *dynamics* requiring decades to centuries to be expressed and spatial scales that range from sub-continental to global in size.

PROBLEMS AND PROCESSES

Many large-scale (including global) phenomena such as increasing concentrations of greenhouse gases, erosion of agricultural lands, and loss of genetic diversity, are mediated by ecological processes that occur at small scales at the surface of the globe (Bowden and Bormann, 1986; Schimel *et al.*, 1986). Some of these problems are significant for the welfare of humans, and all are influenced by 'natural' and managed ecosystem behaviour (reviewed in Smil, 1985; Bolin and Cook, 1983). We currently have considerable information about important ecological processes that influence numerous problems at the small scale of spatial resolution and the short term (less than one year) time scale. However, extrapolation from small scales to larger scales is tenuous at best. Among the most significant problems facing humans are increases in the concentrations of trace gases including CO_2, erosion and sedimentation, and human caused changes in the composition of biotic communities.

Concentrations of numerous biogenically controlled trace gases in the atmosphere are increasing. Among these are several gaseous species of nitrogen which are involved in complex chemical reactions which may ultimately influence the transmission of ultraviolet radiation to the earth's surface, influence atmospheric heating, or become involved in complex tropospheric chemical reactions which may profoundly influence the chemical environment of the biosphere (Crutzen, 1983). The influence can be direct, through absorption by organisms, or indirect, through deposition of chemical substances at the surfaces of ecosystems, resulting in complex reactions that alter our chemical environment (i.e. Strickland and Fitzgerald, 1984; Schindler *et al.*, 1985; Olsen *et al.*, 1985; Bowden, 1986).

Quantities of CO_2 are increasing in the atmosphere because of industrialization and land-management activities (Smil, 1985; Bolin and Cook, 1983; Houghton *et al.*, 1983). A consequence of higher CO_2 levels may be altered patterns of plant productivity, with consequent effects on animal and microbial populations and functioning. Alteration of the biological components of ecosystems subsequently may alter the patterns of soil organic matter accumulation and biogeochemical cycling (Schlesinger, 1984; Schimel *et al.*, 1985a). These system responses may result from either direct enhancement of CO_2 levels on plants or indirect influences on the local environment caused by climate change (Strain, 1985).

Questions that must be addressed in the next decades include: If there is an increased incidence of ultraviolet radiation at the surface of the earth due to the chemical destruction of O_3, as some scientists suspect, can biogenically produced gases such as N_2O be shown to be significantly implicated in that destruction? Will there be a significant impact on the world's agricultural productivity that can be attributed to increased biogenic emissions of trace gases? Will a changing chemical composition of the atmosphere have detectable effects on the physiology of species that are deemed important to humans?

Erosion and sedimentation are linked processes which are extensively modified by man's activity (Schumm, 1977). Productivity of many terrestrial and aquatic ecosystems is being significantly altered by either increasing or decreasing rates of these processes (Schimel *et al.*, 1985b,c). Even though erosion has been studied extensively and our knowledge of small scale dynamics is extensive, our understanding of these linked processes at scales larger than uniform field plots is limited. The public is told that enormous quantities of soil are lost from our farmlands each year, but little attention is paid to the amount of soil that is moved and redeposited within the same landscape. What is the ecological importance of this reworked soil material? What is the importance of differentially deposited soil sediments in aquatic ecosystems? These questions remain unanswered.

The patterns of biotic succession at many sites throughout the world are undergoing rapid change, largely because of major man-caused disturbance and greatly enhanced rates of exchange of propagules among sites (i.e. West *et al.*, 1981; Houghton, *et al.*, 1983; Detwiler, 1986). Implications of these changes include the possibility that large-scale genetic diversity will be significantly reduced and/or exotic species will significantly alter local ecosystem behaviour. Will changes in species composition of major biotic communities influence biogeochemical cycles locally or globally? Will there be a detectable change in the earth's biological productivity that can be attributed to human-caused loss of species?

These problems represent global-scale issues that will require information about ecological processes for solution. I contend that the global-scale problems cannot be solved until we learn to skilfully integrate appropriate smaller-scale information into the context of the defined large-scale concerns.

SEMANTICS

I view global-scale concerns as including two classes of problems. One class I call universal phenomena, and the other I call disjunct phenomena. Universal phenomena are experienced with relative uniformity throughout the globe. Examples of such phenomena are CO_2 and N_2O concentrations in the atmosphere. These gases are distributed throughout the atmosphere. Disjunct phenomena are local or regional, but they are experienced to some degree in many different locations in the world. Examples of disjunct phenomena are groundwater contamination, erosion and sedimentation, and biotic succession. Unfortunately, many important phenomena are expressed along a gradient of scales from global to local. These phenomena include increasing CH_4, NH_3, atmospheric particulate and aerosol concentrations and climate change. These latter phenomena show distinct patchiness around the globe, but they are present to some extent everywhere. We must learn to express what class of phenomena we are interested in, and clearly define problems associated with the phenomena.

The term *ecosystem* is used herein to mean any level in an ecological hierarchy defined as an interconnected system of parts. Definitions of various ecosystems are developed and organized by humans, to evaluate a specific problem or enquiry. The word ecosystem has no distinct meaning itself and must be accompanied by an adjective to become descriptive.

An ecological hierarchy is used here to indicate a ranking or ordering of ecosystems that is useful in describing and analysing a particular complex problem that requires information of varying resolution for evaluation. The concepts of ecosystem and ecological hierarchy are conveniences of the human mind that should be used as tools of logic (O'Neill, 1988).

A useful premise aiding analysis of hierarchical systems is that the understanding of one level in a hierarchy requires a thorough consideration of the level above and the level below (DeWit, 1970; Forrester, 1968). Pragmatically, questions guiding an analysis are developed from the level above the level of interest, and the specific analysis of system components and processes are conducted on the level below. I will develop the concept of ecological hierarchy below using an example of trace gas dynamics in the biosphere and atmosphere.

TRACE GASES

An ecological hierarchy for use as a conceptual device aiding analysis of trace gas dynamics in the terrestrial biosphere and atmosphere is shown in Figure 2.1. A guiding global question requiring analysis is: What is the relative contribution of biogenically produced trace gases in the total global budgets of the gases? Appropriate analysis of this question requires determining flux rates of these gases at their sources, that is, at the location where the biological activity responsible for their formation is prevalent. That location is, in reality, at the soil pore and individual bacterium, bacterial colony or soil enzyme scale of spatial resolution (a few mm^3). However, for pragmatic reasons, most field studies are conceptualized at the *patch* scale of spatial resolution.

16

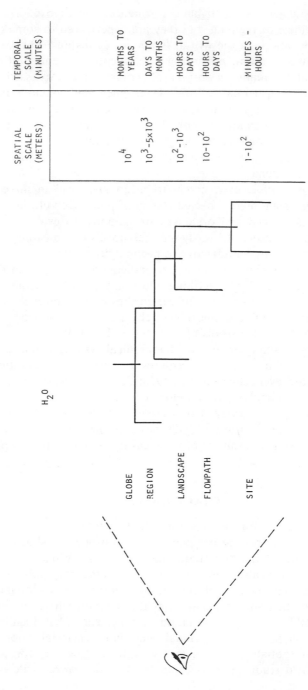

Figure 2.1 An ecological hierarchy useful for conceptualizing the levels of resolution necessary to relate small-scale trace gas emission phenomena to global-scale atmospheric chemistry

A patch in the hierarchy of Figure 2.1 can be an ecosystem defined as a specific biotic community that resides on a definable soil type (Figure 2.2). An appropriate box and arrow representation of a biogeochemical cycle, such as N of a patch ecosystem, is shown in Figure 2.3. The size of a patch is typically a few square metres to a few hectares; it commonly has been described in units of an average square metre or an average hectare. A critical assumption that is made in defining a patch is that all organisms, physical features such as soil pores, soil particles, dead organic matter, nutrients, and water have known distributions within the patch. Thus, any square centimetre or square metre behaves like any other spatial unit within the patch.

Most ecological research has stressed spatial scales of patch size or smaller. Many ecological controlling influences that have been and are currently being addressed illustrate the importance of patch-scale processes in interpreting larger-scale dynamics. Water is probably the most widely studied control of

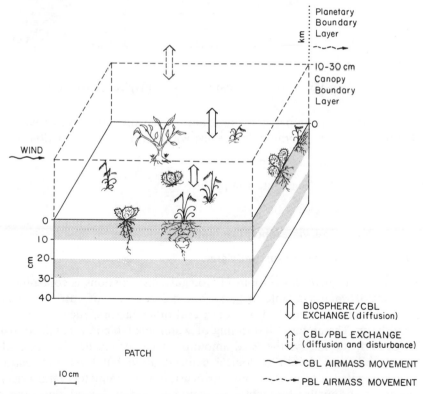

Figure 2.2 Representation of a patch ecosystem. Both the biotic components and soil properties are assumed to be uniformly distributed and describable with ordinary statistical procedures

Comparative Nutrient Cycles of Natural and Agricultural Ecosystems

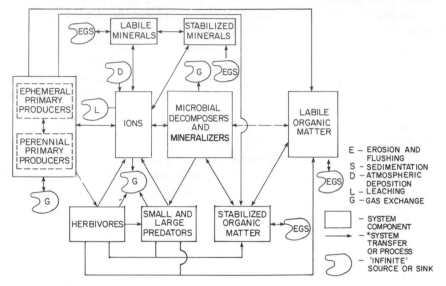

*EVERY TRANSFER IS CONTROLLED BY ONE OR MORE INTRINSIC ENVIRONMENTAL FACTORS.

Figure 2.3   The major components and processes of typical patch ecosystems

trace gas emissions from patches. For example, the amount of water available for biological activity producing trace gas within a patch is a function of:

(1)   the amount that falls as precipitation

(2)   the amount that runs onto the patch

(3)   the amount that runs off

(4)   the quantity lost via evaporation and transpiration

(5)   the availability of subsurface water.

The response of organisms responsible for gaseous emissions is controlled by the amount of water available for growth; the availability of nutrients, which are, in turn, largely controlled by water; and other factors, such as temperature and light. The amount and timing of water available in a patch controls the rates and, ultimately, the total amounts of gas produced. For example, saturated soils may favour biological denitrification but limit nitrification, while well-aerated soils may favour nitrification and completely retard denitrification. Additionally, frequently saturated soils may favour one suite of organisms, while well-aerated soils may support a completely different set of organisms. As we generalize about ecosystem behaviour, it is critical to recognize that adjacent patches often display very different properties, and, thus, very different responses.

The discussion in this section relates appropriately to patch-scale phenomena. Meaningful questions at this scale that are related to the global question above are: What is the contribution of trace gases of each patch (the level of field research on biological processes) to the air mass in the canopy boundary layer? To what extent does water derived from uphill patches control the biogeochemical cycles and consequently gas emissions from downslope patches? Simple generalization from this scale to larger scales is often misleading, if not foolhardy.

A *flowpath* is a connected and interrelated series of patches. At least two concepts of flowpaths are important in consideration of trace gas dynamics. An example of one, a *product flowpath* shown in Figure 2.2, is an air mass that moves along wind pathways within or directly above vegetation canopies of adjacent patches. As the air mass moves along its trajectory, the gases within not only are subject to chemical transformation in the air, but some also are subject to reabsorption by the vegetation, soils, or free water along the ground surface. Simple examples of *control flowpaths* are toposequences or soil catenas, where water or wind are the vectors of transport; small streams; and groundwater channels. More complex examples of flowpaths exist, but they will not be discussed here (see Woodmansee and Adamsen, 1983, for discussion).

The flowpath is a concept that embraces the influences of one patch on another. However, some patches may be isolated and thus are not interactive with other patches. Questions important at the flowpath scale are beginning to receive scientific attention. For example, do patches that occur in downwind positions significantly alter the chemical composition of canopy air masses derived from upwind patches? To what extent are trace gases emitted from one patch absorbed, or chemically transformed, by soils or biota of adjacent or downwind patches? What are the rates of chemical transformations of various trace gases as they move along atmospheric flowpaths?

Few examples exist that can be used to answer these questions about trace gas behaviour in canopy boundary layers. One notable exception is found in Hutchinson and Viets (1969). They showed biogenically evolved gases (NH_3) from one patch to be absorbed or adsorbed by adjacent or downwind patches along product flowpaths. Simple calculations of gas losses from individual patch ecosystems may be inadequate for estimating gaseous inputs to the atmosphere.

Increasing attention is being devoted to the influence of adjacent patches on each other in relation to surface and subsurface drainage pathways or control flowpaths (Correll, 1983). The focus of much of this research is directed towards determining the degree to which the ecological processes in one patch mediate the responses (i.e. nutrient output) of other patches. Recent research results indicate that even though large amounts of fertilizer nutrients may be lost in drainage waters from upslope farm fields, the same nutrients are taken up by forest ecosystems in downslope positions (Correll, 1983). These uptake processes result in the transfer of much higher-quality water out of the watershed than if the slopes are cultivated to the streamside. This example illustrates that knowledge about either of the patch ecosystems alone is insufficient to explain larger-scale system behaviour. Comparable examples need to be developed for atmospheric flowpaths.

Current ecological research is beginning to focus on information that will help explain the responses of *landscape* ecosystems, the scale important to man's physical, economic, or emotional welfare (i.e. pastures, farm fields, and habitat for wildlife). At this scale the focus is on the interaction of flowpaths (and isolated patches) and how they integrate into complex, whole-system responses, such as introduction of chemicals to the planetary boundary layer of the atmosphere, as water quality, beef production, farm-product yields, or sediment yields to waterways. Landscape is a term used by many disciplines, but it has no universal definition. Unfortunately, I know of no word that better describes this level in the ecological hierarchy. I use the term to be a geographic unit that contains complexes of interacting flowpaths that are conceptually related in an analysis of trace gas dynamics, the linked processes of erosion and sedimentation, and biotic community dynamics. The geographic unit may be a watershed, a farm or farms, a lake, or a pasture. Using the trace gas example, an air mass that has a characteristic chemical composition above a mosaic of patches and flowpaths could define the boundaries of the landscape. A critical assumption, using this logic, is that even though emission and absorption rates from patches within the landscape may greatly vary, the mixing processes and chemical reactions within the air mass in the canopy boundary layer are distributed in a relatively uniform way. Neighbouring landscapes would yield air masses with different characteristics. Exchange processes occur between the canopy boundary layer and the planetary boundary layer at this level of resolution. A scientific hope would be that appropriate analyses of regions, the next level above landscapes in the hierarchy, would reveal repeating and predictable patterns of landscapes.

A fundamental need in environmental sciences is to describe better the dynamics of interacting flowpath-scale phenomena, but to do so we need to examine some fundamental facts related to the formation of landscapes. One fact is that the landscape is the level at which the products of biogenic gas formation are admixed into the planetary boundary layer. The landscape is the smallest geographic unit that can pragmatically and routinely be monitored using current remote sensing technology. It is also the level in which control pathways interact and form unique conditions that may favour or inhibit biogenic formation of trace gases.

The landscape-scale concept in the ecological hierarchy discussed here greatly expands the descriptive meaning of the term currently being used by many geographers and, indeed, many ecologists. The meaning here includes the descriptive aspects, but the emphasis is on the processes that occur within landscapes that lead to the exchange of matter, radiation, or information (*sensu* controlling factors for biological and physical processes). I believe that the first two are the vital issues in the context of global change.

A *region* has no consistent biological meaning, since larger geographic units are admixtures of landscapes ranging from urban-industrial to relatively natural. However, regional air masses are formed by the coalescence of air masses derived from landscapes. The boundaries of regions are highly dynamic because of varying wind-directed air mass movement. The chemical constituents of specific regional air masses ultimately mix as they migrate along global circulation pathways. Excellent, but terrifying, hypothetical examples of this type of mixing in the troposphere can be seen in the SCOPE nuclear winter scenarios (Harwell and Hutchinson, 1985; Pittock *et al.*, 1986).

Even though regional ecosystems are heterogeneous and complex arrays of landscapes, they do have many characteristics that distinguish them from one another and from their constituent components. An important question to be asked at the regional level is: Can generalizations about the chemical composition of air masses be made for different regions of the world, or is each region of the world unique because of its own peculiar combination of chemical constituents? The chemical composition of air masses seems to be a function of many interacting sources and processes that typically occur in patch-scale ecosystems. Emissions of gases from one landscape, whether an industrial complex or a grouping of farms, is typically inadequate to explain the chemical nature of the atmosphere above a region. The air mass above a region is a complex mixture of all sources combined with continuous transformations therein. Our scientific hope is that patches and flowpaths within landscapes form definable and repeatable patterns, and thus the landscape can be considered a homogeneous unit. Within this context, we can integrate information from landscape units into broader regional contexts.

Our knowledge of the interaction of patch-, flowpath-, or even landscape-scale ecosystem characteristics with one another to form regional ecosystems concepts is sketchy at best. Likewise, our knowledge of the interactions of regional climates with those of smaller scale is deficient.

ECOLOGICAL PROCESSES

Understanding the behaviour of ecological processes, especially in terrestrial ecosystems, requires a thorough knowledge of the interactions of the hydrologic cycle with dominant biogeochemical cycles, and biotic community composition appropriate to each scale within the ecological hierarchy described above. Because of the importance of these interactions in ecosystem functioning, they are emphasized in the following discussion.

The discussion above emphasized spatial patterning and relationships. Temporal dynamics and scales also are critical, as they relate to hydrologic and other physical processes. Physical attributes of water control the dynamics of important ecological phenomena at many spatial scales. For example, at the time scale of seconds to minutes, erosion and sedimentation are profoundly influenced by the energy associated with moving water. This energy is, in turn, influenced by rainfall intensity (or snowmelt rate), runoff and runon rates, the angle of slope of the flowpath, the size of particles, and surface roughness. If we assume that each patch along the flowpath has its own characteristics with respect to these controlling influences, then each patch displays not only its own response but also contributes to the collective behaviour of the larger flowpath and the even larger landscape during intense erosion and sedimentation events.

Examples of ecological responses appropriately evaluated during time scales of hours to days are the formation, evolution, and transformation of biogenically derived cases, such as NH_3, NO, CH_4, and N_2O. Conditions that lead to the formation of these compounds are frequently linked to wetting, aeration, or freezing cycles of soils.

Whereas the ecological processes that yield numerous trace gases can begin to operate in hours, the combination of processes that lead to plant biomass production generally require days to weeks to become apparent. Water must be present to support root and stem growth and to allow microbial mineralization to supply nutrients required by the plants. Thus, plant production should be envisioned as an integration of many ecological processes that must occur in synchrony to yield significant plant growth (Woodmansee, 1984).

Biotic succession and soil development are complex ecological phenomena that are integrations of myriad biological, chemical, and physical processes that require years to decades, and often centuries, to be expressed. The processes of succession and soil formation may never come to completion because they are often closely correlated to climate and climate changes. A major unanswered question in many patch- to landscape-scale ecosystems is the extent to which the biota and soils are interrelated with climate in a dynamic sense. Superimposed upon this uncertainty is the dramatic influence of humans, both as agents of climatic change and as a major force in disturbance of ecosystems.

Position within the landscape is vital to the expression of ecological responses within patches. These responses are often strongly influenced by water availability, erosive potential, seasonality, and duration. However, the configuration of the landscape itself is often the result of erosion and sedimentation. Water can be the direct geomorphic agent that shapes the landscape; or, in its absence, wind can be the major agent. The time scale for landscape shaping is centuries to millennia.

Thus, landscapes, which are composites of patches linked into flowpaths of varying relationships, form units that can be described and explained and whose behaviour can be predicted with some reasonable degree of confidence. Global scale problems can be addressed if patches and flowpaths can be described and their composite behaviours integrated at the landscape-scale, and if landscapes can be shown to form repeatable patterns. If these requisites can be met, prediction of regional-scale phenomena may be achievable and a goal of global-scale understanding may be within our grasp.

SCIENTISTS: HUMANS OF LIMITED INTELLECTUAL CAPACITY

The discussion above describes examples of the importance of phenomena that occur at small spatial scales, especially the patch, in influencing the behaviour of larger-scale phenomena. The dimension of time is inextricably interwoven with the dimension of space. However, our ability to extend knowledge from small spatial scales and short time scales to evaluation of problems of regional and global importance is not well developed. A major challenge facing humans who perform science in the next decade and beyond is to develop our knowledge base to adequately evaluate large-scale problems.

Limitations on development of knowledge are:

(1) lack of a widely accepted conceptual framework (or frameworks) for integrating information of complex and detailed, but vitally important, small-scale phenomena into relatively simple and tractable models of large-scale system behaviour

(2) lack of attention to basic underlying assumptions

(3) poorly developed communication among scientists of the various disciplines necessary for explaining the complex interactions that lead to large-scale system behaviour

(4) lack of standardized technical and computational facilities so that manipulation of vast amounts of data will be consistent between research groups

(5) lack of long-term records of important transient phenomena

(6) lack of appropriate institutional structures and awareness to ensure that competent and motivated scientists are adequately funded and personally rewarded as they seek to accomplish original research and synthesis of existing knowledge at the regional and global scales.

We know that large-scale system characteristics such as the trace-gas composition of the global atmosphere, the sediment loads of waters at the mouths of major rivers, contamination of groundwater in aquifers, and other phenomena of vital importance to humans, are greatly influenced by patch-, flowpath- and landscape-scale ecosystem behaviour. As scientists, we need a means of organizing our knowledge into comprehensible forms. I contend that a working acceptance of a holistic, systems approach to the evaluation of regional and global problems is necessary for this integration. A hierarchical view of the world is probably necessary. In fact, we probably will need to develop the ability to envision several levels of any hierarchy simultaneously. Rigorous philosophical structures such as 'hierarchy theory' should prove valuable (O'Neill, 1988).

Regardless of the intellectual approach, integration of knowledge across disciplines will be vital because no single individual or discipline holds the key to understanding regional- and global-scale behaviour. Our disciplines have evolved to universally sophisticated states using advanced models of reality, state-of-the-art chemical analyses, and remote-sensing capabilities; and, indeed, most have adopted systems analysis as a powerful research approach. However, most of this evolution has taken place within the disciplines themselves. Our next step must be to cross disciplinary lines and describe the world as being organized around problems rather than disciplinary viewpoints.

The continuing instrumentation, analytical, and computer revolution has enhanced our ability to acquire, manage, manipulate and report data at tremendous rates. Unfortunately, researchers often develop unique data management facilities, thus limiting exchange. As our conceptual models of large-scale ecosystems become more reliable, and as we gain the wisdom to evaluate the environment in the context of human needs, our ability to meaningfully analyze data and formulate information should make great progress, especially if exchange is made easier.

The need for well-conceived and tightly focused long-term research addressing transient phenomena is beginning to be recognized as vital in evaluating ecological systems behaviour (Callahan, 1984). In the past, long-term data gathering or monitoring was viewed with skepticism by many scientists and funding agencies alike. There are many reasons for this skepticism, some reasonable and some not. Regardless of past naiveté, or even abuses and neglect of records, we must establish sites and networks of sites to evaluate long-term trends of ecological phenomena. Without these evaluations we will have no means of placing our ephemeral observations in historical context.

Institutions ranging from large, international and national funding agencies to academic departments and small research units are ill-equipped to organize cross-disciplinary research, even within a single scale of resolution. Many current and future problems of urgent human concern will require not only cross-disciplinary research team building to deal with small-scale concerns, but also research that integrates knowledge from the patch to the global scale. Among the most important problems facing the research community will be the education, legitimization, funding, and rewarding of scientists who devote their energies to research of global change and problem solving.

Educational institutions will need to expand traditional views of disciplinary graduate training to provide the curricula and research opportunities for students to develop broad systems-oriented backgrounds. Additionally, the traditional view of graduate students will need to be expanded to include post graduate retraining in a science that integrates across disciplines.

Academic, public, and private employing institutions need to openly accept participation in team-oriented, integrative research as a legitimate activity at critical junctures during career development, such as promotion, tenure evaluation, and pay rises. Dogmatic adherence to the traditional review criteria of single- or dual-authored papers published in disciplinary based journals, individual achievements in 'grant getting', and over-emphasis on the importance of holding office in scientific organizations all may be counter-productive for the development of this new practice of science.

Funding agencies will need to develop review and funding criteria that can accommodate cross-disciplinary, multi-investigation, and, often, novel research proposals. Clearly, review of such proposals is difficult because few or no individuals will be able to review them competently in their entirety. However, review difficulty will be unacceptable as an excuse for not developing an acceptable review mechanism. Overcoming the obstacles mentioned above will not be easy; however, the current scientific community has some experience with large (e.g. IBP, IGY, WCRP, and others) programme science and imbedded in that experience are some valuable precedents. We are now ready to build on the past and embark on a quest for knowledge about the global environment and how humans can maintain an acceptable existence on earth beyond the next decade.

ACKNOWLEDGEMENTS

Preparation of this chapter was supported by the National Science Foundation Grant #BSR–8114822 for the Central Plains Experimental Range, Long-Term Ecological Research Program at Colorado State University and the Colorado State Experiment Station. I thank Dr D. S. Schimel for his help with the literature survey. Drs D. S. Schimel, C. V. Cole, J. A. Logan and Peter Vitousek made excellent comments about the manuscript.

REFERENCES

Bolin, B., and Cook, R. B., (1983). *The Major Biogeochemical Cycles and Their Interactions*. John Wiley and Sons, New York: 532 pages.

Bowden, W. B. (1986). Gaseous nitrogen emissions from undisturbed terrestrial ecosystems: An assessment of their impacts on local and global nitrogen budgets. *Biogeochemistry*, **2**, 249–280.

Bowden, W. B., and Bormann, F. H., (1986). Transport and loss of nitrous oxide in soil water after forest clearcutting. *Science* **223**, 867–869.

Callahan, J. T. (1984). Long-term ecological research. *Bioscience*, **34**, 363–367.

Correll, D. E. (1983). N and P in soils and runoff in three coastal plain land uses. In Lowrance, R. R., Todd, R. L., Asmussen, L., and Leonard, R. A., (Eds.) *Nutrient Cycling in Agricultural Ecosystems* pp. 207–224. Spec. Publ. No. 23, Univ. Georgia, College of Agric. Exp. Sta., Athens, GA.

Crutzen, P. J. (1983). Atmospheric interactions—homogeneous gas reactions of C, N, and S containing compounds. In Bolin, B., and Cook, R. B. (Eds.), *The Major Biogeochemical Cycles and Their Interactions*. John Wiley and Sons, New York: 532 pages.

Detwiler, R. P. (1986). Land use change and the global carbon cycle: The role of tropical soils. *Biogeochemistry*, **2**, 67–94.

DeWit, C. T. (1970). Dynamic concepts in biology. In Sellik, I. (Ed.) *Prediction and Measurement of Photosynthetic Productivity*, pp. 17–23, Centre for Agric. Publ. Docu. Wageningen, The Netherlands.

Forrester, J. W. (1968). *Principles of Systems*. Wright-Allen Press, Cambridge, MA. 400 pp.

Harwell, M. A., and Hutchinson, T. C. (1985). *Environmental Consequences of Nuclear War: Ecological and Agricultural Effects. SCOPE 28*. John Wiley and Sons, Chichester: 523 pages.

Houghton, R. A., Hobbie, J. E., Melillo, J. M., Moore, B., Peterson, B. J., Shaver, G. R., and Woodwell, G. M. (1983). Changes in the carbon content of terrestrial biota and soils between 1860 and 1980: A net release of CO_2 to the atmosphere. *Ecol. Mongr.*, **53**, 235–262.

Hutchinson, G. L., and Viets, F. G., Jr. (1969). Nitrogen enrichment of surface water by absorption of ammonia volatilized from cattle feedlots. *Science*, **166**, 514–515.

Olsen, R. K., Reiners, W. A., and Lovett, G. M. (1985). Trajectory analysis of forest canopy effects on chemical flux in throughfall. *Biogeochemistry*, **1**, 361–374.

O'Neill, R. V. (1988). Hierarchy theory and global change. (Chapter 3, this volume.)

Pittock, A. B., Akerman, T. P., Crutzen, P. J., MacCracken, M. C., Shapiro, C. S., and Turco, R. P. (1986). *Environmental Consequences of Nuclear War: Physical and Atmospheric Effects. SCOPE 28*. John Wiley and Sons, Chichester: 359 pages.

Schimel, D. S., Coleman, D. C., and Horton, K. A., (1985a). Soil organic matter dynamics in paired rangeland and cropland toposequences in North Dakota. *Geoderma*, **36**, (1985), 201–214.

Schimel, D. S., Stillwell, M. A., and Woodmansee, R. G. (1985b). Biogeochemistry of C, N and P in a soil catena of the shortgrass steppe. *Ecology*, **66**, 276–282.

Schimel, D. S., Kelly, E. F., Yonker, C., and Heil, R. D. (1985c). The effects of erosional processes on nutrient cycling in semiarid landscapes. In *VI International Symposium on Environmental Biogeochemistry*, pp. 571–580. Van Nostrand Reinhold, New York.

Schimel, D. S., Parton, W. J., Adamsen, F. J., Woodmansee, R. G., Senft, R. L. and Stillwell, M. A., (1986). The role of cattle in the volatile loss of nitrogen from a shortgrass steppe. *Biogeochemistry*, **2**, 39–52.

Schindler, D. W., Turner, M. A., and Hesslein, R. H. (1985). Acidification and alkalinization of lakes by experimental addition of nitrogen compounds. *Biogeochemistry*, **1**, 117–134.

Schlesinger, W. H. (1984). Soil organic matter: A source of atmospheric CO_2. In Reichle, D. E. (Ed.). *The Role of Terrestrial Vegetation in the Global Carbon Cycle*. Springer-Verlag, Berlin and New York.

Schumm, S. A. (1977). *The Fluvial System*. John Wiley & Sons, New York: 338 pages.
Smil, V. (1985). Carbon–Nitrogen–Sulfur: *Human Interference in Grand Biospheric Cycles*. Plenum Press, New York: 459 pages.
Strain, B. R. (1985). Physiological and ecological controls on carbon sequestering in ecosystems. *Biogeochemistry*, **1**, 219–232.
Strickland, T. C., and Fitzgerald, J. W., (1984). Formation and mineralization of organic sulfur in forest soils. *Biogeochemistry*, **1**, 79–95.
West, D. C., Shugart, H. H., and Botkin, D. B., (Eds.) (1981). *Forest Succession: Concepts and Application*. Springer-Verlag, Berlin and New York.
Woodmansee, R. G. (1984). Comparative nutrient cycles of natural and agricultural ecosystems: A step toward principles. In Lowrance, R., Stinner, B., and House, G. J., (Eds.) *Agricultural Ecosystems: Unifying Concepts*, pp. 145–156. John Wiley & Sons, New York.
Woodmansee, R. G., and Adamsen, F. J., (1983). Biogeochemical cycles and ecological hierarchies. In Lowrance, R. R., Todd, R. L., Asmussen, L., and Leonard, R. A., (Eds.) *Nutrient Cycling in Agricultural Ecosystems*. Spec. Publ. No. 23, Univ. Georgia, College of Agric. Exp. Sta., Athens, Georgia.

Scales and Global Change
Edited by T. Rosswall, R. G. Woodmansee and P. G. Risser
© 1988 Scientific Committee on Problems of the Environment (SCOPE)
Published by John Wiley & Sons Ltd.

CHAPTER 3

Hierarchy Theory and Global Change[1,2]

ROBERT V. O'NEILL

Environmental Sciences Division,
Oak Ridge National Laboratory,
Oak Ridge, Tennessee 37831

ABSTRACT

Ecological systems are organized across a range of space and time scales. This scaling must be explicitly considered in any attempt to link ecological and atmospheric models. This chapter outlines the elements of Hierarchy Theory which are particularly relevant to this scaling problem. Results are presented as a set of nine criteria. These criteria must be considered before processes, operating on independent scales, can be accurately interfaced to make predictions at the global scale. Hierarchy Theory is proposed as a critical element in developing an understanding of global change.

INTRODUCTION

Environmental systems are complex and multiscaled. Ecologists deal with systems that range from a single organism and its environment all the way up to the biosphere. It is not surprising, therefore, to find that ecologists are beginning to look closely at hierarchy theory to help them deal with this range of systems.

Hierarchy theory was developed by General Systems theorists, notably Koestler (1967, 1969) and Simon (1962, 1969), to deal with complex, multi-scaled systems. Overton (1972, 1974) must be given major credit for introducing ecologists to the theory. The potential utility of the theory has also been pointed out by MacMahon *et al.*, (1978, 1981) and Webster (1979).

---

[1] Research supported by the National Science Foundation's Ecosystem Studies Program under Interagency Agreement SR-8021024 with the United States Department of Energy under Contract No. DE-AC05-840R21400 with Martin Marietta Energy Systems, Inc.
[2] Publication No. 2800, Environmental Sciences Division, ORNL.

Recently, two books have been produced (Allen and Starr, 1982, O'Neill *et al.*, 1986) that provide extensive analysis of the theory and its application to ecological systems. As a result, there is little need to develop a detailed tutorial for purposes of this workshop. Instead, I will focus on applying the theory directly to the problem of global change.

It has long been apparent to ecologists (e.g. Egler, 1942, and Schultz, 1967) that ecological systems are hierarchically structured. As a result, ecologists generally have a positive reaction to hierarchy theory. But while the theory is intuitively appealing as a way of looking at the natural world, ecologists have been frustrated by the lack of application of the concepts. Theoreticians, familiar with the concepts and terminology, seem able to develop the theory but progress has been slow in directing applications to research.

In this paper, therefore, I will avoid expanding the already extensive literature on the basic concepts. Instead I will consider the question at issue: understanding, measuring, and predicting global change. More precisely, I will derive a set of criteria that we can use to address workshop objectives.

Because of my background, I will rely on ecology for examples and illustrations. As I present the criteria, the basic concepts of the theory will emerge and it will become clear that the principles are general and not limited to ecological systems.

FUNDAMENTALS OF THE THEORY

To apply the theory we only need grasp some simple, underlying concepts. The theory asserts that a useful way in which to deal with complex, multiscaled systems is to focus on a single phenomenon and a single time–space scale. By so limiting the problem, it is possible to define it clearly and choose the proper 'system' to emphasize.

The system of interest (Figure 3.1, Level 0) will itself be a component of some higher level (Level + 1). Dynamics of the upper level usually appear as constants or driving variables in a model of Level 0. The behaviour of Level 0 appears to be constrained, bounded, and controlled by the higher level.

The higher level also gives the significance of the phenomena of interest. Let us consider, for example, that the object of study is an individual organism. In studying the organism, we discover reproductive structures and behaviours that are difficult to explain if our attention remains limited to the single organism. It is only by reference to a higher level, the population, that the significance of reproduction can be explained.

The next step in studying the system is to divide Level 0 into components forming the next lower level (Level − 1). The Level − 1 components are then studied to explain the mechanisms operating at Level 0. A mechanistic explanation ordinarily means that a phenomenon is the logical consequence of

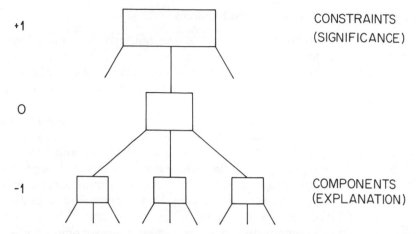

Figure 3.1 Relationships between levels in a hierarchical system. The level of interest is in the centre of the diagram (Level 0). The dynamics of Level 0 will be constrained, bounded, and controlled by the higher level (Level + 1). The higher level also gives the significance of phenomena at Level 0. Level 0 can be divided into components (Level − 1). Interactions among the components provide mechanistic explanations for phenomena at Level 0

the behaviours and interactions of the lower level entities. Another characteristic of the lower level entities is that they appear as state variables in a model of Level 0.

Thus, hierarchy theory dissects a phenomenon out of its complex spatiotemporal context. Our understanding of the phenomenon depends on referencing the next higher and lower scales of resolution. Levels higher than + 1 are too large and slow to be seen at the 0 level and can typically be ignored. Levels lower than − 1 are too small and fast to appear as anything but background noise in observations of Level 0. In this way, the theory focuses attention on a particular subset of behaviour and permits systematic scientific study of very complex systems.

Starting from this introduction, and without drawing out the mathematical and theoretical details, let us see what this theory has to say about studying global change. I will develop the material as a set of criteria that must be satisfied before geophysical and ecological models can be linked.

HIERARCHICAL CRITERIA

The first three criteria deal with scale-dependent biases. These biases are common in applications of hierarchy but they are based on a naive concept of hierarchy that does not stand up to careful analysis.

(1) Searching for the fundamental hierarchy

It is seldom fruitful to go in search of the one and only hierarchy that characterizes the natural world. There is no single, a priori criterion for developing a hierarchy. Instead, a number of different hierarchies may be used to address different problem areas. There is no dogmatic way to determine the superiority of one way of cutting the pie.

To illustrate this point, let us consider the ways one might divide a forest ecosystem (Level 0) into state variables (Level − 1). For a physiological problem, one might divide the system into leaves, boles and roots. This division permits one to emphasize the differences in function among the state variables. For another problem, one might choose a more structural subdivision into canopy, understory, herbaceous, and belowground. This division emphasizes different species strategies within the system. For a problem dealing with species composition, one might choose a taxonomic subdivision that retains the identity of species or groups of species. Any of these subdivisions would be useful for a particular class of problems and no single division stands out as a fundamental. There certainly seems no good reason to force all problems into a single framework.

A similar situation arises in choosing higher levels (Level + 1). Allen *et al.*, (1984) illustrate this point for observations on insect feeding. If one is interested in how this behaviour results in the pollination of plants, the next higher level might be a guild of insects. If one is concerned with the calories of plant being eaten, the next higher level might be a trophic level with the insects aggregated with other consumers.

To summarize, the theory recommends that we set up some hierarchy for studying global change. However, requiring that the hierarchy fit a priori biases is an unnecessary constraint. The bias might force our thinking into a framework that was designed for, and is probably more appropriate for, another problem area.

It may be argued that other natural sciences, such as physics, eventually converged on hierarchical levels that are fundamental in some sense. Nevertheless, physics required several centuries to arrive at this point. During that period, the commonalities between light, magnetism, electricity, and gravity were not evident and each set of phenomena required its own explanatory structure. It suffices to state that at the present stage of development, no absolute, a priori designation of scales is imposed on us by nature in such a way that no other way of looking at the natural world is feasible or useful.

(2) Searching for the fundamental level

It follows from the preceding discussion that it is not fruitful to designate the one and only level to which all other phenomen must be reduced. While most ecologists agree that environmental systems are multi-scaled, some still attempt to reduce all of ecology to a fundamental level such as the population or ecosystem.

In fact, the phenomenon of interest is likely to determine the time and space scales that will be emphasized. There may be as many 'fundamental' scales as there are problems. Once again, it might be argued that progress in physics depended on focusing on the atom as a fundamental level. Similarly, ecology should focus on the organism, the population or the ecosystem. However, we must realize that physics made considerable progress in optics dealing with waves rather than particles and that it eventually dismissed the atom as a fundamental particle.

To illustrate how the scale of observation determines the level of interest, we can consider the study of Sollins *et al.* (1983) on soil organic matter formation at Mt. Shasta. At a time scale of thousands of years, the data show a gradual and continuous accumulation of organic matter. The dynamics causing variation around this curve involve long-term processes such as fire, flood, vulcanism, and earth flow. If one focuses on a shorter time scale, perhaps a hundred years, the long-term accumulation is not evident and the observed dynamics are due to seasonal production and plant succession.

As indicated by the theory, Sollins *et al.* (1983) found it useful to focus on a single level of resolution for each phenomenon and explain dynamics by reference to lower and higher scales of organization. At the same time, adequate explanations of the larger scales were possible without explicit reference to some fundamental level, such as individual organisms or species. In fact, at the largest scales, it was quite irrelevant which plants were involved. It was sufficient to know that a viable plant community existed on the site.

Thus, the theory recommends focusing on a single level but the appropriate level should be based on the problem at hand. It is not useful to force a new problem area into the mold that was appropriate for other problems at other levels.

(3) Translating principles between levels

This brings us to the relationship between levels in a hierarchy. Given that the system is scaled, what can we say about interactions between adjacent levels?

In general, it is not possible to transpose principles developed at one hierarchical level to higher and lower levels. Most concepts and models in ecology have been developed for a single scale. Yet this hidden assumption of scale is often ignored. As a result, some strange errors crop up when a scale-dependent concept is applied at other levels.

Consider the following hypothetical example. Fisheries' managers often take information on rates of reproduction and carrying capacity and calculate an optimum sustained yield. This is the largest continuous harvest that permits the population to persist. On the basis of this calculation, they set limits on the number of fish that can be caught. For the sake of our example, let us assume that 50% of the population can be harvested in a given year. The fishermen, however, operate on a different scale. For example, the fishing fleets may go to Georges Bank in the North Atlantic and harvest almost all of the population over 50% of its total global range.

Now the conditions of the model have been satisfied: 50% of the population has been harvested. However, the recovery rate is no longer proportional to the reproductive rate but to a dispersal rate that can be much slower. The result is that the population begins to drop in numbers. The model was appropriate for a finer scale than was used to harvest the fish. The change to a large spatial scale violated the assumptions of the model because there was a hidden dependence on scale.

Another example of a transposition of scale is provided by our experience with pesticides. If a chemical causes significant mortality in a laboratory population, it seems logical to assume that it will eliminate populations in the field. However, at the larger scale, the additional factor of genetic variability must be considered. Often a small fraction of the total population develops resistance. Instead of decimating the population, as predicted from the fine-scaled model, the population flourishes as the resistant strain expands.

These examples illustrate that it is perilous to transpose principles across hierarchical levels. Constraints imposed at higher levels may dominate, and the overall system behaviour may have little resemblance to the behaviour of isolated components. It is certainly well known among experimentalists that dynamics in the field may be quite different from what is measured in the laboratory.

(4) Be prepared to accept innovative approaches

There is a clear conclusion to be drawn from the first three criteria. We must be prepared to investigate innovative approaches to global measurement and study. A likely conclusion of this workshop and an almost inevitable conclusion of subsequent research is that a new scale of resolution will require new approaches.

It is unlikely that the approaches required to understand global change will reduce to simple repetitions of approaches we have taken in the past. In particular we must avoid the temptation to translate global change into a new justification for old approaches. Obviously, we will not simply dismiss the tremendous resource of manpower and methodology available to study ecological systems at small scales. We still have the obligation to mobilize the existing resources and apply them to the new problem. However, we must also accept that many things will be difficult to transfer to the global scale. We can begin with what we know, but we cannot afford to reduce the global problem to one that fits neatly within existing programmes, approaches, and methodologies.

We must be prepared for new approaches because it is notoriously difficult to measure long-term changes at the scales currently emphasized in ecology. Likens (1983) has concluded that 20 years were needed to describe trends in the water chemistry data at Hubbard Brook. Goldman (1981) indicates that 15 years were required to establish a statistically significant trend in the turbidity of Lake Tahoe. Limiting ourselves to familiar scales of measurement may mean that we cannot even detect a significant trend until it is too late to prevent permanent changes. It is very likely, therefore, that significant measures of global change may require larger scale approaches than we have traditionally dealt with in ecology.

In the short period of this workshop, it is unlikely that we can come up with a definitive list of new approaches. The approaches will emerge as we begin serious research on the global problem. But it may be possible to point to some of the general properties of these new approaches.

It seems likely that the study of global change must take advantage of the technology of remote sensing. Satellite imagery has already proven its capabilities of measuring and predicting agricultural productivity and trends in land use changes. It also seems clear that the new measure of global change will deal with larger landscape units than we have traditionally considered. Thus, the question of landscape indices that can be remotely sensed may be a relevant topic of discussion for the workshop.

A good example of a landscape index has recently been proposed by Krummel *et al.* (1987). They calculated the fractal dimension of land use types on disturbed landscapes. The fractal dimension indicated the complexity of the shape of patches on the landscape. Natural forested areas tended to be small and simple in shape. Disturbed, agricultural areas tended to be large and complex. Based on their results it appears possible to sense remotely a single index that indicates the degree of disturbance to the landscape.

The message of the first four criteria may prove difficult to deal with. The problem of researching global change is unlikely to reduce to aggregating known scales and known ecological principles. This is not to say that we will not take advantage of what we already know, but it does mean that the new problem on the new scale is unlikely to reduce to the familiar and comfortable.

But if it is not possible simply to move concepts and laws across levels, and it is not possible to simply translate this new problem into finer and familiar scales, then how are we to deal with the insights and data that we have available on other scales of resolution? This question brings us to the next two criteria that deal with interlevel relationships.

(5) Effect of a higher level on a lower

One of the most powerful insights of hierarchy theory deals with the concept of constraint. Simply stated, higher levels set constraints or boundary conditions for lower levels. This insight has had widespread influence, for example, on the way engineers control complex systems (Mesarovic *et al.*, 1970).

The dynamics of upper levels are much larger in scale and much slower in time than the level of interest (Level 0). Therefore, over a normal period of observation, the upper levels appear to be constant. Upper level dynamics also appear unaffected by the level of interest, while Level 0 is constrained to follow the dynamics set by the higher level. Thus, upper level dynamics ordinarily appear in a model of Level 0 as forcing functions or driving variables. Thus, the theory has something very definite to say about the effect of the global on the local. Given sufficient difference in scale, it is possible to predict how the higher level will affect the lower.

An excellent example of how higher level constraints can determine system behaviour is provided by the aquatic production relationship worked out by Vollenweider (1975, 1976) and Schindler (1977, 1978). In nutrient-limited, fresh water systems, annual production is closely related to phosphorus loading. It is possible to predict productivity without information about the species of phytoplankton involved in the process. Dynamics can be determined simply by knowing the higher level phosphorus constraints. Detailed data on lower levels are not required.

(6) Predicting the higher level from the lower

While hierarchy theory predicts how higher levels affect lower ones, it is more difficult to move in the opposite direction. Some higher level properties are the sum or integral of lower level dynamics; for example, biomass production may be summed over forest stands. However, this is not always true and serious problems can result.

To illustrate the problem we can consider a recent example worked out by Cohen (1985) for the concept of fitness. It is possible to talk about the fitness of a genotype. If genotype A is more fit than genotype B, individuals with genotype A will produce more offspring. However, given environmental and genetic variability, it is impossible to aggregate this information upward and talk about the fitness of a species. A species with almost all genotype A may or may not be more fit than a species with almost all genotype B.

The dilemma is best illustrated by an example. Cohen (1985) considers the case in which there are small areas in the environment within which the relationship is reversed and genotype B is more fit than genotype A. Even though these areas are small, the reproductive rate of B within these areas may more than compensate for the disadvantage that B experiences elsewhere. As a result, species B does much better than expected. The situation is similar to the pesticide example discussed earlier. The fine-scaled information that almost every mosquito dies when exposed to DDT could not be moved up scale to predict the response of the species.

Stated as a general problem, the influence of lower levels on the higher is known as the 'aggregation problem.' The problem is of real importance to this workshop because the most extensive ecological information is at small scales; for example, physiological responses of organisms and processes operate at the ecosystem scale. The problem is how to make use of the available data and expertise when we deal with larger scales. How can the smaller scale phenomena be aggregated to understand global levels?

The problem is a complex one that is best approached in several stages. Under many circumstances, hierarchy theory has little difficulty with this question. Level -1, the next lower level, forms the components of a model of Level 0. The component dynamics and interactions are the dynamics of Level 0. Thus, there is no problem seeing how the dynamics of Level -1 relate to Level 0.

Under ordinary circumstances, the rapid dynamics of much lower levels can be ignored. They usually appear as averages or assumptions in the model. In fact, it is a good objective to set up the problem so that lower levels do not have to be considered.

When we are dealing with the normal, unperturbed dynamics of the system, hierarchy theory provides a clear explanation of how lower levels affect higher levels. However, there are exceptions and it is sometimes very insightful to seek explanations at much finer scales. Consider, for example, the Monod function for nutrient limitations on production and decomposition processes. Although the function is useful at macro-scales, it is derived from enzyme kinetics at biochemical scales. Thus, although the theory states that it is not necessary to look at much finer scales, it may still be advisable under some circumstances.

But it is the possibility of global catastrophe that introduces the real problem for the workshop. The theory states that finer scales can be ignored as long as the system is behaving normally and stably. But when the system is disrupted, it is the rapid dynamics of the lower levels that break up the constraint system and move the system into a new configuration. In global problems, we are most interested in these catastrophic changes and the problem of aggregation is a real one.

Thus, the problem of aggregation becomes important for three reasons:

(1)   we wish to take advantage of available fine-scaled information

(2)   it is sometimes insightful to seek explanations at much finer scales

(3) we need to understand lower level behaviour to predict when unstable responses will occur. As a result, the aggregation problem may be a serious impediment to understanding and predicting global change.

(7) Interactive state variables

To this point in the discussion, I have focused on the criteria needed to set up the levels to be studied within a single discipline. But the objectives of the workshop require that we interface hierarchical levels in several disciplines. So now we must turn our attention to additional criteria for interfacing different hierarchies.

The seventh criterion is both simple and powerful. You have found a useful scale for interfacing different disciplines (e.g. atmospheric and ecological processes), if and only if the state variables of a model from one discipline appear as state variables in the model of the other discipline.

The simplest way to explain this criterion is to consider an example involving the interfacing of atmospheric and ecological processes. At the scale of a single, isolated tree, moisture content of the air appears as a higher level constraint. The scale of water dynamics in the air is much larger than a single tree. Precipitation constrains the dynamics of the tree which is unable to exert any control back on this driving variable.

Let us now increase the scale and consider a larger forest area. Eventually the water content of the air is significantly affected by moisture released from the vegetation through transpiration. Given sufficient lifting of the air mass due to topographic relief, local precipation results. At this scale it is possible to interface the two hierarchies. The critical element is that precipitation is a function of transpiration and vegetation dynamics is a function of precipitation. At this scale, the vegetation appears as a state variable of the precipitation model and precipitation appears as a state variable of the vegetation model. The loose criterion of 'equivalent' scales can, through the hierarchy theory, be given much more precise definition in terms of interacting state variables.

Of course, the problem of interfacing atmospheric and ecological hierarchies is still not solved. There are any number of levels at which the criterion is satisfied. At a microscopic scale, a single leaf affects, and is affected by, the immediate envelope of air. At continental scales, vegetation cover determines whether desertification processes will take over. So exactly which of these scales is most appropriate for studying global change remains an open question for the workshop. Once we choose a problem area, the theory provides a means for determining a specific scale at which atmospheric and ecological processes can be interfaced.

(8) Seek coherent levels

The discussion under the previous criterion might lead one to believe that any level of resolution can be chosen arbitrarily. This is not the case, and an additional criterion is needed based on the concept of 'coherent levels' (Sugihara, personal communication).

Once again, it is easiest to explain this concept by way of example. It is difficult to predict the behaviour of a single animal. The information required to predict whether it will walk to the right or to the left makes the problem almost unsolvable. However, the problem becomes simpler as we move up scale. It may be possible to predict the movement of a herd of individuals, for example, as they move along a migration route. Thus, there are scales at which one's predictive capability is improved over slightly larger or slightly smaller scales. The scale at which the predictive power is maximized is the coherent level. A coherent level is one that 'makes sense' as an isolated object of study and it is quite likely to correspond to a traditional level of study within a discipline.

While it would be possible to interface disciplines at arbitrary levels of resolution, arbitrary scales do not take advantage of the innate organization in hierarchial systems. It is only when we focus on coherent levels that we are able to take advantage of the information and insights that have developed in each discipline about their own systems.

Together, Criteria 7 and 8 form a guideline for interfacing disciplines. The scales of interest in both disciplines must be coherent levels as well as appearing as state variables in the model from the other hierarchy.

To illustrate the power of Criteria 7 and 8, let us consider a difficult situation which arises in the context of global change. Ecological systems are scaled in a particular way in space and time. Small systems change rapidly and large systems change slowly. But atmospheric phenomena may be arrayed in quite a different manner and large spatial areas may change rapidly. Thus, sub-continental areas show similar seasonal changes in temperature. How can one interface large space, small time changes in temperature with either small space, small time or large space, large time ecological systems?

The answer to the dilemma can be deduced from Criteria 7 and 8. The state variables describing large ecological systems do not change on a seasonal basis and there are no common state variables at the large spatial scale. We must go to the smaller common scale, the rapid time response, to find interactive state variables. However, ecological systems are not organized in large space, small time scales. Therefore we must apply Criterion 8 and find the coherent ecological system which reacts at the common time scale. That is, the relevant point of intersection is the coherent ecological system that responds seasonally, e.g. a forest stand or a grassland patch. Ecological models at the stand/ patch scale can handle changes in gaseous exchange in response to seasonal changes and we have found interactive state variables.

We have found the relevant model. How do we extrapolate the results of the stand model to the large spatial scale of the atmospheric model? The problem reduces to finding the expected value of the small scaled process distributed across space. The expected value is, by definition, the integral of the stand model across the frequency distributions representing the spatial variability of the relevant independent variables. In our example, the independent variables would include seasonal temperature and the biotic parameters of the model. There may be additional sources of variability but these seem to be the most important. Given that this information is available and given that there are no spatial correlations in the response of stands, Monte Carlo simulation of the small scaled model yields an unbiased estimate of the expected value of the process across the region.

Thus, Criteria 7 and 8 provide the necessary background for interfacing phenomena on very different spatiotemporal scales. The positive part of the result is that it is possible to devise a specific approach to answering the question. The negative part of the result is that the information requirement (frequency distributions characterizing the spatial variability of seasonal temperatures and biotic parameters) is rigorous and such data may not be available.

(9) Critical points in parameter space

Once we have picked a scale, we must decide what to measure. Large scale changes occur very slowly and are difficult to detect, as was pointed out earlier. We are once again caught in a dilemma. The need to detect change on the global scale is apparent. Yet long-term changes are very difficult to detect and it may take 20 years before the first significant results can be stated. Monitoring at this scale may be too little too late. Without a way of interpreting changes, we may be able to detect the change only after the catastrophe has begun. So how can hierarchy theory help us determine what it is we want to measure?

A potential solution can be offered if we are not overly concerned with the normal functioning of the system. It is this normal, stable behaviour of the system which is most difficult to monitor. In fact, we are most concerned with the unusual circumstances in which the system will respond unstably to a new perturbation. That is, we are interested in monitoring for the critical points in the behaviour space of the biosphere, the points of bifurcation.

The concept of critical points in parameter space should be familiar to ecologists. For example, we are familiar with the Holdridge classification system. In this system, temperature and moisture parameters are used to array the world's vegetation types. When climate parameters pass certain critical points, the state of the system can change drastically. An example of a critical point would be the annual temperature below which permafrost becomes a feature of the environment. Below this value, the vegetation can change radically. Other examples would be moisture changes that alter forest into grassland or grassland into desert. The ecologist is also familiar with the fact that, over geologic time, it is quite possible for the climate to move across these critical points and change the vegetation.

Movement across these critical points need not be caused by climate or catastrophic perturbations. The radical changes seen in the fossil record can occur through the normal dynamics of a complex nonlinear system.

As an example of how this can happen, let us consider the following hypothetical scenario. At some point in geologic history, diatoms evolved a new enzyme that permitted a silicon shell. The change could have been as simple as a single point locus change. Because none of the predators had silicacious jaws, the diatoms were freed from a constraint and expanded rapidly both in numbers and in taxonomic varieties. Eventually, the diatoms became the dominant organisms on the continental shelves. As a result of breaking a constraint, the well-behaved system became unstable and the diatoms expanded until they hit a new constraint, probably lowered concentrations of silica in sea water. In this scenario, the radical change in the system did not require any radical change in the environment.

Of course, similar hypothetical scenarios could be developed for many of the major changes in the fossil record. Such changes are normal in nonlinear systems. Thus, global monitoring should be designed to indicate whether or not the system is approaching such a point.

Examples of rapid changes illustrate that the aggregation problem is important. The critical change that altered the taxonomic composition of the continental shelves of the world could have been a single mutation, explainable at the subcellular level. The changes we are most interested in monitoring at the global level may be caused by alterations at much lower levels in the hierarchy. This fact significantly complicates what we are trying to do. But the fact remains that the scenario is feasible.

Is the effort to monitor and predict global change a quixotic task, a search for the Holy Grail? Fortunately, we need not draw this conclusion. It may still be possible to monitor for the approach to critical changes such as desertification, glaciation, or the initiation of an erosion or peneplanation cycle. It may be possible to detect the approach of such changes, even though we do not have at our disposal all of the lower level information required to provide a mechanistic explanation for the change.

To see how this prediction is possible, we need to consider some technical details of the theory. The points of critical change are called bifurcation points in the underlying mathematical theory. There is a necessary and sufficient condition for determining when the radical change will occur. The change occurs when the rapid components cease to be stable, that is, the lower level components do not return to normal behaviour following a minor perturbation.

What causes the system to behave normally is that the rapid portions of the system are constrained by higher levels. If the system is perturbed, the rapid components simply return to the slowly changing trajectory. The rate of recovery can be taken as an indicator of the relative stability of the system. It seems reasonable to assume that as the system approaches a bifurcation point, the recovery becomes slower (i.e. the system becomes less stable).

Thus, it should be possible to monitor globally for the approach to a catastrophic change by monitoring the recovery rate of lower levels in the hierarchy. If the response times are increased the system is being moved toward a point of radical self-amplifying change. Thus, even though the actual point of change could be precipitated by fine-scaled changes, the proximity to any point of radical change should be indicated by a change in recovery times.

Perhaps an example of such approaching change would help clarify the concept. Unfortunately, such examples are hard to come by and once again we are limited to a hypothetical experiment. Suppose we were using remote sensing to monitor for land use change in the tropics. In addition to monitoring the square miles that are converted from forest to agriculture, we also measure the shorter term successional recovery rate of a number of sites. As the cleared areas increase, we might find erosion cycles beginning or changes in air moisture and precipitation. In either case, we might be able to measure decreases in the succession rate. At this point, we know not only that a gradual change in a large-scale phenomenon is occurring but also that the change is occurring in the direction of an instability. In this way, a shorter term measurement of stability might permit interpretation of the large-scale change, but would be considerably easier to measure.

The difficulty of aggregating lower level behaviours to higher levels remains and there is a real problem with maximizing the fine-scale understanding we possess. Nevertheless, the theory provides us with an intriguing new approach. In so far as we want to prevent catastrophic changes, measurement of natural rates of recovery could be a key ingredient in any study of global change.

CONCLUSION

To be most useful, the conclusions we draw from this analysis of hierarchy theory must be directly related to the objectives of the workshop. Therefore, I will summarize the criteria by emphasizing the major points and their application to the objectives.

(1) What are the scales in each discipline?

Hierarchy theory cannot tell us which scales of resolution would be most useful to study at the global scale. It can, however, guide us away from a naive concept of hierarchy which would force this new wine into old bottles. The theory recommends that we look at global change as a new and challenging problem area and not try to force it into hierarchies (Criterion 1) or levels (Criterion 2) that are more appropriate for other problems. The theory also warns us not to feel that once the appropriate hierarchy is selected it will be a simple matter of transposing principles developed at finer scales up to the new scale of interest (Criterion 3).

(2) How to integrate scales across disciplines

The theory provides concrete advice for choosing scales that will permit fruitful interfacing between disciplines. We should seek coherent levels (Criterion 8) at which state variables from one discipline appear as state variables in a model for the other discipline (Criterion 7). These criteria are straightforward and it would be well to adhere closely to them during the workshop.

(3) How does the global affect the local?

The theory is particularly helpful in describing how larger scales affect the smaller. Under normal operating conditions, the higher levels appear as constraints or boundary conditions (Criterion 5). When the normal operation of the system is disturbed, many of these constraints disappear as the system changes to a new configuration.

(4) How local determines global

Under normal circumstances, the dynamics of a system are determined by the dynamics of Level -1, the next lower level. Smaller scales operate much too rapidly to be of interest (Criterion 6) and can be ignored. As a result it is ordinarily of little interest to aggregate very fine scales in an attempt to describe large-scale phenomena.

The situation changes dramatically when the system becomes unstable. Now, the fine-scale dynamics are unconstrained and tend to change the system drastically. However, there is no theory available to predict exactly which fine-scaled processes will be most important. The theory is of little help then in deciding what fine-scaled research would be most useful to explain global change. The only guidance comes from the scientific method: once a potential disruption or change is identified, we can formulate and test hypotheses about the significant fine-scale mechanisms.

(5) Research needs

The major challenge of the workshop is to begin to develop a sensible research plan to attack the problems of global change. Hierarchy theory makes it clear that this new research must develop innovative approaches (Criterion 4) even while we attempt to synthesize and apply what we know about ecological systems at smaller scales. It is likely that these new approaches will take advantage of remote sensing and large-scale landscape measures.

It is also clear that the most critical problems of global change will deal with catastrophes (Criterion 9). The theory is particularly valuable here since it suggests measuring recovery rates as an indicator of whether or not gradual changes are leading toward bifurcation points.

As one could anticipate, hierarchy theory is useful in leading us away from naive errors and suggesting new directions to take. However, it is beyond the purview of this or any theory to constrain our deliberations and the theory does not dictate which scales of resolution or which research will be most fruitful. Hopefully the theory does provide a framework which can guide workshop deliberations in the most fruitful directions.

REFERENCES

Allen, T. F. H., and Starr, T. B., (1982). *Hierarchy: Perspective for Ecological Complexity*. University of Chicago Press, Chicago.

Allen, T. F. H., O'Neill, R. V., and Hoekstra, T. W. (1984). *Interlevel Relations in Ecological Research and Management: Some Working Principles from Hierarchy Theory*. General Technical Report RM–110, United States Forest Service, Rocky Mountain Forest and Range Experiment Station, Fort Collins, Colorado.

Cohen, J. E. (1985). Can fitness be aggregated. *Am. Nat.*, **125**, 716–29.

Egler, F. E. (1942). Vegetation as an Object of Study. *Philo. Sci.*, **9**, 245–60.

Goldman, C. R. (1981). Lake Tahoe: Two Decades of Change in a Nitrogen Deficient Oligotrophic Lake. *Int. Ver. Theor. Angew. Limnol. Verh.* **21**, 45–70.

Koestler, A. (1967). *The Ghost in the Machine*. Macmillan, New York.

Koestler, A. (1969). Beyond Atomism and Holism—The Concept of the Holon. In Koestler, A., and Smythies, J. R. (Eds.) *Beyond Reductionism*, pp. 192–232 Hutchinson, London.

Krummel, J. R., Gardner, R. H., Sugihara, G., O'Neill, R. V., and Coleman, P. R. (1987). Landscape patterns in a disturbed environment. Oikos, 48, 321–324. Gardner, R. H., Sugihara, G., O'Neill, R. V., and Coleman, P. R. (1987).

Likens, G. E. (1983). A priority for ecological research. *Bull. Ecol. Soc. Am.*, **64**, 234–43.

MacMahon, J. A., Phillips, D. L., Robinson, J. V., and Schimpf, D. J. (1978). 'Levels of biological organization: an organism-centered approach, *Biosci.*, **28**, 700–4.

MacMahon, J. A., Schimpf, D. J., Andressen, D. C., Smith, K. G., and Bayn, R. L. (1981). 'An organism-centered approach to some community and ecosystem concepts. *J. Theor. Biol.*, **88**, 287–307.

Mesarovic, M. D., Macko, D., and Takahara, Y. (1970). *Theory of Hierarchical Multilevel Systems*. Academic Press, New York.

O'Neill, R. V., De Angelis, D. L., Waide, J. B., and Allen, T. F. H., (1986). *A Hierarchical Concept of Ecosystems*, Princeton University Press, Princeton, New Jersey.

Overton, W. S. (1972). Toward a General Model Structure for Forest Ecosystems. In Franklin, J. F. (Ed.) *Proceedings of the Symposium on Research on Coniferous Forest Ecosystems*. Northwest Forest Range Station, Portland, Oregon.

Overton, W. S. (1974). Decomposability: A unifying Concept? In Levin, S. A. (Ed.), *Ecosystem Analysis and Prediction*, pp. 297–98. Society for industrial and applied mathematics, Philadelphia.

Schindler, D. W. (1977). Evolution of Phosphorus Limitation in Lakes. *Sci.*, **195**, 260–62.

Schindler, D. W. (1978). Factors Regulating Phytoplankton Production in the World's Freshwaters. *Limnol. Oceanog.*, **23**, 478–86.

Schultz, A. M. (1967). The Ecosystem as a Conceptual Tool in the Management of Natural Resources. In Cieriacy-Wantrup, S. V., (Ed.) *Natural Resources: Quality and Quantity*, pp. 139–61. University of California Press, Berkeley.

Simon, H. A. (1962). The Architecture of Complexity. *Proc. Am. Philos. Soc.*, **106**, 467 82.

Simon, H. A. (1969). *The Sciences of the Artificial*. MIT Press, Cambridge.

Sollins, P., Spycher, G., and Topik, C. (1983). Processes of Soil Organic-Matter Accretion at a Mudflow Chronosequence, Mt. Shasta, California. *Ecology*, **64**, 1273–82.

Vollenweider, R. A. (1975). Input–Output Models with Special Reference to the Phosphorus Loading Concept in Limnology. *Schwiz. Z. Hydrol.*, **37**, 53–84.

Vollenweider, R. A. (1976). Advances in Defining Critical Loading Levels for Phosphorus in Lake Eutrophication. *Mem. Ist. Ital. Idrobiol Dott Marco de Marchi Pallanza Italy*, **33**, 53–83.

Webster, J. R. (1979). Hierarchical Organization of Ecosystems. In Halfon, E. (Ed.), *Theoretical Systems Ecology*. pp. 119–31. Academic, New York.

Scales and Global Change
Edited by T. Rosswall, R. G. Woodmansee and P. G. Risser
© 1988 Scientific Committee on Problems of the Environment (SCOPE)
Published by John Wiley & Sons Ltd.

CHAPTER 4

Statistical and Mathematical Approaches to Issues of Scales in Ecology

JOHN N.R. JEFFERS

Institute of Terrestrial Ecology,
Merlewood Research Station,
Grange-over-Sands,
Cumbria LA11 6JV *England*

ABSTRACT

Our current ability to evaluate interactions of biological, chemical, and physical processes is limited by our difficulty of coping meaningfully with small-scale spatial and temporal heterogeneity in terrestrial, freshwater, marine, and atmospheric environments. This paper reviews the problems of measuring interactions in the presence of spatial and temporal heterogeneity from the viewpoint of mathematical statistics, and suggests that the difficulty stems as much from a general failure to plan environmental research adequately as from any supposed properties of the biosphere or geosphere.

Particular attention is focused on the problems of sampling in environmental research and on the measurement of interaction by the use of factorial experiments. It is stressed that the vast improvements which have taken place in data processing and in the modelling of ecological processes cannot be used as a substitute for the detailed and careful planning necessary for the valid estimation of population parameters and interactions. However, there are a number of exciting possibilities in the development of mathematical and statistical techniques which, combined with strict methods of experiment and survey design, will enable us to:

(a) integrate scales, conceptually and practically, between and among disciplines

(b) aggregate local scales meaningfully into larger spatial scales

(c) evaluate how global scale processes influence local processes.

INTRODUCTION

The general objective of this Workshop is to identify and describe the research needed to solve the scaling problems associated with quantifying and interpreting interactions within the biosphere and geosphere. This objective is part of a wider study of spatial and temporal variability of biosphere and geosphere processes, and it is suggested that our current ability to evaluate interactions of biological, chemical, and physical processes is limited by our difficulty of coping meaningfully with small-scale spatial and temporal heterogeneity in terrestrial, freshwater, marine, and atmospheric environments. It is this difficulty that is identified as 'the scaling problem'.

This chapter reviews the problems of measuring interactions in the presence of spatial and temporal heterogeneity from the viewpoint of mathematical statistics, though without embarking on any exercises of formal mathematics. It suggests, using examples drawn from current ecological research programmes, that the difficulty addressed by this Workshop stems as much from a general failure to plan environmental research adequately as from any supposed properties of the biosphere or geosphere. Repeated attempts to aggregate uncoordinated research programmes in the past have failed for reasons which are entirely explained by the logic of the scientific method, and 'the scaling problem' is a reflection of our scientific inadequacy in dealing with the complexity of environmental systems.

STATISTICAL CONCEPT OF HETEROGENEITY

If the biological, physical, and chemical processes of the biosphere all behaved in a perfectly orderly way, we would detect variation in space and time, but we would not regard such variation as 'heterogeneity'. The deterministic models of the Newtonian calculus could be used, with more or less difficulty, to model the changes that take place in time and space, or both together. The Lotka−Volterra equations might be used to describe, for example, the relationships between the numbers of a predatory animal and the numbers of its prey (Maynard Smith, 1974). While we might occasionally be surprised at the magnitude of the changes arising from relatively small induced perturbations in the system—because of the effects of feedback and the non-linearity of the relationships—the same changes would occur each time that the same starting conditions were input to the system. The presence of heterogeneity indicates either that there are parts of the process which are not adequately accounted for in our deterministic relationships or that the system itself is non-deterministic. Indeed, any system which includes living organisms is certain to show some degree of heterogeneity because of the genetic variation which occurs through sexual reproduction. Models of the reaction of some organism to persistent applications of a chemical which do not allow for the possible selection of resistant strains are mere caricatures.

The statistician regards the measurement of variation as being of greater importance than the measurement of central tendency, i.e. means or algebraic relationships, and most of the now extensive theory of mathematical statistics, deal with the measurement of variation and its characterization. Where many scientists concentrate on the average values and regard any apparent variation from those average values as 'error', the statistician accepts the variation as an essential part of the system being studied, and often uses that variation constructively to improve the modelling of the system. The variation can frequently be described by statistical distributions with well-known properties and parameters that are capable of independent estimation. Even where the variation does not follow exactly one of the better known statistical distributions, such a distribution may often be used as an approximation, sometimes by relatively simple transformations of the scale or dimensions of the original measurements. The dispersion of radionuclides in grassland ecosystems, for example, while showing extreme heterogeneity, can be regarded as approximately normal if the measure of radioactivity is first transformed to its reciprocal.

The most important of the statistical distributions is undoubtedly the normal distribution, not because it is 'normal' in the accepted sense of that word, but because its mathematical properties have made it the basis of a very wide range of statistical theory. Many of the other distributions can be shown to be a special case of the normal distribution, and the Central Limit Theory states the very remarkable fact that the distribution of the means of samples from any parent distribution approaches the normal distribution very closely as the size of the sample increases. While it is certainly true that the variability displayed by many environmental systems is markedly non-normal in the statistical sense, this distribution remains at the centre of the development of statistical theory. A relatively recent trend towards the use of the distribution-free methods in statistics has, if anything, served to emphasize the importance of the main core of statistical development (McNeil, 1977).

Nevertheless, many of the other statistical distributions have an important role to play in our description of environmental systems. The randomness of the Poisson distribution is often used as a first test to establish the presence of a spatial or temporal pattern, and there are several alternative distributions to describe 'clumping' in space or time, while retaining a random element in the heterogeneity. Statistical distributions may also be used to detect discontinuity in variability, and are especially useful when that variability is multivariate, so that the variation is represented by a cloud of points in multi-dimensional space. Where the discontinuities exhibit hysteresis, so that a system has an apparently delayed response to a changing stimulus, and the response follows one path when the stimulus increases and another when it decreases, the topological models of catastrophy theory may serve to describe the variation and its apparent heterogeneity (Poston and Stewart, 1978).

In short, variation and heterogeneity are neither unexpected nor unwelcome to the statistician (Bartlett, 1975). Much of his craft has been developed in order to describe, measure and characterize heterogeneity. However, if we regard variation, and hence heterogeneity, as a property of environmental systems, it follows that what we measure can only be used to characterize that variation if our samples are fairly drawn from some defined population. It is the role of fair samples that forms the content of the next section of this chapter.

THE ROLE OF FAIR SAMPLES

The total set of individuals about which inferences are to be made is said by the statistician to constitute a 'population'. Those individuals may be organisms, communities, societies, or whole ecosystems, or indeed any measure or characteristic of these individuals. While such populations will usually be finite in both time and space, they will often be so large as to make it impossible for every member of the population to be investigated, measured, or counted. As scientists, we are usually forced to work with samples drawn from the population, and we will need to make the assumption that those samples are representative of the complete population. Only then can values calculated from the samples be regarded as estimates of the same values of the population. Characteristic values of the population are defined as parameters, in contrast to the corresponding values of samples which may, under certain conditions, be regarded as estimates of those parameters, possibly as constants or coefficients in model equations.

Much of the problem of 'scaling' arises from this distinction. As an example, we may be concerned to make inferences about tropical forests in the world. Ideally, we would like to make valid inferences about all tropical forest, but such forests exhibit marked heterogeneity of almost all of their properties in space or time, or both. Our access to tropical forests may be limited by political boundaries, by difficulties of travel and working in remote areas, by climate and topography, and by the sheer magnitude of the task of measuring, say, biological productivity of a tropical forest ecosystem. Inevitably, we have to make our measurements on sample areas of tropical forest, and those samples have to be selected so that they are representative of the population about which we want to make inferences. Also, because the area we can measure will be limited, our measurement of the heterogeneity of biological production will be related to the scale of the sample areas. In what sense, then, can we regard the samples as justifying inferences about our defined population?

To take another example, there is currently much interest in the effect of acidic inputs to the environment on a range of ecological systems, including agricultural crops, forests, and freshwater. The acidic inputs are derived from point sources of sulphur, oxides of nitrogen, and ozone, widely scattered. In the atmosphere, the original emissions are subjected to complex chemical changes, as well as dispersion, mixing, and concentration by the atmospheric pressure systems. Our measurement of the effects of the resultant pollutant mix on actual ecosystems is necessarily based on sample sites, and on records taken over sample times, and these measurements display considerable heterogeneity. Here, the definition of the population about which we are seeking to make inferences needs particular care. It would be unrealistic to assume that the population was the totality of atmospheric pollution. More likely would be the assumption that our sample records of pollution episodes and their effects were representative of a particular type of forest or freshwater system over a defined period of time. Our ability to make inferences about 'episodes' would also depend on the frequency with which we had recorded the presence of pollutants during the period of observation. Changing our scale of definition requires a change of experimental procedure.

The statistician's requirement for a set of samples to be regarded as 'representative' of some defined population is uncompromising. The samples must be taken by a method which is objective and unbiased. Selection by some form of subjective choice, guided by whatever personal consideration of the representativeness of the samples, will not do—an unfortunate fact which eliminates much of the use of 'case-studies' by economists, sociologists, and ecologists from the statistician's definition of valid inference. Two methods of objective sampling have been traditionally used by statisticians. Systematic sampling has the appeal of simplicity, and, having selected the first individual or location at random, the remaining sample units are taken at a fixed interval. However, unless systematic sampling is repeated, severe problems occur in estimating the precision of estimates derived from the samples, and in characterizing the heterogeneity of the population. For these reasons, statisticians have given most emphasis to random sampling. The simple expedient of ensuring a genuinely random choice at an appropriate part of the sampling procedure guarantees the lack of bias, and also provides a methodology for estimating the heterogeneity of the sampled population, and the precision with which population parameters are estimated from the sample (Green, 1979).

Considerable refinements of methods of random sampling have been developed during the last 60 years—the major period of statistical development. The precision of sample estimates can be greatly increased by stratification of the population, ideally based on knowledge of major differences in heterogeneity of different parts of the population. Alternatively, stratification can be used as a kind of insurance, spreading the coverage of the sample units so that estimates can be made with a defined minimum precision for each part of the population. Multi-stage and cluster sampling provide further adaptations to particular requirements, while retaining the essential properties of randomness. There can be little excuse today for any investigation to be conducted with an inadequate sampling methodology, so diverse is the range of methods available to ensure that valid estimates can be made of population parameters. However, it still needs to be stressed that the design of valid sampling has to be done before any data are collected. It cannot be superimposed on a data collection that has already taken place. Even today, scarce resources are often wasted by inadequate design of the underlying sampling technique, leaving the research worker with a frequently impossible task of trying to unravel results which can never be reconciled with statistical theory. If anything, the use of computers, and the widespread belief that data processing techniques can be found to solve any problem, have only helped to encourage scientists to slide towards post-hoc inadequacies (Jeffers, 1979).

MEASUREMENT OF INTERACTION

The statistical definition of interaction is that it is a measure of the extent to which the effect of one factor varies with changes in the strength, grade, or level of other factors in an experiment. Thus, the response of an organism to the available supplies of some nutrient may be modified by the concentrations of other nutrients or pollutants to which it is exposed. Clearly, interactions may be either very simple, as when only two factors interact linearly, or extremely complex, as when many factors interact, and some of these interactions are non-linear. In both physical and biological systems, interactions may be greatly complicated by the existence of 'feedback', i.e. the carrying back of some effect to modify the factors causing that same effect, either positively or negatively.

As long ago as the mid-20s, R.A. Fisher (1925, 1935) pointed out that interactions could only be investigated experimentally if all of the factors were included in the same experiment. Together with his co-workers at the Rothamsted Experimental Station, he developed the concept of factorial experiments through which some or all of the combinations of factors could be used to determine the strength and character of the interactions. Combined with the analysis of variance, of which regression analysis is a special case, multifactorial experimentation provides an extremely powerful tool for the measurement and characterization of interaction. Like valid sampling, however, the use of factorial designs requires preliminary planning and a rigorous approach to experimentation. Estimation of the higher order interactions requires replication of treatments, together with randomization of treatment combinations in order to ensure valid inferences to the target population.

Some 60 years later, it is sad to record that a significant proportion of the scientific community still does not understand what Fisher was saying in the 1920s. In every country in the world, developed or developing, major experiments are planned with neither replication nor randomization. The excuse is often that it would be too costly, or too difficult, to replicate treatments, and 'not commonsense' to randomize the application of treatments when it is obvious that one treatment would be more suitable for a particular experimental plot. The more costly the research, however, the more important it is that the experimental design should be efficient, and that the results that are obtained are capable of valid interpretation, including the correct identification of interactions (Jeffers, 1978).

Improvements in computing and automatic data collection are often used as an excuse for neglecting the provision of adequate design of sample surveys and experiments. Somehow, it seems to be felt that collecting and processing large numbers of observations can substitute for the features of design which are essential for the valid estimation of the effects of factors and their interactions. 'Within-plot' variation is commonly regarded as a perfectly adequate substitute for a valid estimate of 'experimental error'. Computational techniques, often poorly understood, are used to derive estimates of interactions from unplanned samples which do not cover the full range of admissible values of factors and their combinations. The effects of outliers to the main bulk of observations are frequently neither checked nor observed. 'Modelling' has sometimes been used as a substitute for careful thought about the logic of the experiment or survey which is being conducted (Jeffers, 1980).

MATHEMATICAL MODELS OF VARIATION AT DIFFERENT SCALES

However, I would not wish this paper to be regarded as being totally negative, and there are certainly some exciting possibilities for the understanding of biosphere or geosphere processes which exhibit variation at different scales. Indeed, it is these possibilities which underline the importance of the correct use of procedures of sampling and experimentation if the parameters of the heterogeneity exhibited by mathematical models are to be estimated efficiently. A review of all the possible methods would make this paper too long, however, and it will be sufficient to mention a few which have especially interesting characteristics.

(a) Harmonic analysis

Perhaps the best known of such methods is the harmonic analysis of time series data. Periodic variations of short duration can be superimposed on much longer term variations, with widely different frequencies and amplitudes being combined in the model. In recent years, some confusion has resulted from the competing claims and counterclaims of different schools of time series analysis, but there is now a reasonably unambiguous theoretical base for the analysis of consistently recorded data (Cliff and Ord, 1981). The fitting of harmonics to incomplete data remains both difficult and problematic, and is unlikely to yield useful results to the practical scientist, although theoreticians will undoubtedly continue to 'play' with the theory. As has been emphasized above, no useful statement can be made about variations which have not been sampled.

(b) Markov models

Although there is probably very little justification for the applicability of the Markov model to biosphere or geosphere processes, the model remains remarkably useful as a first approximation to changes taking place from one state to another when only the probabilities of the transitions can be estimated (Kemeny and Snell, 1960). Within major states, the variations in sub-states can also be modelled as a Markov process, resulting in a series of embedded Markov chains. Surprisingly, rather few attempts have so far been made to exploit the properties of Markov models possibly because most of the emphasis in the past has been on 'functional' models, using deterministic differential or difference equations (Usher, 1979).

(c) Fractal geometry

One of the most exciting developments of the last few years has been the use of fractal geometry to reproduce patterns at increasingly smaller scales. Some mathematicians have even hinted at the possibility that all of the heterogeneity which is evident in natural systems might be capable of being represented by this type of geometry (Mandelbrot, 1983). Most of us will probably remain somewhat sceptical until we have more evidence of the usefulness of the formulation in predicting change in environmental processes.

(d) Expert systems

Our experience of modelling biosphere processes has been almost entirely based on the use of procedural algorithms, principally because it was this type of algorithm which was emphasized by the development of the early computers. The more recent development of intelligent knowledge-based systems and declarative programming heuristics has opened up some alternative approaches to the modelling of processes. There are already some examples of data sets which have proved difficult to intrepret by the more conventional statistical models, but which have responded to analysis by rule-based methods (Jeffers, 1985). Computer languages like LISP and PROLOG are already available for the exploration of models incorporating widely varying scales of heterogeneity, but, so far, relatively few ecologists have made the transition from the better-known procedural languages (Conlon, 1985).

CONCLUSIONS

I have argued that much of our difficulty with spatial and temporal scales within the biosphere/geosphere arises from a failure to apply the necessarily strict rules of the logic of the scientific method. That method demands the clear definition of the population about which we will seek to make inferences, and the adoption of valid methods of sampling and experimental design in order to estimate the parameters—including the heterogeneity—of the target population. If my argument is correct, it will, therefore, be useless for us to seek for computational methods which bypass the necessarily laborious and methodologically strict formulation of hypotheses and collection of data in order to:

(a) integrate scales, conceptually and practically, between and among disciplines

(b) aggregate local scales meaningfully into larger spatial scales

(c) evaluate how global scale processes influence local processes.

Each of these goals, separately or in combination, requires the formulation of a priori hypotheses, and the collection of data in carefully prescribed ways so as to test those hypotheses explicitly. The search for a method, computational or conceptual, which will convert a haphazard collection of data into a useful working hypothesis is a delusion!

Elsewhere, I have provided checklists for sampling, experimental design and modelling (Jeffers, 1978, 1979, 1980). The difficulty that the scientist confronts in making a correct use of the scientific method is reflected in the fact that each of those checklists contains some 70 questions which need to be asked of any piece of research. Improvement in our ability to quantify and interpret the interactions of the biosphere and geosphere with a changing global environment depends on our convincing the working scientist—and his administrators—that time spent considering the logic of what he intends to do is not wasted. Indeed, taking sufficient time to plan the investigation with due regard to the difficulties of estimating population parameters, before any data are collected, is the essential pre-requisite for the efficient use of the scarce resource of scientific expertise and equipment.

REFERENCES

Bartlett, M. S. (1975). *The Statistical Analysis of Spatial Pattern*. Cambridge University Press, Cambridge.

Cliff, A. D. and Ord, J. K. (1981). *Spatial Autocorrelation*. Pion, London.

Conlon, T. (1985). *Start Problem Solving with PROLOG* Addison–Wesley, London.

Fisher, R. A. (1925). *Statistical Methods for Research Workers*. Oliver & Boyd, Edinburgh.

Fisher, R. A. (1935). *The Design of Experiments*. Oliver & Boyd, Edinburgh.

Green, R. H. (1979). *Sampling Design and Statistical Methods for Environmental Biologists*. John Wiley & Sons, New York.

Jeffers, J. N. R. (1978). *Design of Experiments. Statistical checklist No. 1*. Institute of Terrestrial Ecology, Huntingdon.

Jeffers, J. N. R. (1979). *Sampling. Statistical checklist No. 2*. Institute of Terrestrial Ecology, Huntingdon.

Jeffers, J. N. R. (1980). *Modelling. Statistical checklist No. 3*. Institute of Terrestrial Ecology, Huntingdon.

Jeffers, J. N. R. (1985). *Ecological Advice through Expert Systems*. Proceedings of First International Expert System Conference. Learned Information, Oxford.

Kemeny, J. G. and Snell, J. L. (1960). *Finite Markov chains*. Van Nostrand, New York.

Mandelbrot, B. B. (1983). *The Fractal Geometry of Nature*. W. H. Freeman & Co., New York.

Maynard Smith, J. (1974). *Models in Ecology*. Cambridge University Press, Cambridge.

McNeil, D. R. (1977). *Interactive Data Analysis*. John Wiley & Sons, New York, London.

Poston, T. and Stewart, I. (1978). *Catastrophe Theory and its Applications*. Pitman Publishing Ltd, London.

Usher, M. B. (1979). Markovian approaches to ecological succession, *J. Anim. Ecol.*, **48**, 413–426.

Scales and Global Change
Edited by T. Rosswall, R. G. Woodmansee and P. G. Risser
© 1988 Scientific Committee on Problems of the Environment (SCOPE)
Published by John Wiley & Sons Ltd.

Chapter 5

Atmospheric Systems and Global Change

ROBERT E. DICKINSON

National Center for Atmospheric Research, [1]
P.O. Box 3000,
Boulder, Co 80307 USA

ABSTRACT

Quantification of atmospheric systems has established processes that occur on a wide range of time and space scales. Numerical models of these processes are used for various applications, such as simulation of mesoscale processes, global weather predictions, and projections of global climate change. The climate system has experienced large changes over geological times, forced by changes in atmospheric composition, location of continents, and earth's orbital parameters. Climate changes on time scales of seasons to centuries are a current focus. The contemporary research agenda includes the questions of seasonal climate prediction from an observed initial state and the projection of decadal climate changes. Treating the wide range of spatial scales of the climate system is difficult, both in the atmosphere and especially at the terrestrial surface. Current societal issues include the question of the gradual warming occurring from increasing concentrations of trace gases and CO_2 and questions of climate change from land-use changes, such as desertification and deforestation.

INTRODUCTION

We all have an intuitive sense of the temporal variability of the atmosphere through our personal experiences as residents in one or more locations. Thus, we see individual clouds float by, sometimes in 5 to 10 minutes or so, and cloud systems build up and dissipate for times of less than an hour to several days. We experience the absence of solar radiation at night and night-time declines

---

[1] The National Center for Atmospheric Research is sponsored by the National Science Foundation.

in temperature whose magnitude increases with increasing distance from oceans, increasing dryness, and absence of clouds.

We are familiar with the various precipitation systems that produce rain or snow, thunderstorms lasting for tens of minutes to a few hours at most, and the extratropical cyclonic storms that can last up to several days. Some of us have experienced nature's severest weather—tornados lasting for a few minutes and hurricanes or severe downslope mountain-lee windstorms. On longer time scales, we notice the variations of temperature and precipitation with seasons. We also sometimes note excesses or deficits in rainfall and abnormally warm or cool temperatures over a month or a season. Such information on the time scales of atmospheric processes has been available to humankind since the dawn of history.

With the availability of instruments, we have been able to extend and quantify our description of atmospheric processes at individual points in the globe. In particular, at the surface we can now measure turbulent fluctuations of wind, temperature, moisture, and trace gases on time scales of seconds to minutes, and through correlation techniques infer the fluxes between surface and atmosphere of momentum, sensible and latent heat, and trace gases. We

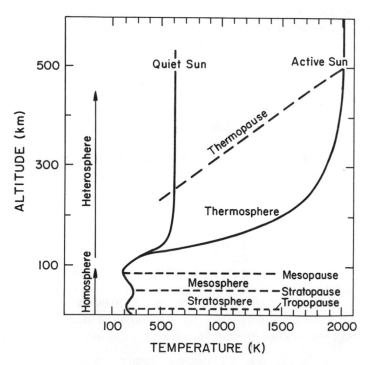

Figure 5.1   Regions of the atmosphere and vertical distribution of temperature (redrafted from Banks and Kockarts, 1973)

can measure how much solar energy is incident and absorbed, as well as the fluxes of thermal infrared energy, and so deduce the balance of energy for various terrestrial surfaces.

By probing the atmospheric column above us using radars, balloons, rockets, and satellites, we see the vertical layering of the atmosphere (Figure 5.1). Most of the humidity and clouds is found in the first several kilometres of the atmosphere. The cumulonimbus clouds of thunderstorms, frequently extending above 10 km, are exceptions. Cirrus ice-crystal clouds also can extend above 10 km. Over the first 10 to 15 km or so in the troposphere, temperatures decrease with altitude, typically by about 5 to 7 $^\circ$C km^{-1} (Figure 5.2). None of the systems that produces our 'weather' reach very far above this

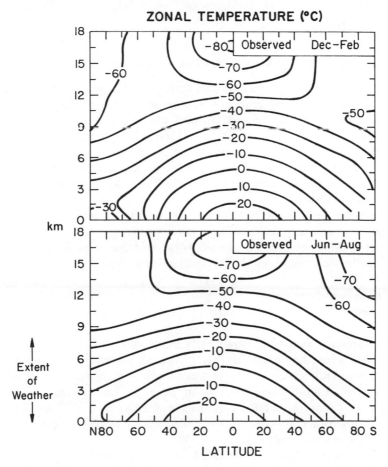

Figure 5.2 Longitudinal average atmospheric temperatures (Newell *et al.*, 1972) (Copyright 1972 MIT Press)

region. Above and extending to about 50 km is the stratosphere, where temperatures increase with altitude. The stratosphere has significant radiative, dynamic, and photochemical connections to the lower atmosphere. Perhaps of greatest practical significance is its storage of most of the atmosphere's ozone, hence controlling the levels of ultraviolet radiation incident at the earth's surface. Higher layers of the atmosphere contain many interesting processes, but significant connections to the lower atmosphere have not been identified and are not discussed further here.

Indeed, rapid vertical turbulent fluxes of various properties are largely confined to the planetary boundary layer, typically about 1 km thick, but extending up several kilometres or more during summer continental conditions and shrinking to less than 100 m over land during night-time and polar conditions.

SPATIAL SCALES, OBSERVING SYSTEMS, AND MODELS

Most atmospheric systems have spatial scales that are larger than the extent of human vision. For many centuries, sailors and geographers have noted the average patterns of surface winds and temperatures over the globe, in particular, the trade easterlies and mid-latitude westerlies, and the much stronger seasonal variation of temperature over continents than over oceans (Figure 5.3). Establishing the spatial patterns and scales of the day-to-day weather superimposed on the mean patterns has required spatially distributed systems of instruments.

The initial weather observation networks developed in the 1850s linked by telegraph the surface measurements of temperature, pressure, and incidence of rain over continental areas. These networks and the maps drawn from their data described and projected the generally eastward movement of mid-latitude cyclonic storm systems and anticyclones with spatial scales of, typically, 1000 km. About 50 years ago, meteorologists began to link data from upward balloon ascents to describe the vertical extent and structure of these weather systems.

Procedures for analysing maps nearly immediately after the acquisition of these data were developed in the 1950s, along with the initial digital computers and numerical weather forecast models for projecting the measured fields up to several days forward in time. These systems of synoptic weather–data acquisition and analysis have since then been greatly improved, now utilizing other sources of data that allow global coverage, such as satellite-temperature soundings.

The numerical models have likewise been refined to be integrated on a global mesh and to provide useful weather forecasts up to more than a week in advance. The initial thrust of these models was the solution by numerical methods of the three-dimensional, time-dependent equations for atmospheric hydrodynamics. The determination of atmospheric temperature is also crucial in these solutions, since winds are driven by pressure gradients whose vertical variation depends on temperature.

Atmospheric temperatures over the time and space scales of cyclonic disturbances are largely adiabatic, that is, they change primarily because of air movement and pressure variation. On longer time scales, atmospheric 'diabatic' heating terms become crucial for establishing temperature. These terms include release of latent heat during condensation of rainfall, atmospheric radiational processes, and convective heating from the earth's surface. General Circulation Models (GCMs) of atmospheric processes were first developed in the 1960s. These models began with the equations of the weather prediction models, but included equations for atmospheric water vapour, its evaporation, precipitation, and movement within the atmosphere, equations for atmospheric radiation, and frictional and thermal coupling to the earth's surface. The intent of these models was to simulate the average statistics of atmospheric circulation systems.

In the 1970s, the atmospheric science community, concerned with the question of climate change resulting from increasing concentrations of atmospheric carbon dioxide (CO_2) and other trace gases, first developed simple energy–balance and radiative–convective equilibrium models to study this problem. Previous GCMs had used fixed sea-surface temperatures taken from observations, but by assuming that oceans acted as wet surfaces and were locally in energy balance, Manabe and Wetherald (1975) applied a version of their GCM to study climate change from increasing CO_2. A decade of further such studies (e.g. as reviewed most recently by Bolin *et al.*, 1986) has clarified the difficulties confounding detailed precise projections of future climate change from increasing trace gases. These difficulties largely involve problems of matching global climate change to other scales and to interfaces with other disciplines.

Many of the severest weather systems occur on the mesoscale, ranging from a few kilometres to a few hundred kiolometres in horizontal extent (Figure 5.4). At the smaller end of this range are the individual thunderstorm cells; at the larger end are frontal phenomena, including rainbands, and mesoscale convective complexes. The conventional radiosonde observing systems are too coarse to capture these features. Weather radars have been deployed for two decades to provide early warning of dangerous mesoscale systems. At present, considerable effort is devoted to developing and deploying a variety of new radars, automatic weather stations, and other new wind measurement techniques, as well as fine-resolution, limited-area numerical models for simulating mesoscale motions over a time scale of a day or less. Such research and operational developments are needed to better describe the mesoscale atmospheric processes (e.g. Anthes, 1983).

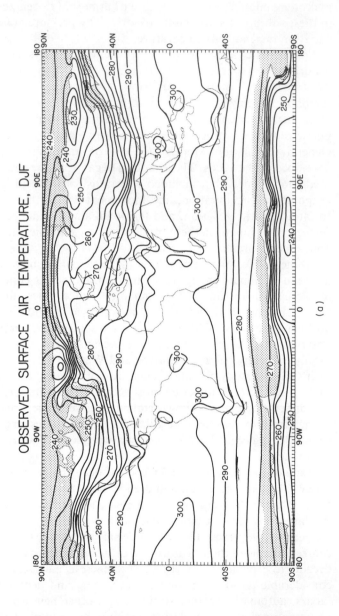

OBSERVED SURFACE AIR TEMPERATURE, DJF

(a)

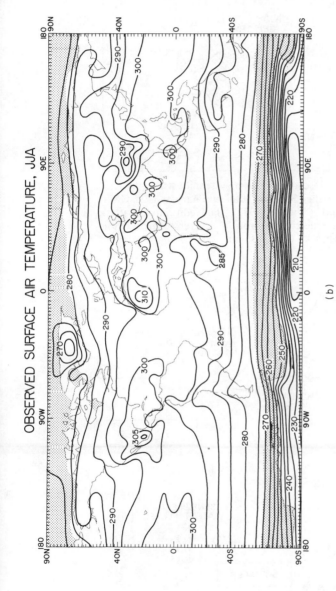

Figure 5.3 Geographical distribution of winter and summer average ((a) December–February and (b) June–August) global surface temperatures (Washington and Meehl, 1984)

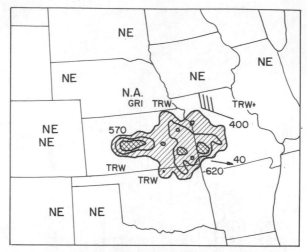

Figure 5.4   Mesoscale weather system illustrated by U.S.
National Weather Service radar summary at 1435 GMT,
22 June 1981 (Anthes, 1983). The contoured pattern
shows large reflectivities from a severe storm over the
states of Nebraska and Missouri

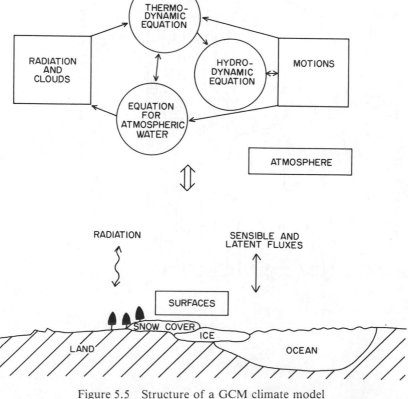

Figure 5.5   Structure of a GCM climate model

Global forecast models and the observing systems that provide them with data are also being improved. With improvements in computer power, the spatial resolution of global forecast models is shrinking to about 100 km or less. Weather forecasts can now provide useful information up to at least 10 days into the future, and some centres, e.g. the European Centre for Medium Range Weather Forecasts, are actually disseminating forecasts that far ahead. With the extension of global forecast models to finer space scales and to larger time scales, inclusion of many of the physical processes previously treated in climate GCMs (Figure 5.5) is also becoming important for forecast models. Likewise, considerable emphasis is placed on upgrading the treatment of physical processes in mesoscale models. In sum, to a certain extent, mesoscale, forecast, and climate models are converging to similar descriptions of atmospheric physical processes, and their primary distinction becomes their spatial resolution and time span of integration.

GLOBAL-SCALE CLIMATE CHANGES

Consideration of climate change and variations in the time frame of human experience benefits from the perspective of longer time-scale variations revealed only in the geologic record.

Climate change on geological time scales

Geological processes range over time scales of thousands to billions of years. Several intriguing questions over this range of time scales have been emphasized in recent investigations. On the billion-year scale, the atmosphere, oceans, and continents themselves evolved. The primary observation for climate theorists on this time scale is that climate must not have been so drastically different from that of today as to preclude the development and evolution of life. On the other hand, at the time of our planet's formation, four and a half billion years ago, the output of energy from the sun was about 30% less than it is today (Newkirk, 1983). With the composition of today's atmosphere but with 30% less solar radiation, land and oceans would have become covered with ice and snow, and the resulting high albedos would have maintained this condition even with current fluxes of solar radiation. Thus, CO_2 and possibly other gaseous absorbers must have been more abundant. At least 100 times the current concentrations of CO_2 would have been required, if it were the only factor preventing such an 'ice-covered earth' catastrophe. Such concentrations could have been supplied by interactions between geological and climatic processes (e.g. Walker *et al.*, 1981).

On time scales of 100 million years, sea floors are created from spreading centres and destroyed by subduction, rafting continents together and apart, and modifying their climate by changing their latitude. The Cretaceous period,

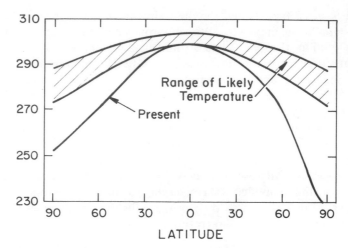

Figure 5.6   Estimated excess of Cretaceous temperature over present (Barron and Washington, 1984)

centred about 100 million year ago, was much warmer than today, especially in high latitudes (Figure 5.6). GCM climate modelling studies (e.g. Barron and Washington, 1984) indicate that this warmth could not have been maintained only by differences in the size, shape, and location of continents. Rather, there must have been some large change in the concentration of the atmosphere's radiatively active gases; most likely, CO_2 concentrations were 4 to 10 times greater than at present—as also suggested by evidence from geochemical cycling (Budyko and Efimova, 1981; Berner *et al.*, 1983).

Subsequent to the Cretaceous, the global climate cooled until it entered the sequence of ice ages experienced over the last one and a half million years, with large fluctuations in temperature and ice cover over 100-, 40-, and 20-thousand-year periods. These periodic fluctuations, apparently, are largely driven by variations of the earth's orbital parameters (e.g. Imbrie *et al.*, 1984) and accompanied by fluctuations in atmospheric CO_2 levels. For example, during the peak of the last ice age 20 000 years ago, CO_2 levels were less than 60% of their current levels. These changes in CO_2 levels may have been controlled by variations in ocean-uptake rates driven by biological processes (Knox and McElroy, 1984). On continental areas in the tropics, far from the ice sheets, the cover of tropical forests was greatly diminished (e.g. Ab'Saber, 1982; Haffer, 1982). Whether this decline in forest cover resulted mostly from more arid conditions as indicated or represented, in part, by a response to the different CO_2 levels is not known. Evidence for tropical ocean temperatures during the last ice age is conflicting (e.g. Rind and Peteet, 1985).

At the end of the last ice age, 10 thousand years ago, the earth passed closest to the sun in August. At present, it does so in January. Consequently, the

Northern Hemisphere absorbed about 7% more solar radiation in July than it does today and about that much less in January. Summer temperatures in the Northern Hemisphere were warmer then than now and monsoon circulations were more pronounced (Kutzbach and Guetter, 1984), as has been simulated with GCM integrations.

The contemporary climate research agenda

Of greatest interest for human affairs are climate changes on time scales of seasons to centuries. The current World Climate Research Programme divides research over this range of time scales into three 'streams' as follows (WCRP, 1984):

> The *first stream* aims at ... weather anomalies on time scales of one to two months, ... observing the initial value of the ocean surface temperature and sea ice fields, and making progress in the ability to predict the ... amount of stored soil water and the rate of evaporation, ... precipitation and extended clouds, and in the formulation of radiative transfer in the presence of such clouds.
>
> The *second stream* aims at ... variations of the global climate over periods up to several years ... particularly evident in the tropical regions. The largest contribution to the variations of the global atmosphere which may be predictable on interannual time scales is ... the influence of the oceans and especially, the tropical oceans in which large-scale circulation and temperature anomalies can be forced by remote atmospheric events and propagate along the Equator. ... Stream 2 is ... modelling the coupled atmosphere—ocean system using a truncated version of oceanic dynamics restricted to the tropical part of the ... oceans.
>
> The *third stream* aims at variations of climate over periods of several decades and assessing the response of climate to either natural or man-made influences.

In other words, the first stream is concerned with extending the time span of current numerical weather prediction approaches out to several weeks or longer by improving GCMs used for forecasting to better incorporate fluctuations in the hydrological cycle, radiation, and surface fluxes. On a time scale of seasons to years, much of the natural climate variability, especially in the tropics, is coupled with variations in ocean-surface temperatures. The most dramatic example of such coupling has been the El Niño-Southern Oscillation climate fluctuations that occur every 3 to 5 years or so and last for about a year (Wright, 1985). These fluctuations encompass most of the tropical Pacific and extend into the Indian and Atlantic Oceans with teleconnections to extra-tropical latitudes. Thus, the second stream emphasizes coupling of atmosphere, sea ice, and ocean dynamic models.

On a time scale of decades or longer, the largest projected climate changes are those resulting from the addition of CO_2 and other trace gases to the atmosphere. CO_2 has increased from concentrations less than 280 ppm a few hundred years ago (Neftel *et al.*, 1985; Oeschger, 1987) to present values of nearly 350 ppm. Of most concern besides CO_2 are the chlorofluorocarbons CFC-12 (CCl_2F_2) and CFC-11 (CCl_3F) (Figure 5.7) and methane (CH_4) (Figure 5.8), tropospheric ozone (O_3) and nitrous oxide (N_2O), listed in order of their probable relative importance for future climate change.

The climate effect of future increases of the CFCs alone may be about 70% of that of CO_2, assuming no serious attempt is made to further regulate their production (e.g. Dickinson and Cicerone, 1986). The warming from CO_2 and other trace gases will probably exceed that implied by a doubling of CO_2 within the next 50 years.

The question of steady-state climate change from large increases of atmospheric CO_2 has been considered with the most advanced GCMs suitable for such study, including realistic continents and a seasonal cycle of solar radiation (Manabe and Stouffer, 1980; Hansen *et al.*, 1984; Washington and

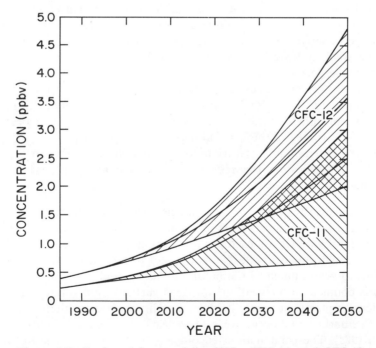

Figure 5.7 Projected future of the CFCs, CFC-11, and CFC-12. Upper, lower, and middle scenarios are shown for each gas as inferred from an economic analysis. (Dickinson and Cicerone, 1986). (Reprinted by permission from *Nature*, **319**, No. 6049, 109–115, copyright © 1986 Macmillan Magazines Limited.)

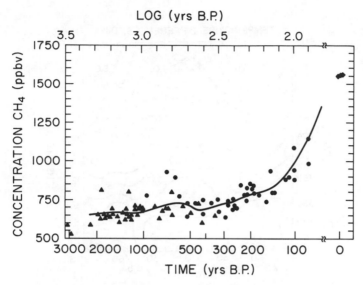

Figure 5.8 Past history of CH_4. The circles refer to data from Greenland ice cores, the triangles to South Pole ice core data, the diamonds to current concentrations. (Rasmussen and Khalil, 1984.)

Meehl, 1984; Wetherald and Manabe, 1986). These model studies indicate a steady-state global warming of from 2.0 to 4.8°C global-average temperature for a doubling of CO_2. The range of possible outcomes depends heavily on the extent of positive feedbacks from cloud radiative effects, which are now modelled with extreme uncertainty. It also depends significantly on the modelling of changes in the seasonally varying sea-ice distribution, in the vertical distribution of atmospheric water vapour, and in vertical lapse rates.

Global-average temperature change is constrained by global-energy balance and is thus simplest to estimate approximately. However, there is no reason to expect a uniform change, and not only large regional variations in temperature change but also geographical and seasonal shifts in rainfall patterns are likely. Figure 5.9 illustrates the latitudinal variation of temperature calculated from one GCM simulation of climate change from CO_2 doubling. All GCM simulations have indicated a considerably larger surface-temperature increase in high than in low latitudes, especially in winter, apparently a result of the highly stable temperature stratification and the ice-snow albedo feedbacks. Figure 5.10 shows relative changes in soil moisture from another such model that suggests a strong summertime mid-latitude, mid-continental increasing dryness. However, not all modelling groups obtain this result. All models for climate change from increasing CO_2 have used a very simplistic treatment of soil moisture, e.g. the Budyko bucket method as illustrated in Figure 5.11.

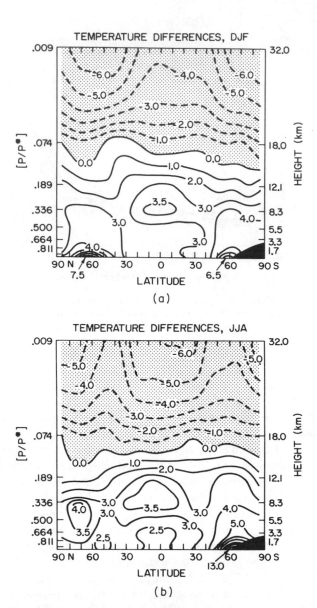

Figure 5.9   Zonal average annual average temperature change from CO_2 according to Washington and Meehl (1984). Top frame shows values for December–February; bottom shows those for June–August. (P/P^* = pressure/surface pressure)

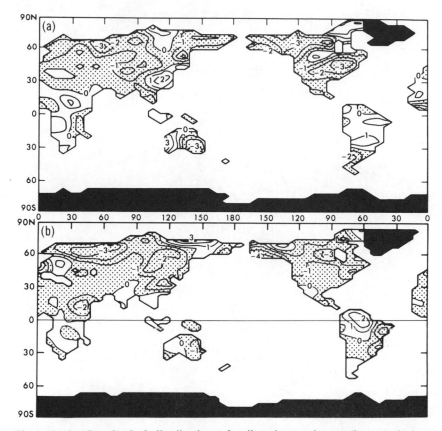

Figure 5.10 Geophysical distribution of soil-moisture change (in cm) during June–August for two model simulations. The upper frame is for relatively coarse resolution and the lower frame for a finer resolution. (Manabe *et al.*, 1981)

One of the difficulties in projecting future climate change from increasing concentrations of trace gases is that the oceans require decades to centuries to equilibrate with changes in global-energy balance. There has already been a considerable increase of atmospheric trace gases since 1900, but the thermal inertia of the oceans is thought now to be reducing the realized warming to about half of that implied by a steady-state assumption. Still, global temperatures should have increased from between 0.3 and 1.0 °C over those of 1900. Indeed, global temperatures are now the warmest ever recorded and are about 0.5 °C warmer than in 1900. However, temperatures were nearly as warm in 1940 as at present (Figure 5.12). Evidently, natural fluctuations in global temperature occur with up to 0.5 °C range and obscure the signal from increasing trace gases.

BUDYKO BUCKET MODEL

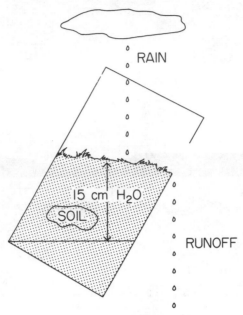

Figure 5.11 Budyko bucket model for
GCM soil moisture

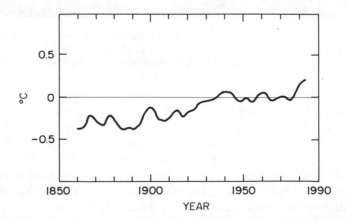

Figure 5.12   Record of smoothed global-average surface temperature (heavy line) relative to 1950–1979 mean value (light
line) (Jones *et al.*, 1986). (Reprinted by permission from
Nature, **322**, 430–434, Copyright © 1986 Macmillan Magazines Limited.)

OUTSTANDING SCALE-DEPENDENT PROBLEMS

The atmospheric sciences now have an excellent conceptual and practical frame-work for considering the global atmosphere over time scales of days to years. We are learning to include in GCMs details of the diurnal cycle and to carry out simulations for up to several decades of time. Mesoscale models simulate the smaller-scale atmospheric processes out to a day or two. Some of the most serious current difficulties relate to time- and space-scale mismatches to this framework, involving either processes internal to the atmosphere or at the interface with other disciplines.

Scale problems internal to the atmosphere

The severest difficulties for modelling the atmosphere come from the need to parameterize many subgrid-scale atmospheric processes, that is, the net effect of processes occurring on scales of e.g., 10 m to 100 km, must be included in global models that resolve typical minimum horizontal scales of at least several hundred kilometres. Especially significant are the descriptions of the radiative properties of clouds and of the vertical transfer of heat, moisture, and momentum by moist and dry convection.

Cloud-property changes may be important for climate change on time scales of years or longer, but even the sign of the net effect on radiation is still difficult to establish. Subgrid-scale vertical fluxes of water vapour, sensible heat, and momentum are generally described by simple one-dimensional, conceptual models whose relationship to real three-dimensional processes is poorly known. Thus, the sign of these fluxes is generally expected to be correct, but their magnitudes for given atmospheric profiles may be far from correct. Considerable compensation and negative feedbacks may make the net effect of errors in subgrid-scale vertical flux descriptions less serious than they might otherwise be. Errors in the treatment of horizontal momentum fluxes from grid scale to subgrid scale may also degrade atmospheric simulations.

The above discussion has taken the perspective of a global-scale modeller frustrated by the difficulty in treating linkages to smaller scales. However, severer problems may be encountered by modellers on smaller scales attempting to link their models to larger-scale, up to global, processes. An example is Idso's difficulties in extrapolating a good understanding of local micro-meteorological processes to global climate problems (Idso, 1980; CO_2/Climate Review Panel, 1982). Most small-scale investigators of atmospheric problems are more timid and prefer to assume that larger scales are fixed. Since the coupling of the small-scale processes to larger scales may be their most significant aspect, the practical output of such research tends to be primarily support for improving parameterizations of larger-scale models.

Scale problems at the terrestrial interface

From the viewpoint of the atmosphere, the terrestrial surface has a simple role—its main functions are to determine how much solar radiation is absorbed versus reflected and how absorbed energy is partitioned into latent and sensible heat versus storage. Longwave radiation is also absorbed and emitted but varies less and is less sensitive to details of surface processes. However, when the surface processes actually involved in energy transfers are examined, they are found to be very complex with considerable small-scale variability in space and time.

For example, a tree is not a simple flat surface but is a collection of a large number of surfaces, often idealized as plates or cylinders. These surfaces absorb solar radiation, depending on their orientations, shading from over-lying elements, and transfer of radiation through canopy elements (e.g. Dickinson, 1983). Such elements individually reach equilibrium temperatures that balance their solar heating with sensible and latent heat losses. These losses, in turn, depend not only on element temperature but also on the rate of air movements past the element. If the element is a photosynthetic surface, it loses water by transpiration through its stomates. Stomatal resistance depends on the plant species, but varies strongly with environmental variables, e.g. incident solar radiation, water-vapour pressure deficit, temperature, and leaf-water potential, which depends on the availability of water to the roots. Incident solar radiation, CO_2 mixing ratio, wind flow, and stomatal resistance, in turn, determine gross carbon uptake by the canopy.

Thus, one major modelling issue is the development of simple but effective and reliable parameterizations for the detailed energy-transfer processes within vegetation canopies that yield satisfactory descriptions of the exchange of heat and moisture with the atmosphere. A second question is how to include root resistance, that is, the limits to the rate at which water can be transferred from soil to leaves depending on soil properties and moisture (Dickinson, 1984; Bolin, 1988).

Soil properties can be included in one-dimensional soil models with reasonable accuracy. However, these properties vary widely on almost all horizontal scales and within the vertical column as well. To model the properties of soil, we must specify statistical properties of soils on the model grid and obtain the required global data. Holes by termites and worms can be important in many areas, and removal of vegetation may lead to compaction and pore closure, both by the effects of raindrops and of travel by animals and equipment.

Another issue of concern is the modelling of surface runoff, which depends not only on the vertical soil-column physics but also on steepness and extent of slopes, on vegetation including debris cover, and on the intensity of the rainfall. Rainfall intensities, especially in convective situations, can be much more intense locally than they are averaged over a model grid square. Thus,

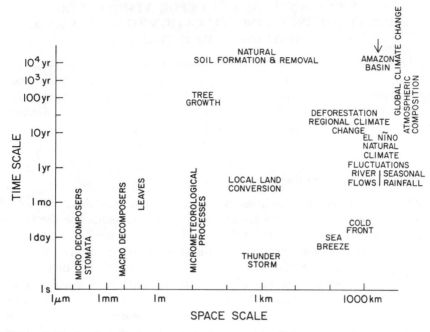

Figure 5.13 Scales of the atmospheric processes linked to surface processes (Dickinson, 1987)

how to include the subgrid-square distributions of soil properties, slopes, and rainfall intensities is a key issue for representing land processes in global models.

Trace-gas emissions, e.g. those of methane and nitrous oxide, presumably also have considerable small-scale variability as a consequence of the mosaic structure of land surfaces. Figure 5.13 illustrates the range of time and space scales involved in connecting the surface to the atmosphere.

The smallest-scale processes above the molecular level that determine the fluxes of moisture and gases to the atmosphere occur on the spatial scales of cells that are of the order of tens of microns; stomatal closure controls the fluxes through leaf surfaces, and microdecomposers, in metabolizing carbon, release various gases. Trees develop over spatial scales of tens of metres but have lifetimes of up to hundreds of years. In studying climate change and variations, we are concerned with spatial scales of at least several hundred kilometres and with time scales of seasons out to a century. Land surfaces are linked to this climate change through micrometeorological processes, which have individual eddies with 10 to 100 m spatial scales. One possible approach to linking large-scale model calculations to local ecosystems is that suggested by Gates (1985).

SOME KEY SOCIETAL ISSUES FOR ATMOSPHERIC MODELLING REQUIRING ATTENTION TO SCALES AND INTERDISCIPLINARY INTERFACES

We have already discussed the development of atmospheric models from mesoscale to global and the difficulties in accounting for the wide range of scales of phenomena internal to the atmosphere and at the terrestrial surface. Some current societal issues are also of considerable scientific interest in forcing us to come to grips with the wide range of scales.

Climate warming from trace gas increases

Past and future increases in atmospheric trace gases may drive the climate system within a century to a warmer state than it has realized in the last ten million years (Dickinson and Cicerone, 1986). It is difficult to imagine the implementation of practical controls on these increases, since the global distribution of the trace gases will translate sources in one location to climate impacts in another location. In particular, the industrialized countries control most of the emissions, whereas the less-developed countries—who have many more immediate concerns—are likely to suffer the greatest costs. If societal adjustments are the primary recourse to this event (Kellogg and Schware, 1981), such adjustment should be less painful if we anticipate decades in advance of the likely climate changes. Approaches to ameliorating the trace-gas climate change should be developed in the context of all the other environmental stresses from modern industrialization (Clark, 1986).

As already discussed, projections of future climate change for a given trace-gas scenario are limited by questions involving coupling between scales. How will cloud radiative properties change in step with such climate change? What shifts in regional climate patterns including rainfall will accompany the overall global change? What will be the contribution of decreases in sea ice and snow cover to amplifying the warming? How will shifts in surface-energy balance couple with the dynamics of soil-moisture and plant-water usage to modify the growth patterns of natural vegetation and cultivated crops? Attempts have been made to treat all these questions, but the answers can yet be viewed with little confidence.

Desertification and droughts

Periods of rainfall deficit are one of the known consequences of natural climate variability. Furthermore, for given rainfall, vegetation and soils can be degraded such that runoff is amplified and moisture storage reduced. What is less obvious is whether changes of land cover can shift mean rainfall or change its seasonal and stochastic variability. We know that changing an isolated small-scale patch of land can have little effect. The water for rainfall over this patch comes from elsewhere, and such a change cannot significantly modify the patterns of atmospheric vertical motion. On the other hand, changing all continental land surfaces from wet to dry condition leads to large reductions in rainfall over land areas and widespread shifts in climatic patterns (Shukla and Mintz, 1982; Mintz, 1984). Up to half of the rainfall over continental areas is derived from evapotranspiration of land surfaces, and widespread changes in surface-energy processes necessarily affect the dynamics of the overlying atmosphere. On a smaller scale, irrigation over 100 km scale areas apparently enhances warm-season precipitation in the presence of atmospheric uplift (Barnston and Schickedanz, 1984).

However, we cannot describe with any confidence the role of such actual changes of the land surface on rainfall patterns and amounts. The Sahel region of Africa is a dramatic example of a large area where both vegetation degradation and a major drought have evolved over the last several decades (e.g. Nicholson, 1983; Lamb, 1983; Dennett *et al.*, 1985). The connections between the vegetation degradation and the drought span a wide range of space and time scales. Equally convincing arguments can be made that the drought is entirely natural or that it is largely man-made. Large and widespread albedo increases can promote drought (Charney *et al.*, 1977), but Gornitz and NASA (1985) find that albedo over West Africa has increased by, at most, 0.5%. We must greatly improve our descriptions of the land-surface changes and their links to the atmosphere in parallel with improving our ability to model atmospheric response to such changes. Only by doing so can we hope to quantify the relative contributions to such droughts of natural fluctuation and environmental degradation.

Deforestation

Tropical forests are being degraded and destroyed with alarming rapidity and serious ecological damage. However, for reasons described above, at present we can just begin to evaluate what effects deforestation might have on local, regional, and global climate (e.g. Henderson-Sellers, 1987).

Sioli (1984) has speculated that, when too much forest is destroyed, water cycles will be interrupted by lack of sufficient water vapour, total annual rainfall will decrease, and the length of dry season will increase, in turn affecting the structure of the forest. He suggests that 'once a critical point has been reached an irreversible chain reaction will start.' What are the most sensitive aspects of tropical forest ecosystems to climate change? Mori and Prance (1987) suggest that pollination and seed-dispersal processes may be disrupted. Fires may also cause considerable damage. How will the survival of seedlings be modified?

CONCLUSIONS

Atmospheric systems span a wide range of time and space scales. Changes in climate are studied using models that couple hydrodynamic and physical processes in the atmosphere and link these processes to the land and ocean systems. Improving descriptions of the links to land will require determining what factors best describe the fluxes of water and energy between land surfaces and the atmosphere. Establishing such factors will likely involve coping with the wide range of spatial scales of land processes and especially those that are small compared to the grid resolution of the atmospheric models. Treating the two-way interactions between climate change and ecosystems will also demand better descriptions of the influence of the atmospheric systems on the ecosystems.

REFERENCES

Ab'Saber, A. N. (1982). The paleoclimate and paleoecology of Brazilian Amazonia. In Prance, G. T. (Ed.) *Biological Diversification in the Tropics*. Columbia University Press, New York 741.

Anthes, R. A. (Ed.) (1983). *The National STORM Program—Scientific and Technological Bases and Major Objectives*. UCAR Report, University Corporation for Atmospheric Research, Boulder, CO 520.

Banks, P. M., and Kockarts, G. (1973). *Aeronomy Part A*. Academic Press, New York 430.

Barnston, A. G., and Schickedanz, P. T. (1984). The effect of irrigation on warm season precipitation in the southern Great Plains. *J. Climate Appl. Meteor.*, **23**, 865–888.

Barron, E. J., and Washington, W. M. (1984). The role of geographic variables in explaining paleoclimates: Results from Cretaceous climate model sensitivity studies. *J. Geophys. Res.*, **89**, 1267–1279.

Berner, R. A., Lasaga, A. C., and Garrels, R. M. (1983). The carbonate–silicate geochemical cycle and its effect on atmospheric carbon dioxide over the last million years. *Amer. J. Sci.*, **283**, 641–683.

Bolin, B. (1988). Linking terrestrial ecosystem process models to climate models. (Chapter 7, this volume).

Bolin, B., Döös, B. R., Jäger, J., and Warrick, R. A. (Eds.) (1986). *The Greenhouse Effect, Climatic Changes, and Ecosystems. A Synthesis of Present Knowledge*. John Wiley and Sons, Chichester, London 541 pages.

Budyko, M. I., and Efimova, M. A. (1981). The influence of carbon dioxide on climate. *Meteorol. Hydrol.* **2**, 15–17.

Charney, J., Quirk, W. J., Chow, S.-H., and Kornfield, J. (1977). A comparative study of the effects of albedo change on drought in semiarid regions. *J. Atmos. Sci.*, **34**, 1366–1385.

Clark, W. C. (1986). Sustainable development of the biosphere: Themes for a research program. In Clark, W. C., and Munn, R. E. (Eds.) *Sustainable Development of the Biosphere*, pp. 5–48. International Institute for Applied Systems Analysis, Laxenburg, Austria.

CO₂/Climate Review Panel (1982). *Carbon Dioxide and Climate: A Second Assessment*. National Research Council, National Academy Press, Washington, DC: 72 pages.

Dennett, M. D., Elston, J., and Rodgers, J. A. (1985). A reappraisal of rainfall trends in the Sahel. *J. Climat.*, **5**, 353–361.

Dickinson, R. E. (1983). Land–surface processes and climate–surface albedos and energy balance. *Advances in Geophysics.*, **25.**, 305–353.

Dickinson, R. E. (1984). Modelling evapotranspiration for three-dimensional global climate models. In *Climate Processes and Climate Sensitivity, Maurice Ewing Series 5*, pp. 58–72. American Geophysical Union, Washington, DC.

Dickinson, R. E. (Ed.) (1987). *The Geophysiology of Amazonia. Vegetation and Climate Interactions*. John Wiley & Sons, Inc., New York: 526 pages.

Dickinson, R. E. and Cicerone, R. J. (1986). Future global warming from atmospheric trace gases. *Nature*, **319**, No. 6049, 109–115.

Gates, W. L. (1985). The use of general circulation models in the analysis of the ecosystem impacts of climatic change. *Cimatic Change*, **7**, 267–284.

Gornitz, V., and NASA (1985). A survey of anthropogenic vegetation changes in West Africa during the last century—Climatic implications. *Climatic Change*, **7**, 285–325.

Haffer, J. (1982). General aspects of the refuge theory. In Prance, G. T. (Ed.) *Biological Diversification in the Tropics*. Columbia University Press, New York: 714 pages.

Hansen, J., Lacis, A., Rind, D., Russell, G., Stone, P., Fung, I., Ruedy, R., and Lerner, J. (1984). Climate sensitivity: Analysis of feedback mechanisms. In Hansen, J. E., and Takahashi, T. (Eds.) *Climate Processes and Climate Sensitivity, Maurice Ewing Series 5*, American Geophysical Union, Washington, DC: 368 pages.

Henderson-Sellers, A. (1987). Effects of change in land use on climate in the humid tropics. In Dickinson, R. E. (Ed.). *The Geophysiology of Amazonia. Vegetation and Climate Interactions*. pp. 463–496. John Wiley & Sons, Inc., New York.

Idso, S. B. (1980). The climatological significance of a doubling of the earth's atmospheric carbon dioxide concentration. *Science*, **207**, 1462–1463.

Imbrie, J., Hays, J. D., Martinson, D. G., McIntyre, A., Mix, A. C., Morley, J. J., Pisias, N. G., Prell, W. L., and Shackleton, N. J. (1984). The orbital theory of pleistocene climate: Support from a revised chronology of the marine $\delta^{18}O$ record. In Berger, A. L., Imbrie, J., Hays, J. Kukla, G., and Saltzman, B. (Eds.) *Milankovitch and Climate*, Part 1. pp. 269–305.

Jones, P. D., Wigley, T. M. L., and Wright, P. B. (1986). Global temperature variations between 1861 and 1984. *Nature*, **322**, 430–434.

Kellogg, W. W., and Schware, R. (1981). *Climate Change and Society. Consequences of Increasing Atmospheric Carbon Dioxide*. Westview Press, Boulder, CO: 178 pages.

Knox, F., and McElroy, M. B. (1984). Changes in atmospheric CO_2: Influence of the marine biota at high latitudes. *J. Geophys. Res.*, **89**, 4629–4637.

Kutzbach, J. E., and Guetter, P. J. (1984). The sensitivity of monsoon climates to orbital parameter changes for 9,000 years BP: Experiments with the NCAR general circulation model. In Berger, A. L., Imbrie, J., Hays, J., Kukla, G., and Saltzman, B. (Eds.) *Milankovitch and Climate*, Part 2, 801–820.

Lamb, P. J. (1983). Sub-Saharan rainfall update for 1982: Continued drought. *J. Climat.*, **3**, 419–422.

Manabe, S., and Stouffer, R. J. (1980). Sensitivity of a global climate model to an increase of CO_2 concentration in the atmosphere. *J. Geophys. Res.*, **85**, 5529–5554.

Manabe, S., and Wetherald, R. T. (1975). The effects of doubling the CO_2 concentration on the climate of a general circulation model. *J. Atmos. Sci.*, **32**, 3–15.

Manabe, S., Wetherald, R. T., and Stouffer, R. J. (1981). Summer dryness due to an increase of atmospheric CO_2 concentration. *Climatic Change*, **3**, 347–386.

Mintz, Y. (1984). The sensitivity of numerically simulated climates to land-surface boundary conditions. In Houghton, J. T. (Ed.). *The Global Climate*, pp. 79–105. Cambridge University Press, Cambridge.

Mori, S. A., and Prance, G. T. (1987). Species diversity, phenology, plant–animal interactions, and their correlation with climate as illustrated by the Brazil nut family (lecythidaceae). In Dickinson, R. E. (Ed.) *The Geophysiology of Amazonia. Vegetation and Climate Interactions*, pp. 69–89. John Wiley & Sons, Inc., New York.

Neftel, A., Moor, E., Oeschger, H., and Stauffer, B. (1985). Evidence from polar ice cores for the increase of atmospheric CO_2 in the last two centuries *Nature*, **315**, 45–57.

Newell, R. E., Kidson, J. W., Vincent, D. G., and Boer, G. J. (1972). *The General Circulation of the Tropical Atmosphere, Vol. 1*. MIT Press, Cambridge, MA: 258 pages.

Newkirk, G., Jr. (1983). Variations in solar luminosity. *Ann. Rev. Astron. Astrophys.*, **21**, 429–467.

Nicholson, S. E. (1983). Sub-Saharan rainfall in the years 1976–80: Evidence of continued drought. *Mon. Wea. Rev.*, **111**, 1646–1654.

Oeschger, H. (1988). The ocean system—ocean/climate and ocean/CO_2 interaction. (Chapter 15, this volume.)

Rasmussen, R. A., and Khalil, M. A. K. (1984). Atmospheric methane in the recent and ancient atmospheres: Concentrations, trends, and interhemispheric gradient. *J. Geophys. Res.*, **89**, 11,599–11,605.

Rind, D. and Peteet, D. (1985). Terrestrial conditions at the last glacial maximum and CLIMAP sea-surface temperature estimates: Are they consistent? *Quart. Res.*, **24**, 1–22.

Shukla, J., and Mintz, Y. (1982). Influence of land-surface evapo-transpiration on the earth's climate. *Science*, **215**, 1498–1501.

Sioli, H. (1984). Former and recent utilizations of Amazonia and their impact on the environment. In Sioli, H. (Ed.) *The Amazon: Limnology and Landscape Ecology of a Mighty Tropical River and its Basin*. W. Junk, Dordrecht: 763 pages.

Walker, J. C. J., Hays, P. B., and Kasting, J. F. (1981). A negative feedback mechanism for the long term stabilization of the earth's surace temperature. *J. Geophys. Res.*, **86**, 9776–9782.

Washington, W. M., and Meehl, G. A. (1984). Seasonal cycle experiment on the climate sensitivity due to a doubling of CO_2 with an atmospheric general circulation model coupled to a simple mixed layer ocean model. *J. Geophys. Res.*, **89**, 9475–9503.

Wetherald, R. T., and Manabe, S. (1986). An investigation of cloud cover change in response to thermal forcing. *Climatic Change*, **8**, 5–23.

Wright, P. B. (1985). The Southern Oscillation: An ocean-atmosphere feedback system? *Bull. Amer. Meteor. Soc.*, **66**, 398–412.

WCRP (1984). *Scientific Plan for the World Climate Research Programme*. WMO-ICSU Joint Scientific Committee, WCR Publication No. 2: 95 pages.

Scales and Global Change
Edited by T. Rosswall, R. G. Woodmansee and P. G. Risser
© 1988 Scientific Committee on Problems of the Environment (SCOPE)
Published by John Wiley & Sons Ltd.

CHAPTER 6

Variability in Atmospheric–Chemical Systems

PAUL J. CRUTZEN

Division of Atmospheric Chemistry,
Max-Planck-Institute for Chemistry,
P.O. Box 3060, D–6500 Mainz, F.R.G.

ABSTRACT

The spatial and temporal variability of the atmospheric concentrations of individual gases is determined by the location of their production sources and strongly related to their photochemical lifetimes, which in many cases depends on reactions with hydroxyl radicals. Most chemical compounds are emitted into the atmosphere at very discrete locations. The photochemically most reactive ones are largely processed within the lowest levels (1–2 km) of the troposphere under variable meteorological conditions, resulting in a spatially and temporally very inhomogeneous system that is extremely difficult to model and explore experimentally. Fortunately, the photochemistry of the background free troposphere is much simpler. It is substantially determined by the far less variable compounds O_3, CO and CH_4, which have atmospheric lifetimes of the order of months to years. Here, the greatest difficulty is the correct modelling of the chemical transformations and transport of NO and NO_2 which act as powerful catalysts in the photochemistry of the atmosphere. As the photochemistry of the background atmosphere is changing substantially due to growths in the atmospheric burdens and emissions of the important gases CH_4, CO, O_3, and NO_x, the development of global photochemical–meteorological models is an important enterprise. It is bound to be an essential ingredient in the future 'Global Change' programme.

Although the development of high-resolution, global photochemical transport models may become feasible some time in the future through increasing computer resources, 'off-line' low resolution models which utilize the essential information extracted from high resolution, large-scale meteorological models probably will be the most useful tools for a long time to come.

INTRODUCTION

The chemical composition of the atmosphere is the result of a great variety of processes and the field of atmospheric chemistry intersects with such broad and diverse scientific disciplines as meteorology, optical physics, chemistry, and biology. It requires high technology for measurements of many ultra-minor constituents in an uncontrollable and greatly variable environment.

The most abundant atmospheric gases, molecular nitrogen and oxygen, may be considered to represent the equilibrium state of global biogeochemical processes that have operated on time scales of tens of thousands of years. Because of their very long chemical lifetimes, the volume mixing ratios of these gases are extremely constant in the atmosphere below 100 km altitude. Only above this height, photochemical processes driven by solar ultraviolet radiation become fast enough to affect the distribution of molecular oxygen. For molecular nitrogen photodissociation becomes effective at even higher altitudes. Gravitational separation becomes more important than meteorological mixing above about 100 km.

There is a general rule that the shorter the lifetime, the higher the variability of a trace gas. For this, Junge (1974) has derived a useful approximate empirical equation relating the spatial and temporal, relative standard deviation σ' (the ratio between standard deviation and arithmetic average) of the volume mixing ratio of a chemical constituent to its atmospheric lifetime (t_c/s):

$$\sigma' = 0.02/t_c,$$

which is especially useful for compounds with average lifetimes of the order of a year or more, because it allows the mean atmospheric residence times of chemical compounds to be roughly estimated from the measured variability of their concentrations. With this information their emission rates into the atmosphere and their integrated atmospheric sinks can be approximately determined.

Other atmospheric chemical compounds are far less abundant than N_2 and O_2. The main reason for this, however, is not much smaller emissions, but much shorter atmospheric lifetimes, leading to much lower abundances. For instance, the net fluxes of CO_2 and O_2 at the earth's surface due to biological (and combustion) processes are roughly equal and opposite in sign. However, the total mass of atmospheric CO_2 is more than a thousand times less than that of O_2. Consequently, human activities affect the distribution of CO_2 much more than that of O_2. For many other chemical compounds the atmosphere is even more strongly affected by human activities.

The shorter-lived and less abundant trace gases are, however, not only important as sensitive indicators of environmental change. Their short atmospheric residence times also indicate, of course, that they are chemically very active. Because of their high reactivity many compounds are only present in the atmosphere at extremely low concentrations. In fact, several of these have so low concentrations that they can never be measured, such as electronically excited atomic oxygen. Many radical species (i.e. molecules with an odd number of electrons) are exceedingly important, because they serve as catalysts in photochemical chain reactions in the atmosphere. The most important among these is hydroxyl (OH). Despite the fact that the average volume mixing ratio of OH in the troposphere is only 3×10^{-14}, it is this radical, and not abundant molecular oxygen, which is responsible for the first critical step in the oxidation and removal of most compounds in the atmosphere. In its turn, because hydroxyl is born from photochemical interactions between ozone and water vapour, ozone too is clearly a very important tropospheric gas.

Most important chemical constituents are emitted into the atmosphere at the earth's surface by natural, mostly biological, processes and increasingly also by anthropogenic activities, whereby the role of man is not only restricted to technological, but also agricultural activities. Generations of humans, all over the world, have been changing the earth's biosphere over millenia, e.g. through forest removal. Tropical agricultural activities produce substantial quantities of important trace gases.

The results of human activities are clearly visible in the atmosphere, not only regionally but also globally. Of substantial importance from the point of view of global atmospheric photochemistry are the changes that are taking place in the abundance of methane, carbon monoxide, and the oxides of nitrogen. It is likely that this is leading to changes in the concentrations of tropospheric ozone and hydroxyl, and the cycling of many compounds in the background atmosphere. As we will discuss, a gradual shift of photochemical activity from clean to more polluted atmospheric environments may be the result of the these changes. It is, therefore, extremely important that long-term measurements are performed to document the changes that are occurring in ozone at various locations in the background troposphere. Field experiments to test atmospheric photochemical chemistry are likewise of major importance. Such efforts can only be carried out fruitfully if appropriate photochemical–meteorological models are also developed. This is exceedingly difficult, especially for the lower troposphere, because emission sources of most photochemically active constituents are distributed very heterogeneously at the earth's surface and occur during variable meteorological conditions. Fortunately, the modelling of the photochemistry of the background atmosphere is substantially dependent on the distributions of the more long-lived gases, carbon monoxide, methane, ozone, and water vapour, which are therefore more uniformly distributed and relatively well known from observations. The main difficulty is the appropriate modelling of the highly-variable, short-lived NO_x (NO and NO_2) gases.

In the following we will mainly discuss global scale atmospheric chemical processes, especially how they might be changing because of various anthropogenic activities and how they might be modelled. From the start it should be clear of course that the complexity of atmospheric chemistry and meteorology is so large that any modelling activity can at the best be only a gross approximation of reality. Fortunately, some recent efforts to develop global scale models for medium and long range weather forecasting at some meteorological research centres have become increasingly successful. The extraction of the essential meteorological data from such models for the purpose of photochemical modelling, therefore, presents itself as a promising approach.

PHOTOCHEMISTRY OF THE BACKGROUND TROPOSPHERE

The main actors: OH, O_3, NO_x, CO, and CH_4

Although only about 10% of all atmospheric ozone is located in the troposphere, the lowest 10–17 km of the atmosphere, this small amount of ozone, with volume mixing ratios of 20–100 ppbv (1 ppbv = 10^{-9}), is nevertheless of fundamental importance for the composition of the earth's atmosphere. The reason for this is the generation of electronically excited atomic oxygen through the absorption of ultraviolet radiation by ozone, and its reaction with water vapour:

$$O_3 + h\upsilon \rightarrow O(^1D) + O_2 \, (<310 \text{ nm}) \qquad \text{(R1)}$$
$$O(^1D) + H_2O \rightarrow 2OH \qquad \text{(R2)}$$

It is the attack by OH that initiates the oxidation of most trace gases in the atmosphere (Levy, 1971; McConnell *et al.*, 1971). This is of fundamental importance, because only a few of the many gases that are produced at the earth's surface can be removed by rainfall and none by reaction with molecular oxygen. According to current estimates, the average volume mixing ratio of hydroxyl radicals in the troposphere is only about 3×10^{-14} (e.g. Crutzen, 1982). Therefore, although the atmosphere contains almost 21% molecular oxygen, it is the ultraminor constituent OH which starts almost all oxidation processes. In the background troposphere about two thirds of the OH radicals react with CO, and one third with CH_4. Smaller fractions react with various other gases. Both methane and carbon monoxide are increasing in the atmosphere (Rasmussen and Khalil, 1981; Rinsland *et al.*, 1985; Blake *et al.*, 1982; Bolle *et al.*, 1985; Rinsland and Levine, 1985). This can clearly influence the background tropospheric concentrations of hydroxyl and, therefore, also those of many other important atmospheric constituents. It is, therefore, clear that understanding of tropospheric chemistry and estimations of the future impact of human activities require detailed knowledge of the photochemical reactions affecting ozone, carbon monoxide, and methane. We will also show how the concentrations of ozone and hydroxyl are strongly affected by catalytic reactions that depend on NO and NO_2.

Although OH reacts overwhelmingly with CO and CH_4 in the background troposphere, these reactions do not necessarily lead to its removal from the atmosphere. They are merely the starting points for various chains of reactions, which may compensate for the initial OH loss and which have important implications for the chemical composition of the troposphere. For instance, in the presence of sufficiently large concentrations of nitric oxide, the oxidation of carbon monoxide will lead to formation of tropospheric ozone, without loss of the catalysts OH, HO_2, NO, and NO_2, through the reaction chain:

$$CO + OH \rightarrow H + CO_2 \tag{R3}$$
$$H + O_2 + M \rightarrow HO_2 + M \tag{R4}$$
$$HO_2 + NO \rightarrow OH + NO_2 \tag{R5}$$
$$NO_2 + h\nu \rightarrow NO + O \ (\leq 400 \ nm) \tag{R6}$$
$$O + O_2 + M \rightarrow O_3 + M \tag{R7}$$

net: $\quad CO + 2O_2 \rightarrow CO_2 + O_3$

In reactions (R4) and (R7), and in the following, the symbol M denotes a third molecule (mostly O_2 and N_2) which serves to stabilize the product which is formed by the association reaction.

A competing chain of reactions, leading to ozone destruction, which dominates in NO-poor environments:

$$CO + OH \rightarrow H + CO_2 \tag{R3}$$
$$H + O_2 + M \rightarrow HO_2 + M \tag{R4}$$
$$HO_2 + O_3 \rightarrow OH + 2O_2 \tag{R8}$$

net: $\quad CO + O_3 \rightarrow CO_2 + O_2$

likewise does not lead to the loss of OH and HO_2 radicals. The second reaction sequence is more important than the first one whenever the ratio of the concentrations of NO and O_3 is less than 2×10^{-4}. With ozone volume mixing ratios increasing from about 20×10^{-9} (20 ppbv) at the earth's surface to 100 ppbv at the tropopause, the break-even point between the reaction chains (R3–R7) and (R3, R4, R8) is attained at nitric oxide volume mixing ratios of 4×10^{-12} (4 pttv) at the ground and 2 pptv at the tropopause. Although these are very low concentrations, they may, nevertheless, not be exceeded in extensive regions of the troposphere in view of the very short residence times of NO and NO_2 due to the rapid formation of highly water-soluble and photochemically quite inactive nitric acid via:

$$NO + O_3 \rightarrow NO_2 + O_2 \tag{R9}$$
$$NO_2 + OH(+M) \rightarrow HNO_3(+M) \tag{R10}$$

during daytime and:

$$NO_2 + O_3 \rightarrow NO_3 + O_2 \qquad\qquad (R11)$$
$$NO_3 + NO_2(+M) \rightarrow N_2O_5(+M) \qquad\qquad (R12)$$

followed by the deposition of NO_3 and N_2O_5 on cloud drops and wetted aerosol during night-time. During daytime, the NO_3 which is formed through reaction R11 is immediately photolysed to reproduce the original reactants NO_2 and O_3.

Because of the very short lifetime of NO_x in the troposphere, we may expect appreciable concentrations of NO_x only within, at most, a few weeks travel distance from where they are produced, i.e. mostly near the highly industrialized regions at mid-latitudes in the Northern Hemisphere, and the tropical upper troposphere, where significant amounts of NO are formed by lightning. In agreement with these thoughts. McFarland *et al.* (1979) have indeed measured background volume mixing ratios of NO of less than 10 pptv in the marine boundary layer of the tropical Pacific. Furthermore, Noxon (1981) has reported a mixing ratio for NO_x of 30 pptv at 3 km altitude at Hawaii.

Besides reacting with NO and O_3 (R5 and R8), HO_2 can also react with itself, especially in NO-poor environments, leading to the production of H_2O_2, which plays an important role in aqueous oxidation chemistry (e.g. Chameides and Davis, 1982; Graedel and Wechsler, 1981):

$$CO + OH \rightarrow H + CO_2 \qquad\qquad 2 \times \;(R3)$$
$$H + O_2 + M \rightarrow HO_2 + M \qquad\qquad 2 \times \;(R4)$$
$$HO_2 + HO_2 \rightarrow H_2O_2 + O_2 \qquad\qquad (R13)$$
$$H_2O_2 + h\nu \rightarrow 2OH \;(\lambda \leq 350 \text{ nm}) \qquad\qquad (R14)$$

net: $2CO + O_2 \rightarrow 2CO_2$

This reaction sequence leads to the oxidation of carbon monoxide without affecting tropospheric ozone. Because highly water-soluble H_2O_2 can be removed efficiently by uptake in clouds and by precipitation, these processes are important sinks for perhydroxyl (HO_2), and indirectly for OH. Furthermore, H_2O_2 is also involved in the reaction pair:

$$HO_2 + HO_2 \rightarrow H_2O_2 + O_2 \qquad\qquad (R13)$$
$$H_2O_2 + OH \rightarrow HO_2 + H_2O \qquad\qquad (R15)$$

net: $OH + HO_2 \rightarrow H_2O + O_2$

which removes OH and HO_2. The oxidation of carbon monoxide in NO-poor environments, therefore, most likely leads to a loss of OH and O_3. In NO-rich environments ozone formation is strongly favoured (Crutzen, 1973).

The oxidation of methane is likewise of very large importance in tropospheric photochemistry. In the first place, about 30–40% of the hydroxyl radicals react with CH_4. Secondly, the oxidation chains of methane strongly affect the atmospheric budgets of hydroxyl and ozone (Crutzen, 1973). Again, nitric oxide plays an important role in determining the oxidation pathways. In NO-rich environments, almost certainly, rapid formation of formaldehyde (CH_2O) occurs, following the initial reaction between OH and CH_4:

$$CH_4 + OH \rightarrow CH_3 + H_2O \qquad (R16)$$
$$CH_3 + O_2 + M \rightarrow CH_3O_2 + M \qquad (R17)$$
$$CH_3O_2 + NO \rightarrow CH_3O + NO_2 \qquad (R18)$$
$$CH_3O + O_2 \rightarrow CH_2O + HO_2 \qquad (R19)$$
$$HO_2 + NO \rightarrow OH + NO_2 \qquad (R5)$$
$$NO_2 + h\nu \rightarrow NO + O \ (\leq 400 \text{ nm}) \qquad 2 \times (R6)$$
$$O + O_2 + M \rightarrow O_3 + M \qquad 2 \times (R6)$$

net: $$CH_4 + 4O_2 \rightarrow CH_2O + H_2O + 2O_3$$

Most important, this sequence of reactions leads to the net production of two ozone molecules with various intermediates, in particular NO an NO_2, serving as catalysts.

In NO-poor environments, the CH_4 oxidation steps may follow several pathways that may also lead to CH_2O formation.

$$CH_4 + OH \rightarrow CH_3 + H_2O \qquad (R16)$$
$$CH_3 + O_2 + M \rightarrow CH_3O_2 + M \qquad (R17)$$
$$CH_3O_2 + HO_2 \rightarrow CH_3O_2H + O_2 \qquad (R20)$$
$$CH_3O_2H + h\nu \rightarrow CH_3O + OH \ (< 330 \text{ nm}) \qquad (R21)$$
$$CH_3O + O_2 \rightarrow CH_2O + HO_2 \qquad (R19)$$

net: $$CH_4 + O_2 \rightarrow CH_2O + H_2O$$

However, the photolysis of methylhydroperoxide (CH_3O_2H) is slow, resulting in a relatively long mean atmospheric residence time for this compound of about a week. Consequently, methylhydroperoxide may be removed from the atmosphere by rainfall, or deposition on cloud droplets, aerosol particles, or on the earth's surface. Wherever this occurs, the oxidation of CH_4 will have led to a loss of two odd hydrogen radicals (OH and HO_2).

Another reaction mechanism which occurs in NO-poor environments and which leads to the net loss of OH and HO_2, is

$$CH_4 + OH \rightarrow CH_3 + H_2O \qquad (R16)$$
$$CH_3 + O_2 + M \rightarrow CH_3O_2 + M \qquad (R17)$$
$$CH_3O_2 + HO_2 \rightarrow CH_3O_2H + O_2 \qquad (R20)$$
$$CH_3O_2H + OH \rightarrow CH_2O + H_2O + OH \qquad (R22a)$$

net: $$CH_4 + OH + HO_2 \rightarrow CH_2O + 2H_2O$$

Finally, the catalytic pair of reactions

$$CH_3O_2 + HO_2 \rightarrow CH_3O_2H + O_2 \tag{R20}$$
$$CH_3O_2H + OH \rightarrow CH_3O_2 + H_2O \tag{R22b}$$

net: $OH + HO_2 \rightarrow H_2O + O_2$

is particularly important, because it destroys both OH and HO$_2$.

The oxidation pathways of methane in NO-poor environments are, therefore, far less predictable than in NO-rich environments and do not always lead to production of CH$_2$O. It is, however, likely that hydroxyl and perhydroxyl radicals are being destroyed rather efficiently.

Three pathways lead to the oxidation of formaldehyde to carbon monoxide with the overall net production of 0.8 HO$_2$ radicals per CH$_2$O molecule:

$$CH_2O + h\nu \rightarrow H + CHO \ (\leq 350 \text{ nm}) \tag{R23a}$$
$$H + O_2 + M \rightarrow HO_2 + M \tag{R4}$$
$$CHO + O_2 \rightarrow CO + HO_2 \tag{R24}$$

net: $CH_2O + 2O_2 \rightarrow CO + 2HO_2$

or

$$CH_2O + h\nu \rightarrow CO + H_2 (\leq 350 \text{ nm}) \tag{R23b}$$

or

$$CH_2O + OH \rightarrow CHO + H_2O \tag{R25}$$
$$CHO + O_2 \rightarrow CO + HO_2 \tag{R24}$$

net: $CH_2O + OH + O_2 \rightarrow CO + H_2O + HO_2$

Because the average photochemical lifetime of CH$_2$O in the troposphere is only a few hours, the likelihood of precipitation scavenging of formaldehyde is rather low and formation of carbon monoxide efficient.

Changes in background atmospheric photochemistry due to human activities

As HO$_2$ is rapidly converted into OH by reactions R5 or R8, leading to net gain or net loss of ozone, respectively, the oxidation sequences leading from CH$_4$ to CO, and from there to CO$_2$, have a strong effect on the abundance of both hydroxyl and ozone concentrations in the background troposphere.

Depending on the ambient concentrations of NO, the oxidation of one molecule of methane to one molecule of carbon dioxide yields the following astonishing net results:

(a) in NO-poor environments, i.e. especially in the unpolluted background troposphere: a net loss of 2–3.5 hydroxyl radicals and 0–1.7 ozone molecules.

(b) in NO-rich environments, i.e. especially in the highly industrialized, middle and high latitude regions of the northern hemisphere: net gain of 0.5 OH radicals and 3.7 ozone molecules.

The amount of methane in the atmosphere has been increasing for considerable time, in agreement with the growth of various human activities, which are summarized in Table 6.1. Long-term observations during the past three decades indicate an average yearly increase by about 1.1% (Rasmussen and Khalil, 1981; Rinsland *et al.*, 1985; Blake *et al.,* 1982; Bolle *et al.,* 1985). Furthermore, analyses of air trapped in ice cores have shown that the atmospheric methane content before 1650 was 2–3 times lower than at present (Rasmussen and Khalil, 1984; Craig and Chou, 1982). These global increases in methane have, therefore, probably caused substantial increases in ozone concentrations in NO-rich environments and decreases in hydroxyl concentrations in NO-poor environments. Over the past decades, the production of ozone at Northern hemisphere mid-latitudes has been strongly enhanced by concomittant increases in NO emissions from industrial activities (Table 6.2), and by the increases in carbon monoxide concentrations in the free troposphere by about 2% per year that have been deduced from solar photographic spectra taken on the Jungfraujoch in Switzerland during the past 30 years (Rinsland and Levine, 1985). Elsewhere, no long-term CO trends have yet been analysed, although air trapped in ice cores may contain relevant information. The observed rise in CO may have led to a decrease of OH also at mid-latitudes in the Northern Hemisphere through reaction R3. However, enhanced ozone and nitric oxide formation and the effects of reaction R5 may have counteracted this.

In the NO-poor atmospheric environments, the observed increase in background CH_4 concentrations have, however, most likely led to lower hydroxyl concentrations. Also in these regions, in agreement with photochemical theory, increases in carbon monoxide concentrations may have occurred, although no data are available. Through reaction R3, this would have led to a further reduction in the hydroxyl concentrations, causing a strong positive feedback. In fact, this raises questions about the stability of the chemistry of the background troposphere. Altogether, we conclude that, as time goes on, lesser quantities of industrial and natural gases may become oxidized in the tropics by reactions with hydroxyl, leading to a build-up of various important trace gases in the troposphere. Gradually, these gases are increasingly oxidized in the temperate latitude zones of the Northern Hemisphere, especially from April to October, leading to enhanced production of tropospheric ozone in these regions, which obtain much NO from fossil fuel combustion processes. Model calculations indicate the possibility of a doubling of average ozone concentrations in the lower background troposphere at mid-latitudes since the start of the industrial period (Crutzen and Gidel, 1983). There are indeed

Table 6.1 Tropospheric sources of CO, CH$_4$ and C$_5$H$_8$ + C$_{10}$H$_{16}$, their average residence times, typical transport distances, and atmospheric concentration ranges. Derived from information in Crutzen et al. (1986), Crutzen (1983), Bolle et al. (1985), Seiler (1974), Zimmerman et al. (1978), and Bingemer and Crutzen (1986). Another significant source of methane may come from northern peatlands (Harriss et al., 1985). Note that most emission estimates are uncertain by at least a factor of two

Gas	Direct source/year Source identification ($\times 10^{14}$ g/yr)	Secondary source/year Source identification ($\times 10^{14}$ g/yr)	Atmospheric lifetime	Transport distances in East–West, North–South and vertical direction; Range of typical volume mixing ratios in unpolluted troposphere.
CO	4–16 : Biomass burning 6.4 : Fossil fuel combustion 0.2–2 : Vegetation	3.7–9.3 : methane oxidation 4–13 : C$_5$H$_8$, C$_{10}$H$_{16}$ oxidation	2 months	4000, 2500, 10 km 50–200 $\times 10^{-9}$
CH$_4$	0.8–1.6 : Rice fields 0.7–1.4 : Natural wet-lands 0.8 : Ruminants 0.1–0.3 : Termites 0.4–0.8 : Biomass burning 0.3–0.4 : Gas leakage 0.3–0.4 : Coal mining 0.2–0.4 : Sanitary landfills		$\approx$ 10 years	Global 1.5–2.0 $\times 10^{-6}$
C$_5$H$_8$ + C$_{10}$H$_{16}$	6–10 : Forests		10 hours	400, 200, 1 km 0–10 $\times 10^{-9}$

Table 6.2 Tropospheric sources, average residence times, typical transport distances, and atmospheric concentrations ranges of NO_x. Derived from information in Crutzen (1983), Galbally and Roy (1978), and Borucki and Chameides (1984). Note the large uncertainties in estimates which must be resolved by future research

Direct source/year Source identification ($\times 10^{12}$ g N/yr)	Secondary source/year Source indentification ($\times 10^{12}$ g N/yr)	Atmospheric lifetime	Transport distances in East–West, North–South and vertical directions; Range of typical volume mixing ratios in unpolluted troposphere
20 : Fossil fuel combustion	1–1.5 : Oxidation of N_2O	1.5 days	1500, 400, 1.0 km
3–10 : Biomass burning			$1–100 \times 10^{-12}$
2–10 : Lightning			
5–10 : Soils			
0.15 : Ocean			
0.25 : Jet aircraft			

several indications of upward trends in tropospheric ozone. Angell and Korshover (1983) report increases in ozone concentrations by 20% in the middle troposphere (2–8 km) at north temperate latitudes over the time period 1967–1979. Similar increases in free tropospheric ozone at Hohenpeissenberg in Southern Germany have also been reported by Attmannspacher *et al.* (1984). Observations of surface ozone at the clean air stations of Mauna Loa, Hawaii and Point Barrow, Alaska, may likewise suggest upward trends by ≈ 1% per year over the past decade (Harris and Nickerson, 1984). Several more examples are given by Logan (1985).

Geographical distribution of sources and sinks of CO and CH_4

So far only very few successful hyroxyl concentration measurements have been made in the troposphere, so that much of the current theory of atmospheric photochemistry is untested by observations. Indirectly, this can, however, be achieved by comparison of calculated and observed concentrations of trace gases that react with hydroxyl. Using photochemical–meteorological models, it is possible to estimate the global distribution of hydroxyl radicals in the background troposphere. These calculations require knowledge of the global distributions of O_3, H_2O, CO and CH_4, which are quite well known from observations. The calculated OH distributions are, however, also strongly determined by the global NO and NO_2 distributions, which are expected to be very variable of which very few observations have so far been made in the 'background' troposphere. Fortunately, the atmospheric sources of NO are much better known (see Table 6.2), so that these can be introduced at the appropriate locations in models. Photochemical reactions and atmospheric transport determine the distribution of NO and NO_2. The global distributions of NO and NO_2 (and associated products), and of OH can, therefore, in principle be calculated. An example of a possible distribution of OH concentrations, calculated with a two-dimensional (latitude and height dependent) model, is shown in Figure 6.1. We note that the highest OH concentrations appear in the tropics, which is to be expected, because it is there that the flux of solar ultraviolet radiation peaks. From these calculations, the average concentration of hydroxyl in the troposphere is estimated to be above 6×10^5 molecules cm^{-3}. Few reliable direct observations of hydroxyl radical concentration are yet available, although efforts to improve the situation are under way. The calculated OH concentration distribution shown in Figure 6.1 is, however, roughly consistent with the global observatons of methylchloroform (CH_3CCl_3), which is removed from the atmosphere by reactions with OH and which has no other sources than the rather well known industrial emissions at mid-latitudes (Crutzen, 1982; Crutzen and Gidel, 1983: Zimmermann, 1984). Methylchloroform is, therefore, a particularly suitable gas to validate tropospheric photochemical–meteorological models. Comparison between obser-

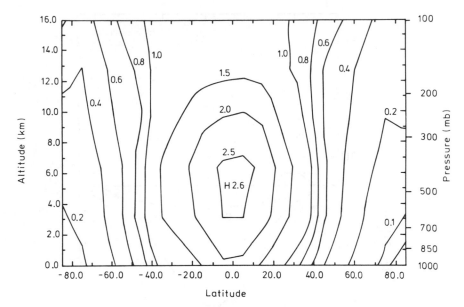

Figure 6.1   Calculated mean annual distribution of average daytime hydroxyl concentrations (units: 10^6 molecules/cm^3). Note the occurrence of maximum concentrations in the tropics. This is due to a minimum in total stratospheric ozone in equatorial regions, which allows substantial penetration of photochemically active solar ultraviolet radiation (Crutzen, 1982. Reproduced by permission of S. Bernhard)

vations and model calculations is facilitated by the fact that the atmospheric lifetime of CH_3CCl_3 is about 10 years, so that its distribution is mainly latitude dependent (Prinn *et al.*, 1983). Of critical importance for a successful simulation of the methylchloroform distribution are the calculated hydroxyl concentrations in the tropics, and the exchange of air masses between the Northern to the Southern Hemisphere. Judging from comparisons between calculated and observed distributions of $CFCl_3$ and CF_2Cl_2, interhemispheric transport is likewise described very well (Zimmermann, 1984; Crutzen and Gidel, 1983). Both gases are produced in rather well known quantities by industry, but in contrast to CH_3CCl_3 they do not react with OH.

We must caution that the apparent success of the adopted model to simulate the global distributions of CH_3CCl_3, $CFCl_3$, and CF_2Cl_2 may have been quite fortuitous. As we will note in the following section, the design of an appropriate photochemical–meteorological model is an extremely difficult task, requiring the modelling of many processes that occur on space and time scales that are far smaller than can be resolved by models. The transport of NO_x is particularly affected by this, so that calculated NO_x and OH distributions should be very model dependent. Nevertheless, it is likely that the solution of Figure 6.1 contains the main features of the global OH distribution

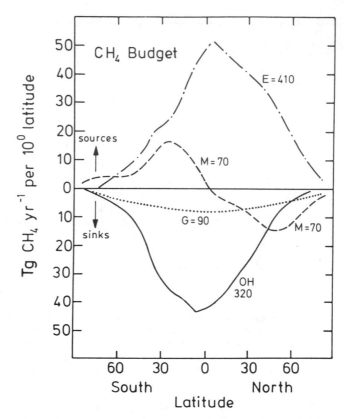

Figure 6.2   Calculated and estimated sources and sinks of methane in 10^{12} g per year (Crutzen and Gidel, 1983). Note the importance of the tropics. The various sources of CH_4 are identified in Table 1. OH = destruction by OH (320×10^{12} g); M = transport from northern to southern hemispher (70×10^{12} g);   G = annual   growth   (90×10^{12} g);   E = total source (410×10^{12} g)

in the background troposphere and roughly the right numerical values. This information is extremely useful, because it allows estimation, for example, of the global distribution of the sources and sinks of CH_4 and CO, which are removed from the atmosphere by reaction with OH. The results are shown in Figures 6.2 and 6.3. It appears that the tropical regions do contribute strongly to the destruction and production of CO and CH_4. The most important tropical sources for carbon monoxide are biomass burning and the oxidation of hydrocarbons, especially isoprene (C_5H_8) and terpenes ($C_{10}H_{16}$), that are emitted from tropical vegetation, as indicated in Figure 6.4. Most biomass burning activities are related to tropical agriculture (forest clearing, shifting

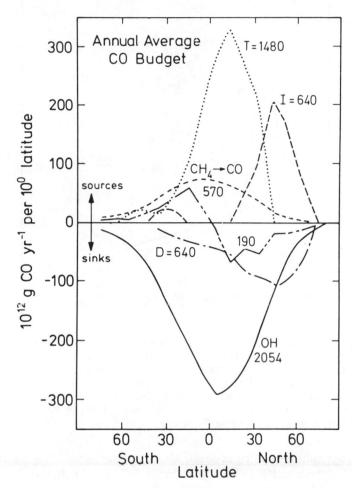

Figure 6.3 Calculated and estimated sources and sinks of carbon monoxide in 10^{12} g per year (Crutzen and Gidel, 1983). Note the importance of the tropics. The various sources of CO are identified in Table 6.1. OH = Destruction by OH radicals $(2054 \times 10^{12}$ g); D = destruction by microbial action in soils $(640 \times 10^{12}$ g); I = fossil fuel burning $(640 \times 20^{12}$ g). $CH_4 \rightarrow CO$ = oxidation of CH_4 to CO (570×10^{12}); T = tropical sources of CO $(1480 \times 10^{12}$ g). The transport of CO from the northern to the southern hemisphere is about equal to 190×10^{12} g

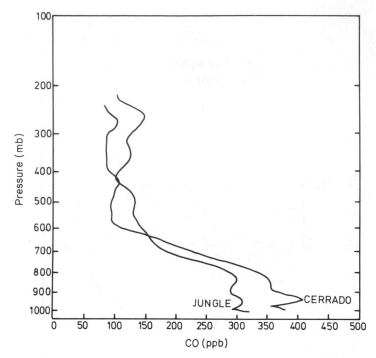

Figure 6.4   Measured average CO volume mixing ratios over the tropical forest (jungle) and savanna (cerrado) regions of Brazil during the dry season (8 vertical profiles over cerrado, 10 profiles over the forests). The concentrations measured in the lower troposphere are as high as in polluted industrial regions in the northern hemisphere (Crutzen *et al*, 1985. Reproduced by permission of D. Reidel Publishing Company)

agriculture, agricultural waste burning). Most methane is probably formed by the decay of organic matter in the anaerobic sediments of natural marshlands and rice fields. The estimated source strengths of the various processes that contribute to the production of CH_4 and CO are shown in Table 6.1. Most of these are still very uncertain. Much research is necessary to establish them with greater certainty.

Sources and sinks of background tropospheric ozone

The term in the tropospheric ozone budget that can be estimated with the greatest confidence is the global ozone loss by the reactions (R1) and (R2), which on average amounts to about 8×10^{10} molecules $cm^{-2} s^{-1}$. The ozone loss rate at the ground is of similar magnitude (Galbally and Roy, 1980). The estimated downward transport of stratospheric ozone by meteorological processes is about equal to 6×10^{10} molecules $cm^{-2}s^{-1}$ (e.g. Galbally and Roy, 1980; Gidel and Shapiro, 1980; Fabian and Pruchniewicz (1977). It is, therefore, clear that the potential production of significant amounts of ozone in the troposphere must also be considered. As discussed earlier, catalytic reactions that result in ozone production in the background troposphere take place during the oxidation of CH_4 and CO in environments with sufficient concentrations of NO. According to the estimates presented in Figures 6.2. and 6.3, the average tropospheric destruction rates of CO and CH_4 by reaction with OH are about equal to 3×10^{11} molecules $cm^{-2}s^{-1}$ and 8×10^{10} molecules $cm^{-2}s^{-1}$, respectively. If all oxidation of CO and CH_4 would occur in NO-rich environments, yielding one ozone molecule for each carbon monoxide, and 2.7 ozone molecules for each methane molecule oxidized, the average, global net ozone production rate would be equal to about 5×10^{11} moelcules cm^2s^{-1}. This is far larger than the ozone that can be destroyed by reactions R1 and R2 and by destruction at the earth's surface. Consequently, a substantial fraction of the background troposphere must contain so little NO_x that ozone is not produced but destroyed. Altogether, we may estimate that about half of the oxidation of CH_4 and CO in the background free troposphere occurs in NO-poor, the other half in NO-rich, environments. As a substantial fraction of NO is produced by anthropogenic activities (see Table 6.2), it is likely that substantial increases in tropospheric ozone may have occurred during the industrial era at northern mid-latitudes.

The potential for tropospheric ozone production

An approximate upper limit to the ozone production rates that could occur in the troposphere can be derived by assuming that there is always and everywhere enough NO in the troposphere, so that, as with methane, there are 2–3 ozone molecules produced per hydrocarbon molecule which is oxidized to carbon monoxide (Crutzen, 1973). Because most of these hydrocarbons, especially isoprene, are produced by tropical vegetation and very short-lived, their oxidation to carbon monoxide occurs mainly in lowest 2–3 km above the tropical forests, as is clearly shown in Figure 6.4. Making use of the information on the sources of CO, and non-methane hydrocarbons (Table 6.1), an average global, tropospheric ozone production of more than 10^{12} molecules $cm^{-2}s^{-1}$ would be possible. As we have seen, the actual tropospheric ozone production rate could only be at most 10% of this amount. This indicates that most oxidation of the reactive hydrocarbons emitted from tropical vegetation must take place without the production of ozone. One may therefore guess that most air masses above the tropical forests contains subcritical NO concentrations. At the same time it is, however, also clear that future expansions of industrial and biomass burning activities in the tropics (and elsewhere), leading to larger NO_x concentrations, could cause large

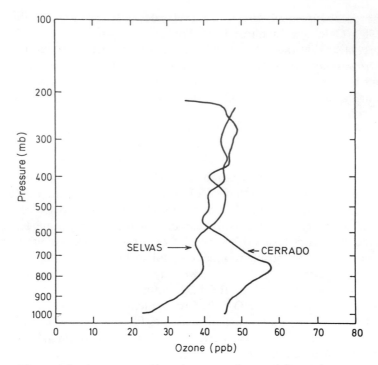

Figure 6.5   Average profiles of ozone volume mixing ratios over the tropical humid forest (selvas) and savanna (cerrado) regions of Brazil during the dry season (8 vertical profiles over cerrado, 10 profiles over the forests). Note the much higher ozone concentrations over the savanna regions. This is due to the photochemical smog created by biomass burning effluents, especially NO (Crutzen *et al,* 1985. Reproduced by permission of D. Reidel Publishing Company)

increases in tropospheric ozone, especially in the boundary layer above the tropical forests in which most of the oxidation of isoprene and terpenes to CO occurs. An example of ozone production in polluted tropical air masses is presented in Figure 6.5. In the this case, the necessary nitric oxide to cause the photochemical production of ozone was supplied by the burning of biomass in the savanna regions of Brazil during the dry season.

THE MODELLING OF ATMOSPHERIC-CHEMICAL PROCESSES

The problem of spatial and temporal variability

The distribution of chemical compounds in the atmosphere is determined by complex photochemical and meteorological processes that occur over a wide range of time and space scales. The compounds that are produced at the earth's surface by biological and by anthropogenic processes are from the start not evenly distributed in the atmosphere. For example, the oxides of nitrogen NO and NO_2 are mostly produced from such discrete sources as power plants, highways, or industrial centres and cities, which together cover at most about one percent of the continents of the earth. As the average atmospheric lifetime of NO_x is only a few days, it is clear that atmospheric motions can not disperse these compounds evenly through a major portion of the atmosphere. The same holds for many other short-lived compounds, such as the reactive hydrocarbons which are released to the atmosphere mostly by forests, but also by industrial processes. As the sources of many short-lived, photochemically active atmospheric constituents are so discrete and often do not coincide in space, efforts to model the atmospheric chemical systems are faced with the immense problem to resolve or approximate the temporal and spatial discreteness of the distributions and sources of short-lived, reactive chemical compounds. They are also very variable in space and time, so that the overlap of chemical plumes can be highly variable, depending on the particular meteorological situation. The discontinuity of emission patterns and photochemical interactions between various classes of compounds is, therefore, a major obstacle in the design of appropriate photochemical–meteorological models that describe the transport and photochemistry of reactive compounds that are released to the atmosphere. It is, therefore, not surprising that up to now no high-resolution models have been developed that take care of these difficulties. In fact, the question may be asked as to what type of models are most appropriate to describe the highly space and time variable photochemistry that occurs in regions of intensive releases of trace constituents to the atmosphere. A good representation of these processes is also important for the correct description of the photochemistry of the background troposphere because this depends critically on the ambient concentrations of short-lived NO_x, which is produced only over a very small fraction of the earth's surface.

Coupling between boundary layer and free tropospheric chemistry

As discussed in section 2, the photochemistry of the background troposphere mainly involves water vapour and the other chemical constituents O_3, CO, CH_4, NO_x, and their photochemical reaction products, such as OH and HO_2. Fortunately, the spatial and temporal variability, especially of CH_4, CO and O_3, is much less than for most other gases that are primarily released from the earth's surface. The explanation for this is that CH_4, CO and O_3 have relatively long tropospheric lifetimes, ranging from months (CO and O_3) to more than ten years (CH_4). The main problem in photochemical modelling of the background troposphere is, therefore, caused by the high variability in the distribution of the NO_x gases and the randomness of the removal of highly water-soluble compounds, expecially HNO_3, H_2O_2, and CH_3O_2H by precipitation processes. The latter requires the appropriate simulation of the effects of cloud processes. However, clouds not only are responsible for the removal of highly water-soluble compounds from the troposphere, but may also efficiently transport less water-soluble gases rapidly from the lower to the middle and upper troposphere. Most gases that are emitted at the earth's surface by natural or anthropogenic processes are insufficiently water-soluble to be affected by precipitation removal. This is e.g. also the case for NO and NO_2. For all those gases, efficient upward transport may occur in cloudy regions, especially in convective clouds or along frontal slopes. The more short-lived the gas, the more important is this transport.

To demonstrate the importance of the proper meteorological modelling of cloud transport processes, we compare in Figure 6.6 the calculated vertical distributions of SO_2 in the tropical marine atmosphere that are obtained with two very different types of models. In this part of the atmosphere, the release of dimethylsulphide (DMS) from the ocean and its oxidation, following

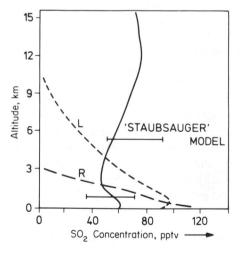

Figure 6.6 Calculated distributions of average distributions of SO_2 in the tropical oceanic troposphere. The curves marked L and R are typical for results obtained with eddy diffusion models. A detailed meteorological model of the tropical circulations which includes rapid cloud transport, the so-called 'Staubsauger' model (Chatfield and Crutzen, 1984) gives totally different results that agree much better with the range of observations that are indicated with horizontal bars (Maroulis *et al,* 1980)

reaction with hydroxyl, is the main source of SO_2 (Andreae and Raemdonck, 1983; Barnard *et al.,* 1982; Cox and Sheppard, 1980; Lovelock *et al.,* 1974). A typical conversion time from DMS to SO_2 is less than one day. The two profiles that show a rapid decrease in volume mixing ratios with altitude are calculated with so-called eddy-diffusion models, which until recently have been used most commonly for global atmospheric chemistry modelling, especially in the stratosphere. In analogy with molecular diffusion theory, in these models the net vertical flux of a gas is set proportional to the vertical gradient of its volume mixing ratio. As we note, with these assumptions, the calculated volume mixing ratios of SO_2 in the middle and upper troposphere are much less than those observed (Maroulis *el al.,* 1980). The reason for this dissatisfac-

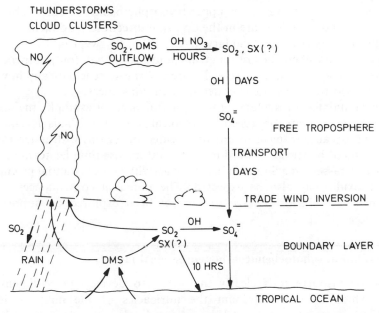

Figure 6.7   A picture of the cycling of the most important sulphur compounds, dimethyl-sulphide (DMS), sulphur dioxide, and particulate sulphate, in the tropical marine atmosphere. Violent upward motions in thunderstorm cells rapidly deposit short-lived DMS to the tropical middle and upper troposphere, where it is oxidized to SO_2, which in turn is oxidized to H_2SO_4 and condensed to sulphate aerosol. Other sulphur compounds (SX) may also be involved. An important role in these oxidation processes is played by hydroxyl during daytime and by NO, during night-time. The latter is formed from the oxidation of NO which is produced by lightning (Chatfield and Crutzen, 1984)

tory behaviour is that the eddy diffusion formulation cannot resolve the effect of brief episodes of vigorous vertical exchange.

On the other hand, very different results, which agree much better with the observations, are obtained with the so-called 'Staubsauger' (the German word for 'vacuum cleaner') model, in which the meteorological exchange processes that are typical for the tropical marine atmosphere are much better resolved in time and space (Chatfield and Crutzen, 1984). In this formulation the tropical troposphere is divided into five regions, ranging from a very turbulent central region with very strong upward motions occurring in intense thunderstorm cloud clusters over only a few percent of the tropics, to a stable outer region, where downwind motions dominate. These regions correspond most closely to the intertropical convergence zone and the horse latitudes. Zonally and temporally averaged, they constitute the well-known Hadley circulation of the tropics. The intense upward motions in the thunderstorms of the center region are fed by strong trade winds, which blow equatorwards in the lowest 1–2 km of the marine atmosphere. In the upper troposphere and the outer subsidence regions the wind directions are in the opposite directions, producing a closed cycle. In this way tropical boundary layer air can be swept up to the middle and upper troposphere in less than an hour, too brief for photochemical processes to play a significant role. Figure 6.7 shows schematically how SO_2 formation and transport occur under such circumstances.

Other candidates for the 'Staubsauger' action are, for example, the reactive hydrocarbons that are produced by tropical vegetation. This is especially interesting, because tropical forests are located in regions with much rainfall which is caused by strong upward motions and intense thunderstorm activity. In these air masses a maximum in lightning activity and substantial production of nitric oxide may also be expected. The potential consequences of this for tropical air chemistry have so far not been explored by atmospheric observations.

High resolution, photochemical–meteorological models

For the troposphere it is clearly important to develop three-dimensional models which take into account the intricacies of the many scales of atmospheric motions. Such models have been developed in the meteorological community for weather predictions and for climate studies and have been rather successful in describing (parameterizing) rather discrete meteorological processes, e.g. moist convection. On the other hand, such models are not faced with anything like the spatial discreteness of the sources of the chemical constituents and their variable lifetimes. Several efforts are indeed under way to introduce chemistry in elaborate meterological models. For example, ambitious efforts take place at the US National Center of Atmospheric Research (NCAR, 1985) to introduce photochemistry in an existing mesoscale meterogical model (originally developed by Anthes and Warner, 1978) with parameterized subgrid scale cloud processes and detailed simulation of the boundary layer processes, both of which are of critical importance for a correct description of the long range transport of atmospheric pollutants. The vertical extent of the model covers the troposphere in 15 layers, while the horizontal domain may be chosen to cover the Western US or parts thereof. In fact, the model is generalized so that it can be adapted to simulate air pollution problems at any other location. For a typical version of the model, covering the Western US, the minimum horizontal grid is equal to 60 km × 60 km, which allows for up to a week simulation of pollution episodes with the most advanced computers that are currently available. A simplified reaction scheme describes the photochemistry of O_3, various classes of hydrocarbons, NO_x, SO_2, and their most important reaction products. The principal aim of the effort is to develop a regional 'acid deposition' model, which is suitable to assess source–receptor relationships and transport across national borders (NCAR, 1985). No doubt, the development of this model is a very interesting effort, which for the first time brings together expertise in such diverse atmospheric science subdisciplines as boundary layer and mesoscale meteorology, cloud physics, and atmospheric chemistry. The task is clearly enormous and it will be especially difficult to test both the photochemical and meteorological subsections of the models sufficiently so that undisputed advice can be given regarding choices between regulatory measures vis-à-vis a multitude of pollutant gases, such as SO_2, NO_x and hydrocarbons that often emanate from different industrial activities and nations. So far, in connection with the 'acid rain' problem, most regulations have been concerned with the emissions of SO_2. No doubt, NO_x and hydrocarbons will come increasingly in the forefront of interest. It should be noted here that for these gases not all emissions are anthropogenic. Nitric oxide is also produced by lightning discharges in quantities that are still very uncertain. Of even greater importance compared to anthropogenic inputs may be the emissions of reactive hydrocarbons, such as isoprene and terpenes, from forests (Zimmerman, 1977).

Finally we may note that even a rather tight grid spacing of 60 km × 60 km can of course not resolve the extreme heterogeneity of emission patterns at the earth's surface, because only the largest metropolitan areas occupy a major fraction of such a grid box. To resolve this problem, higher resolution, so-called nested-grid models, covering important subsections of the larger domain may be designed. Still, even such models cannot resolve the enormously complex patterns of emissions, ranging from point to city size sources.

Low-resolution, large scale models

From the previous discussion it is clear that the design of even a limited, subcontinental-size chemical–meteorological model constitutes an utterly complicated task which requires major team work and computer resources.

The future of this kind of mesoscale, atmospheric modelling is hard to predict. One obvious way to proceed is to build on ever increasing computer power and to go for higher resolution. However, I believe that there is considerable room for alternative approaches, such as the development of models that allow for appreciable stochastic behaviour. In fact, the development of such models may be the only hope of arriving at a comprehensive global model of atmospheric chemistry for the foreseeable future.

Efforts which have been carried out so far in developing global scale models have been restricted to the modelling of some long-lived gases such as $CFCl_3$, CF_2Cl_2, and N_2O, mainly in order to test the dynamic behaviour of the three-dimensional model (Zimmermann, 1984). Although the development of more elaborate models might appear immature, simplified approaches to global scale modelling may, nevertheless, prove useful. For instance, in the previous section we have shown results from a two-dimensional (latitude, height dependent) model with which an average global OH distribution was calculated. Somewhat fortuitously, the calculated OH distribution was verified rather satisfactorily by the globally observed CH_3CCl_3 distribution. This OH distribution was subsequently used to derive important information regarding the global distribution of sources and sinks of methane and carbon monoxide. The future extension of such a model to three dimensions may, therefore, prove fruitful for some atmospheric-chemical applications, allowing, for example, the effects of continental emissions to be separated from those over marine areas. However, in such models it will be necessary to include even rougher parameterizations of the exchange processes between the boundary layer and the middle and upper troposphere than in mesoscale models. Fortunately, the requirements for global models are different than those for mesoscale models, because one is never interested the know the detailed distribution of chemical compounds around the globe at a particular time, but merely in the average distribution, and in variations thereof. Most promising in the future development of three-dimensional models may, therefore, be the introduction of the stochastic simulation of boundary layer and convective processes, which may be derived from a thorough, statistical analysis of the results obtained by detailed large scale, meteorological models. Considering the strongly increasing capabilities of such global scale models in predicting global weather over periods approaching a week, there is considerable hope that the necessary parameters for such stochastic modelling efforts may be derived, so that global modelling with such large grid spacings as, for example, 500 km × 500 km can prove useful. In practice, the meteorological models are supplied with up-to-date information every six hours, so that appropriate statistics should be extractable for photochemical applications.

CONCLUSIONS

Atmospheric chemistry is the interplay between a large variety of environmental processes, such as industrial and agricultural activities, microbial activities in soils and waters, all sorts of meteorological phenomena, and the photochemistry of many trace gases.

Substantial atmospheric composition changes are occurring which affect the overall functioning of tropospheric chemistry. According to current photochemical theory, a gradual build-up of various trace gases that react with hydroxyl should be occurring in the global atmosphere. Because the observed increasing atmospheric concentrations of methane in the background atmosphere probably induce lower hydroxyl concentrations, these in turn would cause higher concentrations of methane and carbon monoxide, implying a potentially strong positive feedback. In fact, the current status of knowledge of atmospheric photochemistry allows a legitimate question about the overall stability of the chemical composition of the background atmosphere. Furthermore, as a result of the processes that are discussed in this paper, a gradual shift of the oxidizing power from the clean background atmosphere to more polluted regions can be postulated, leading especially to enhanced ozone concentrations at mid-latitudes. It is clear that these largely theoretical thoughts should be urgently tested by chemical observations in the background atmosphere and by the high-quality monitoring of various critical compounds at a sufficient number of stations around globe. At the same time, better estimates must be made of the sources of trace consitutents at the earth's surface. Biological emissions are by far most difficult to estimate.

The design of appropriate models is very important in order to study the combined effect of the many environmental factors that contribute to atmospheric chemistry. The development of such photochemical models is made exceedingly difficult by the great variability in the atmospheric concentrations of the most reactive gases, which are caused by the strong temporal and spatial heterogeneity of the meteorological processes and the distribution of the sources of trace gases at the earth's surface. Fortunately, rather successful models of the large-scale meteorological processes have been developed over the past decade especially for the purpose of medium and long range weather predictions. The extraction of the essential meterological data from such models for atmospheric chemistry applications, therefore, presents itself as most promising venue. Because for global tropospheric chemistry modelling, one is not interested in the three-dimensional chemical distribution at any particular time, but more on average distributions and variability, the development of global models containing a substantial element of stochasticity seems most appropriate.

REFERENCES

Andreae, M. O., and Raemdonck, H. (1983). Dimethylsulfide in the surface ocean and the marine atmosphere: a global view. *Science,* **221**, 744–745.
Anthes, R. A., and Warner, T. T. (1978). Development of hydrodynamic models

suitable for air pollution and other mesometeorological studies. *Mon. Wea. Rev.,* **106**, 1045–1078.

Angell, J. K., and Korshover, J. (1983). Global variation in total ozone and layer mean ozone: An update through 1981. *J. Climate Appl. Meteorol.,* **22**, 1611–1626.

Attmannspacher, W., Hartmannsgruber, R., and Lang, P. (1984). Langzeittendenzen des Ozons der Atmosphäre aufgrund der 1967 begonnenen Ozonmeßreihen am Meteorolgischen Observatorium Hohenpeißenberg. *Meteorol. Rdsch.,* **37**, 193–199.

Barnard, W. R., Andreae, M. O., Watkins, W. E., Bingemer, H., and Georgii, H. W. (1982). The flux of dimethyl sulfide from the oceans to the atmosphere. *J. Geophys. Res.,* **87**, 8787–8794.

Bingemer, H. and Crutzen, P. J. (1986). The production of methane from solid wates. *J. Geophys. Res.,* **92** (D2), 2181–2187.

Blake, D. R., Mayer, E. W., Tyler, S. C. Montague, D. C., Makide, Y., and Rowland, F. S. (1982). Global increase in atmospheric methane concentration between 1978 and 1980. *Geophys. Res. Lett,* **9**, 477–480.

Bolle, H. J., Seiler, W., and Bolin, B. (1985). Other greenhouse gases and aerosols. Assessing their role for atmospheric radiative transfer. In: *WMO/ICSU/UNEP International Assessment of the Role of Carbon Dioxide and Other Radiatively Active Constituents in Climate Variation and Associated Impacts.*

Borucki, W. J., and Chameides, W. L. (1984). Lightning: Estimates of the rates of energy dissipation of nitrogen fixation. *Rev. Geophys. Space Phys.,* **22**, 363–372.

Chameides, W. L., and Davis, D. D. (1982). The free radical chemistry of cloud droplets and its impact upon the composition of rain. *J. Geopyhs. Res.,* **87**, 4863–4877.

Chatfield, R. B., and Crutzen, P. J. (1984). Sulfur dioxide in remote oceanic air: cloud transport of reactive precursors. *J. Geophys. Res.,* **89**(D5), 7111–7132.

Cox, R. A., and Sheppard, D. (1980). Reactions of OH radicals with gaseous sulphur compounds. *Nature, 284,* 330–331.

Craig, H., and Chou, C. C. (1982). Methane: The record in polar ice cores. *Geophys. Res. Lett.,* **9**, 1221–1224.

Crutzen, P. J. (1973). A discussion of the chemistry of some minor constituents in the stratosphere and troposphere. *Pure Appl. Geophys.,* **106–108**, 1385–1399.

Crutzen, P. J. (1982). The global distribution of hydroxyl. In Goldberg, E. D. (Ed.) *Atmospheric Chemistry,* pp. 313–328. Dahlem Konferenzen, Springer-Verlag, Berlin.

Crutzen, P. J. (1983). Atmospheric interactions—homogeneous gas reactions of C, N, and S containing compounds. In Bolin, B., and Cook, R. B. (Eds.) *The Major Biogeochemical Cycles and Their Interactions, SCOPE* 24, pp. 68–112. John Wiley and Sons, New York.

Crutzen, P. J., and Gidel, L. T. (1983). A two-dimensional photochemical model of the atmosphere. 2: The tropospheric budgets of the anthropogenic chlorocarbons, CO, CH_4, CH_3Cl and the effect of various NO_x sources on tropospheric ozone. *J. Geophys. Res,* **88**, 6641–6661.

Crutzen, P. J., Delany, A. C., Greenberg, J., Haagenson, P., Heidt, L., Lueb, R., Pollock, W., Seiler, W., Wartburg, A., and Zimmerman, P. (1985). Tropospheric chemical composition measurements in Brazil during the dry season. *J. Atmos. Chem., 2,* 233–256.

Crutzen, P. J., Aselmann, I., and Seiler, W. (1986). *Methane production by domestic animals, wild ruminants, other herbivorous fauna, and humans. Tellus,* 38B, 271–284.

Fabian, P., and P. G. Pruchniewicz (1977). Meridional distribution of ozone in the troposphere and its seasonal variations. *J. Geophys. Res.,* **82**, 2063–2073.

Galbally, I. E., and Roy, C. R. (1978). Loss of fixed nitrogren from soils by nitric oxide exhalation. *Nature,* **275,** 734–735.

Galbally, I. E., and Roy, C. R. (1980). Destruction of ozone of the earth's surface. *Q. J. R. Meteorol. Soc.,* **106,** 599–620.

Gidel, L. T., and Shapiro, M. (1980). General circulation model of the net vertical flux of ozone in the lower stratosphere and the implications for the tropospheric ozone budget. *J. Geophys. Res.,* **85,** 4049–4058.

Graedel, T. E., and Wechsler, C. J. (1981). Chemistry in aqueous atmospheric aerosols and raindrops. *Rev. Geophys. Space Phys.,* **19,** 505–539.

Harris, J. M., and Nickerson, E. C. (Eds.) (1984). *Geophysical Monitoring for Climate Change No 12.* Summary Report 1983, NOAA, Boulder, Colorado, 184 pp.

Harriss, R. C., Gorham, E., Sebacher, D. I., Bartlett, K. B., and Flebbe, P. A. (1985). Methane flux from northern peatlands. *Nature,* **315,** 652–653.

Junge, C. E. (1974). Residence time and variability of tropospheric trace gases. *Tellus,* **26,** 477–488.

Levy, H., II (1971). Normal atmosphere: Large radical and formaldehyde concentrations predicted *Science,* **173,** 141–143.

Levy, H., II. (1974). Photochemistry of the troposphere. *Advances in Photochemistry,* 370–524.

Logan, J. (1985). Tropospheric ozone: seasonal behavior, trends and anthropogenic influence *J. Geophys. Res.,* **90** (D6), 10463–10482.

Lovelock, J. E., Maggs, R. J., and Rasmussen, R. A. (1974). Atmospheric dimethyl sulfide and the natural sulphur cycle. *Nature,* **248,** 625–626.

Maroulis, P. J., Torres, A. L., Goldberg, A. B., and Bandy, R. A. (1980). Atmospheric SO_2 measurements on Project GAMETAG. *J. Geophys. Res.,* **85,** 7345–7349.

McConnell, J. C., McElroy, M. B., and Wofsy, S. C. (1971). Natural sources of atmospheric CO. *Nature,* **233,** 187–188.

McFarland, M., Kley, D., Drummond, J. W., Schmeltekopf, A. L., and Winkler, R. H. (1979). Nitric oxide measurements in the equatorial Pacific. *Geophys. Res. Let.,* **6,** 605–608.

NCAR. (1985). *The NCAR Eulerian Regional Acid Deposition Model.* Technical Note NCAR/TN-256 + STR, National Center for Atmospheric Research, Boulder, CO 80307, USA: 178 pages.

Noxon, J. F. (1981). NO_x in the mid-Pacific troposphere, *Geophys. Res. Lett.,* **7,** 1223–1226.

Prinn, R. G., Rasmussen, R. A., Simmonds, P. G., Alyea, F. N., Cunnold, D. M., Lane, B. C., Cardelino, C. A., and Crawford, A. J. (1983). The atmospheric lifetime experiment. 5: Results for CH_3CCl_3 based on three years of data. *J. Geophys. Res.,* **88,** 8415–8426.

Rasmussen, R. A., and Khalil, M. A. K. (1981). Atmospheric methane (CH_4): trends and seasonal cycles. *J. Geophys. Res.,* **86,** 9826–9832.

Rasmussen, A. N., and Khalil, M. A. K. (1984). Atmospheric methane in the recent and ancient atmospheres: concentrations, trends and interhemispheric gradient. *J. Geophys. Res.,* **89** (D7), 11599–11605.

Rinsland, C. P., Levine, J. S., and Miles, T. (1985). Concentration of methane in the troposphere deduced from 1951 infrared solar spectra. *Nature,* **318,** 245–249.

Rinsland, C. P., and Levine, J. S. (1985). Free tropospheric carbon monoxide concentrations in 1950 and 1951 deduced from infrared total column amount measurements. *Nature,* **318,** 250–254.

Seiler, W. (1974). The cycle of atmospheric CO. *Tellus,* **26,** 117–135.

Stockwell, W. R. (1985). A Homogeneous gas phase mechanism for use in a regional acid deposition model. *Atmos. Environ.* (submitted).

Zimmerman, P. R. (1977). *Tampa Bay photochemical oxidant study, Rep.* EPA 904/9–77–028, Environmental Protection Agency, Research Triangle Park, N. C. 1977.

Zimmerman, P. R., Chatfield, R. B., Fishman, J., Crutzen, P. J., and Hanst, P. L. (1978). Estimates on the production of CO and H_2 from the oxidation of hydrocarbon emissions from vegetation. *Geophys. Res. Lett., 5*, 679–682.

Zimmermann, P. H. (1984). *Ein dreidimensionales numerisches Transportmodell für atmosphärische Spurenstoffe.* Dissertation, University of Mainz, F. R. G.

Scales and Global Change
Edited by T. Rosswall, R. G. Woodmansee and P. G. Risser
© 1988 Scientific Committee on Problems of the Environment (SCOPE)
Published by John Wiley & Sons Ltd.

CHAPTER 7

Linking Terrestrial Ecosystem Process Models to Climate Models

BERT BOLIN

Department of Meteorology,
University of Stockholm,
Arrhenius Laboratory,
S–106 91 Stockholm,
Sweden

ABSTRACT

Climate models, particularly General Circulation Models (GCMs), could serve as a basis for the construction of models for biosphere–geosphere interactions. Such an approach requires the development of methods for how to describe ecosystem processes on the scale of 100×100 km, i.e. how the processes on a smaller scale could be parameterized. This fundamental problem in model building has been addressed in the development of parameterization schemes for physical processes in climate models. Present methods used for describing convection and cloud formation and for considering surface boundary layer processes are summarized. The problem of how to incorporate the role of terrestrial biota for the exchange of heat, moisture, and chemical compounds between the atmosphere and the ground in global models is discussed on the basis of existing ecosystem models that have been designed for analysis of local problems.

INTRODUCTION

The original definition of the term 'biosphere' referred to the part of the earth system in which there is life, i.e. the lower atmosphere, the top layers of the soils, the oceans, and the uppermost layers of the ocean sediments. It is interesting to note that meteorologists, physical oceanographers, and climatologists today call this thin shell around the earth the climatic system. In their pursuit to develop climatic models for this system, chemical and biological processes have been almost completely ignored. Climate models rather describe the thermodynamics of the atmosphere and the oceans as they are being driven by the spatially varying distribution of solar radiation and with due regard to the importance of the rotation of the earth. *Climate models describe the physical behaviour of the biosphere.*

With the major advances in our understanding of how the motions of the atmosphere and the oceans are being maintained as a background, we now ask ourselves: How do the life processes on land and in the sea, i.e. at the interfaces between the atmosphere and land and between the atmosphere and the sea, modulate the climate on earth? To answer this fundamental question we must broaden our approach and consider the *dynamics of the biosphere in its entirety, i.e. the interplay between physical, chemical, and biological processes on earth.*

It is well-known that present conditions for life on earth are, to a considerable degree, the result of the evolution of life on earth during hundreds of millions of years. This evolution is a truly global change of the biosphere, but a slow one. On shorter time scales we know about the variations between glacial and interglacial times that have occurred during the last one million years. We know particularly well the change from the last period of glaciation to the present interglacial period that took place 8000 to 15 000 years ago. Undoubtedly a close analysis of this major climatic change is of great interest in the study of biosphere dynamics as defined above.

The concept of 'Global Change' now is becoming a major issue in the environmental sciences due to a concern about the present rapid change of the environment. The focus is on the problem of biosphere dynamics on time scales from a decade to a few centuries. Although we certainly need to learn about natural variations of the biosphere in the past to understand the changes that are taking place today, we shall here primarily address the problem of physical, chemical, biological interactions on this short time scale. We shall assume that the basic geological setting is not changing except for possible changes of the topsoil by direct exploitation of natural resources by man (e.g. deforestation and changing land use) and changes due the natural response of ecosystems to changing abiotic conditions such as climate.

MODEL HIERARCHY AND SUBGRID SCALE PARAMETERIZATION

Some basic principles

O'Neill (1988) has discussed in some detail theories of hierarchies of ecological systems. He emphasizes that it is most important in developing models for ecosystem behaviour to define carefully the particular problem to be addressed and the scales in space and time that will be considered explicitly. Although processes on smaller scales may be essential for the behaviour of the system on the scale of direct concern, it is usually not feasible to proceed by considering the role of these subgrid scale processes by modelling them through successively increasing the scale (decreasing resolution) starting from the smallest observable scale. We should instead attempt a direct formulation of the role of subgrid scale processes for the processes dealt with explicitly in our model. Usually much of what happens at smaller scales is of no importance in this context and should accordingly be disregarded. Our objective must not be to analyse all aspects of the processes on a smaller scale but merely those of direct importance for the problem being considered.

We are concerned with global processes in the present context. To develop models for biosphere dynamics on this scale it is important to make use of the experiences gained in climate modelling. The most advanced models of the climate system (so-called general circulation models, GCMs, cf. e.g. Manabe and Stouffer, 1980) describe processes in the atmosphere and the sea by resolving features on scales larger than a few hundred kilometres in the horizontal directions and one or a few kilometres in the vertical. Altogether some 100 000 gridpoints are used, in the most detailed computations with such models, and even the advent of the next generation of computers will not permit better horizontal resolution than about 100 km in the case of long-term integrations. As a matter of fact, much cruder models of the climatic system are often used, particularly for principle studies aimed at a better understanding of the interplay and relative importance of different fundamental processes. It is accordingly necessary to develop methods for how to consider ecological processes (at the atmosphere–land/ocean interface) as well as the transfer of important chemical compounds within the atmosphere and the sea by using models with a resolution of a few hundred kilometres in the horizontal directions. Climate models resolve, on the other hand, processes in time with much detail. Changes are computed by forecasting the changing weather from day to day but the analysis of the results is restricted to the consideration of the statistical manifestation of weather such as the seasonal or annual mean temperature, precipitation and cloudiness, the characteristic variability of weather phenomena as well as the motions of the sea and their transfer of heat, momentum, and water in the atmosphere and the sea. It is therefore possible in the model to relate statistically temporal variations of ecosystem behaviour to meteorological variations if required.

Our problem thus becomes how to consider statistically the importance of those biological and chemical processes on smaller spatial scales that influence the large-scale features of the system. Before doing so it may be of some interest as an illustration to describe the approach used when parameterizing physical sub-grid scale processes in climate models. We note for example that the formation of clouds and precipitation, particularly in statically unstable air masses where convective clouds are formed, show important features on scales of the order of kilometres. Also, vertical exchange of heat and momentum both in the atmosphere and the sea next to the atmosphere/land and atmosphere/sea interfaces is accomplished by small scale turbulence, with characteristic eddy dimensions of a few hundred metres.

It is first important to recognize that climate models basically treat the *flux of energy, momentum, and water*. Therefore, when parameterizing the role of sub-grid scale processes we need primarily to consider their role in this regard. If we are successful in formulating models for the large scale behaviour of the biosphere in this way, we may use these deduced features as given when attempting to analyse how small scale characteristics of the biosphere are generated and maintained.

Parameterization of cloud formation and precipitation

Clouds form when the humidity reaches 100%. For precipitation to occur, however, ice nuclei or a broad size spectrum of cloud water droplets with increased probability for coalescence are required. These conditions apply to the microscale and since the moisture field in the atmosphere shows large sub-grid scale variations they cannot be applied directly by using mean values for temperature and humidity on the scale of resolution of the climate model. A crude description of clouds and some conditions for precipitation release is, however, obviously necessary, not only because these phenomena themselves are important meteorologically, but also because cloud formation implies transformation of latent heat into sensible heat, which may mean a drastic change of the vertical static stability of the air. This in turn changes the conditions for vertical transfer of energy, momentum and water, which is of prime importance for the large-scale dynamical behaviour of the system. Let us limit ourselves to considering the formation of convective clouds.

Convection occurs when the vertical temperature gradient (lapse rate) exceeds the temperature decrease that a rising air parcel experiences, which is about $1\,^\circ\mathrm{C\,km^{-1}}$ in non-saturated air (dry-adiabatic lapse rate) and $0.4\text{--}1.0\,^\circ\mathrm{C\,km^{-1}}$ in saturated air (moist-adiabatic lapse rate), the value being smaller the higher the temperature is. Such vertical instability is due to the changing temperature distribution brought about by the large-scale motions of the atmosphere, or by heating or cooling due to radiative processes, or heat exchange with the underlying surface. Convection is a rapid process compared with the rate of change of large-scale systems and one may therefore, to a first approximation, assume that the process of vertical equilibration of a convectively unstable air column takes place instantaneously. The new state thereby created is characterized by neutral static stability and complete mixing within the vertical layer, where the instability develops and implies vertical transfer of heat, moisture and momentum when this vertical mixing is accomplished.

Convective clouds have a horizontal dimension of a few kilometres, which is small compared to the resolution of the global model. Only a part of the grid area is therefore usually covered by clouds, and downward motion takes place in the clear areas between the clouds. Obviously the average relative humidity on the scale of the grid is therefore less than 100%. In an area with deep convection it may merely be about 80%, while it is closer to saturation in the case of shallow convection. To parameterize the conditions for initiation of convection we must determine empirically the grid scale features (e.g. relative humidity) that should prevail in order for convection to occur. Reasonably satisfactory statistical formulations have been developed in this way and been tested by extensive model experimentation (see e.g. Ogura, 1985).

Atmospheric boundary layer modelling

The role of the atmospheric surface boundary layer for the large-scale motion of the atmosphere is primarily associated with its vertical transfer of momentum, heat, and moisture between the surface of the earth and the free atmosphere due to small scale turbulence. (Transfer of biologically important constituents through the boundary layer also takes place and will be considered later.) We are not interested in the detailed structure of the turbulent eddies that bring about the transfer, but merely need to account for the transfer as accurately as possible. One way is, for example, to make use of Monin and Oboukhov similarity theory (cf. Panofsky, 1985) which, on dimensional grounds, permits a reduction of all possible quasi-steady state vertical profiles of wind and temperature to two universal profiles. With the aid of these, knowledge about the characteristic features of the earth's surface (roughness, temperature, and wetness), and wind, temperature and humidity in the atmosphere above the boundary layer, the vertical fluxes can be determined. A simple parameterization of transfer through the boundary layer is possible in this way. Considerably more complex parameterization schemes are, however, often used (cf. e.g. Panofsky, 1985).

The land surface within an area of the size of the grid used for large-scale modelling is seldom homogeneous. Since we are only able to describe the average vertical wind and temperature profiles in a grid and also need to assign average characteristic values for surface roughness, surface temperature and wetness the question arises how these should be determined. Presumably they are not simple arithmetic averages over the area considered but probably depend non-linearly on more detailed features of the surface and the bulk values for wind and temperature in the free atmosphere. Not even the most advanced climate models include any sophisticated treatment of such features. As a matter of fact roughness has often been assumed not to vary at all over land in most experiments with climate models. Similarly soil moisture has been modelled very approximately without much consideration of physical and biological process of importance for evapotranspiration. Still, to a first approximation these models describe the role of boundary layer processes for the large-scale atmospheric motion systems reasonably well. The problem will be further considered in the next section.

A closer comparison between the present distribution of climate on earth and model simulations using the most advanced climate models shows significant discrepancies. It is difficult to determine to what extent these are due to inadequate treatment of the internal dynamics of atmospheric or oceanographic processes or to improper treatment of the transfer processes between the atmosphere and the underlying surface, particularly the irregular land surface. It seems plausible that the latter may be of some significance. We need to analyse present climate models with regard to their sensitivity to the treatment of exchange processes at the earth's surface and presumably develop more realistic models for their description, including the role of biological processes, to find out if exchange processes between the atmosphere and the land surface are adequately treated.

It is important to emphasize that the reverse problem, i.e. the determination of how detailed features of climatic conditions in the boundary layer are created by sub-grid scale irregularities of the land surface, requires a more cumbersome analysis (Gates, 1985). It is, however, obviously necessary that mean conditions on the scale of the grid of the climate model agree with observed climatic conditions to expect that successful simulations of subgrid scale features can be made.

MODELLING BIOTIC PROCESSES IN GLOBAL MODELS OF THE BIOSPHERE

Principal considerations

Let us, in the following, limit ourselves to terrestrial processes. It has already been emphasized that it is not feasible to develop a hierarchy of models starting from the species level and proceeding step-wise to the scale of concern which is required to treat large-scale dynamics of ecosystems. The scientific challenge is, rather, to attempt the formulation of simple hypotheses and try to verify to what extent they may be applicable. Failures in such an approach still means increased insight.

We shall use the experiences gained in developing climatic models as the basis when treating biological and chemical processes. In addition to the consideration of transfer of energy, momentum, and moisture, as now included in climatic models, we also need to consider

—the modulation of water flux between land and the atmosphere due to the presence of terrestrial biota
—the role of soil processes for transfer between the atmosphere and the land surface.

—the role of terrestrial biota for the generation and destruction of chemical constituents in the atmosphere
—the transfer of chemical constituents in the atmosphere
—chemical, often photochemical, interactions within the atmosphere

The prime objective is: How to model the relevant processes and how to develop appropriate parameterization with due regard to the scales we wish to resolve, i.e. those larger than a few hundred kilometres

The problem of parameterizing fluxes does not only concern the development of methods to average spatially over inhomogeneous land surfaces, but also to capture the importance of *long-term changes of biota and soils* which often in reality are obscured by the variations that occur on short time scales i.e. those associated with the daily cycle, weather variations, and the annual cycle. The role of such changes of the characteristics of biota and soils for example due to slow changes of climate has seldom been considered. Important feed-back mechanisms may thereby have been excluded in the analysis of long-term changes of climate. In the following sections we shall consider the transfer processes between the atmosphere and land with particular consideration of the role of the biota. A number of microclimatological models dealing with the detailed transfer problem have been developed in recent years. We shall discuss the global modelling problem with these local models as a starting point. It is appropriate first to consider the exchanges of energy and water, which are closely coupled.

Energy and water transfer between the land surface and the atmosphere

For detailed analyses of energy and water transfer between the land surface and the atmosphere reference is made to Jones (1983), Dickinson (1984), Dickinson and Hanson (1984), Verstraete (1985) and Sellers, Mintz, and Sud (1985). We shall not reproduce their results here, but merely emphasize some principally important points. The fundamental processes to be considered are illustrated in Figure 7.1. The heat balance of a thin surface layer of soil may be written (Verstraete, 1985)

$$R_n - LH - H - S_h + G_n = 0 \qquad (1)$$

where

R_n = net energy gain due to radiative flux
H = sensible heat flux from the land surface to the atmosphere
LH = latent heat flux from the land surface to the atmosphere
S_h = storage of heat in the top soil layer
G_n = heat flux from below into the top soil layer

Similarly we can formulate the moisture balance of this top layer as follows

$$P - E - R_w - S_w + G_w = 0 \qquad (2)$$

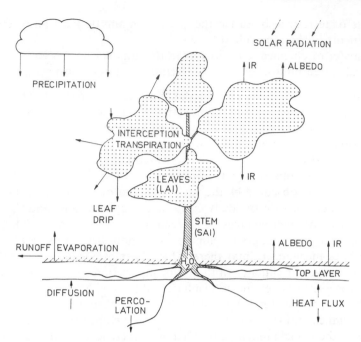

Figure 7.1   The fluxes of water (on the left side) and energy (on the right side) in the atmospheric surface boundary layer and the topsoil as modulated by the presence of a vegetation cover (after Dickinson, 1984)

where

P = precipitation rate
E = evaporation rate
R_w = runoff rate
S_w = storage of water in the top soil layer
G_w = water flux from below into the top soil layer

Within the atmosphere water vapour flux (E) and heat flux $(LE + H)$ are maintained by turbulent processes which are included in present climate models. In the soil, water is transferred by water diffusion, which depends on the soil water potential and which in turn is a function of soil porosity. Heat flux is primarily a molecular diffusion process, but to some extent also dependent on the water transfer in the soil. In addition, however, water is transferred by the plants extending their roots into the ground and their above-soil parts into the atmosphere. If the ground is covered by vegetation most of the water flux (E) is due to evapotranspiration from plants, which obviously also affects the heat transfer between the ground and the atmosphere. The water flux due to the presence of plants depends on plant processes, particularly the stomatal conductance g_s, and the water uptake by the roots. Following Jones (1983) we may write.

$$g_s = g_o + g_r F_R F_S F_P F_T S \qquad (3)$$

where

g_o = minimum conductance
g_r = maximum less minimum conductance
F_R = influence of solar radiation
F_S = specific humidity factor
F_P = water potential factor
F_T = temperature factor
S = the ability of the root system to supply the water that the leaves would be able to evaporate under prevailing conditions

All the latter five factors are between zero and unity and their magnitudes depend on atmospheric conditions (temperature, humidity, solar radiation) and on soil and biota response to such abiotic conditions. Similarly, we need to formulate the transfer of water from the soils to the roots which depends on the potential for water transfer into the root, and on a root geometry factor which takes into account: root length per unit volume of soil, distance between roots, root radii, and root depth.

Rather than considering these processes in some detail as outlined above, most climate models treat the air-land surface exchange as well as transfer processes within the soil in a very simplified manner usually not including biological processes at all. Evapotranspiration is simulated by evaporation from a soil moisture reservoir which has a prescribed maximum storage capacity. Overflow is considered as runoff and the rate of evaporation depends on the degree to which this soil moisture reservoir is filled. Accordingly, $G_w = G_n = 0$ in equations (1) and (2) and no further considerations of soil and biota processes are required.

A more elaborate treatment would be to prescribe characteristics of the soil (e.g. porosity, heat conductivity) and of the biota (e.g. minimum and maximum water conductance, their sensitivity to temperature, humidity etc., root geometry and water uptake capability). A simple model for transfer processes in the soil below the top soil layer is then required to describe the exchange of heat and moisture with deeper soil layers by molecular diffusion.

Microclimatological models of this kind have been developed (cf. Verstraete, 1985). Considerably more computing capacity is, however, required to include such a description in a general circulation model and we may legitimately ask if soil and biota characteristics are known well enough to justify such an effort. It should, however, be done for exploratory purposes. The most advanced model of this kind, which has been developed so far is due to Dickinson (1984).

The role of ecosystem processes in global models

The crucial question to address concerns the mutual interplay between abiotic factors that describe atmospheric conditions, i.e. weather and climate (solar radiation, temperature, humidity, cloudines, precipitation, etc.) and the inorganic substrate (soil conditions) on one hand, and the characteristics of the biota including the organic component of the soil on the other. Ecological models of the kind described briefly above have been reasonably successful for the local scale. Dickinson (1984), Dickinson and Hanson (1984) and Sellers, Mintz, and Sud (1985) now propose to use similar models for the scale of the General Circulation Model grid. We should then consider them as conceptual models, but cannot expect that functional relationships and rate coefficients that have been derived for small-scale processes can be directly adopted when concerned with the scales of the order of 10^4 km^2 (100 × 100 km). This difference between local models and means for parameterization of such processes for the scale of a grid square in a global model is in principle similar to our earlier observation that cloud formation and precipitation formation occur at a mean relative humidity in a grid squre well below 100%.

A serious problem then arises: How do we test our models? Testing against real data is necessary, but it is immediately obvious that such attempts consitute major observational efforts particularly because we need to consider spatially very heterogeneous land surfaces. The planning of an International Geosphere Biosphere Programme (IGBP) should attempt to define the observational design that would be optimal. Both satellite and ground observations are obviously required.

It is important to recognize, however, that it would also be desirable to develop testing procedures of the global model directly related to their use diagnostically, including biological and chemical processes as sketched above. We may assume as a first approximation that the present (or preferably the preindustrial) ecosystem distribution is (or was) in equilibrium with prevailing climate and ecosystem distribution. How well are we able to simulate these distributions when including the consideration of biotic processes? How sensitive are our results to the choice of functional relationships and rate constants that describe the ecosystem behaviour in the model? In this way we are, however, only able to deduce to what extent a given model is compatible with observed conditions but not to exclude the possibility that other model formulations might be superior. The approach is one of trial and error, but the testing of models in this way would undoubtedly increase our understanding of large-scale ecosystem dynamics.

Such an experimental approach must proceed step wise. It is advisable to start with very simple models, perhaps even simpler than the one outlined in the previous section. Some such analyses have been attempted and might be of interest in the present context.

(i) Dependance of surface temperature maximum on stomatal dynamics

It is well known that the daily variations of atmospheric temperature and moisture largely determine the rhythmn of stomata opening and closure. Since evapotranspiration thereby is modified, there is a direct feedback from the biota on moisture and temperature variations in the atmosphere. The weather forecasting model, developed at the European Centre for Medium Range Weather Forecasts, has been tested using different formulations of the process of evapotranspiration (L. Bengtsson, private communication). In one case a simple 'bucket' formulation was used in which availability of water for evaporation remained essentially the same throughout the day and the rate of evaporation was determined by the flux of water vapour through the atmospheric boundary layer. In another case an attempt was made to prescribe roughly the daily variation of the stomatal conductance being comparatively large early during the day and decreasing as the temperature increases and humidity decreases towards noon and during the afternoon. Since the time step of integration of a general circulation model is considerably less than an hour, the daily cycle of meteorological variables is obtained with good resolution. Experiments show that the maximum daily temperature is significantly higher in the latter case, because of decreasing evapotranspiration during the day caused by decreasing stomatal conductance. Because the ecosystem model is part of a large scale three-dimensional forecast model, the horizontal advection of air and other changes of weather is implicitly accounted for. If we accept that the daily variation of temperature in addition to being influenced by changing weather conditions is primarily dependant on the variations of stomatal conductance we can determine an optimum formulation for the dependance of stomata on atmospheric temperature and moisture by trying to find out how the best possible agreement can be obtained between daily temperature variations in the model and in reality.

(ii) The dependance of the monthly mean temperature on availability of water for evapotranspiration

Also, the monthly mean temperature in a region is greatly influenced by availability of soil moisture for evapotranspiration, which in turn, of course, depends not only on precipitation but also on the characteristics of the plant cover (e.g. species composition). An ongoing climate change may well change these and thereby significantly modify the ratio of evapotranspiration to runoff. The rapidity with which this takes place depends on the rate of renewal of species in a plant community. A few experiments which have bearing on this problem have been carried out by Shukla and Mintz (1982). With the aid of a general circulation model for climatic studies they have determined the global surface temperature distribution in July that would prevail in the mean for two extreme cases:

a) constantly wet soils, i.e. enough water would be supplied for evaporation even if precipitation were inadequate

b) constantly dry soils, i.e. all precipitation would run off or penetrate into the soil and not be available for evapotranspiration.

Figure 7.2a and b show the the computed July temperatures for these two cases assuming the present distribution of solar radiation. The difference is shown in Figure 7.2c. The modulation of surface temperature due to the presence or absence of water in the soil is very significant, the difference being 20–30 °C in the interior of the major continents. At high latitudes present climate is close to that of wet soils, but in the subtropical arid and semi-arid regions of Europe and North Africa the climate is 10–20 °C hotter than in the wet case.

With these two extreme cases as starting points it would be most essential to explore in which way simple ecosystem models for different types of biomes would modify the climatic regimes and in this way get a first idea about the response characteristics between abiotic and biotic factors in the climatic

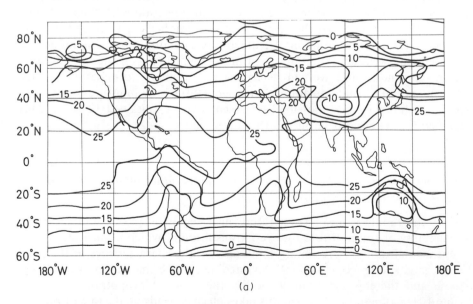

Figure 7.2   July surface temperature (0 °C) as simulated by a General Circulation Model assuming present distribution of solar radiation, ocean surface temperature, and albedo of the earth's surface; (a) the soil is always wet, (b) the soil is always dry and (c) shows the difference between (b) and (a) (after Shukla and Mintz, 1982). (Copyright 1982 by the AAAS)

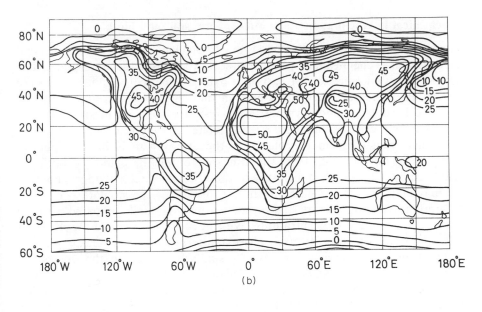

(b)

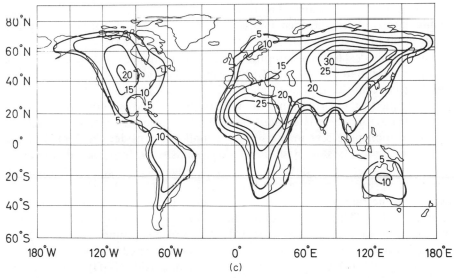

(c)

system. It is particularly important to relate ecosystem characteristics (e.g. biomass, soil properties) to the regional characteristics of the hydrological cycle. Attempts in this direction have been made although not by combining ecosystem models and climatic models but merely considering the local scale. A comparison with climatic changes during the last 15 000 years has also been made (Solomon *et al.*, 1984).

Biochemical interactions between atmosphere, soil, and biota

The composition of the atmosphere is due to biochemical processes in soil and biota as well as chemical and photochemical processes in the atmosphere. The atmospheric concentration of carbon dioxide is markedly influenced by the terrestrial and marine ecosystems and, in the long term perspective, also by geological processes. Photosynthesis of terrestrial plants, and accordingly CO_2 variations in the atmosphere, are directly related to evapotranspiration in that the water transfer from the plants to the atmosphere is in the opposite direction of the CO_2 flux from the atmosphere into the plant and both are regulated by the opening and closure of the stomata. To a first approximation we may assume that the humidity in the interior of the stomata is close to saturation, while the CO_2 partial pressure usually is small compared with the exterior atmospheric concentrations. We can accordingly relate the fluxes of CO_2 and water between the atmosphere and terrestrial ecosystems.

The transfer of some other gases is similarly related to the process of photosynthesis. Nitric oxide (NO) behaves, however, differently. If atmospheric concentrations are above an equilibrium concentration (compensation point) there is a transfer of NO into the plant, while the opposite is true for low atmospheric concentrations. The transfer processes are also dependent on the availability of fixed nitrogen in the soil and accordingly influenced by the application of fertilizers (cf. Johansson and Granat, 1984).

There is also exchange of sulphur in gaseous form between the atmosphere and plants (Hällgren *et al.*, 1982). SO_2 enters the stomata, but for increasing ambient concentrations the total flux of sulphur into the plant decreases, possibly due to the formation and emission of reduced sulphur compounds.

Another set of gases is emitted from the terrestrial ecosystems because of decomposition processes in the soil. We shall not consider the details of the biochemical processes but wish merely to emphasize the importance of relating them to the abiotic, physical processes that are largely governed by atmospheric processes. We need, for example, to find ways of how to determine the rate of their emissions as dependant on precipitation and soil characteristics. There is a profound difference between conditions of dry and wet soils. In the former case primarily oxidized compounds (e.g. CO_2) are emitted, while in the latter case reduced gases are formed (e.g. CH_4 and from basic soils, NH_3). How is the emission of nitrous oxide (N_2O) dependent on the climate? The problem becomes difficult because release and transfer of gases from the soil into the atmosphere is an intermittent process very much governed by the wetness of the soil. We need to establish conditions for how such emissions are related to the large scale parameters of a global model, probably in a satistical sense, in order to be able to analyse the regional and global biogeochemical cycling of elements and compounds. Criteria must be established using the large-scale variables to determine how one or the other process on the small scale is switched on or off. This is not simple and work of this kind has merely begun.

CONCLUDING REMARKS

The global biosphere is a complex system. To gain an understanding of its structure and dynamic features, it is necessary to not only increase our knowledge about the detailed processes, but also to develop models of how global interactions take place. Our attempts to analyse the detailed physical, chemical, and biological processes need, in this context, be guided by an advancement of our understanding of the latter. It is necessary to develop a strategy of data gathering that serves both these purposes. Climate research during the last decade may serve as a useful example of how to approach this difficult problem in a systematic way. While realizing the necessity of a systematic and long lasting effort of observing the atmosphere, the oceans, land, and life on earth, such programmes must remain flexible enough to permit those modifications and even sometimes improvisations that are necessary to maintain a viable programme.

REFERENCES

Dickinson, R. E. (1984). Modeling evapotranspiration for three-dimensional global climate models. In *Climate Processes and Climate Sensitivity*, pp. 58–72. Geophys. Monograph 29, Vol 5., Am. Geophys. Soc.

Dickinson, R. E., and Hanson, B. (1984). Vegetation-albedo feedbacks. In *Climate Processes and Climate Sensitivity*, pp. 180–186. Geophys. Monogr. 29, Vol 5. Amer. Geophys. Soc.

Gates, W. L. (1985). The use of general circulation models in the analysis of the ecosystem impacts on climatic change. *Climatic Change*, 7, 267–284.

Hällgren, J. E., Linder, S., Richter, A., Troeng, E., and Granat, L. (1982). Uptake of SO_2 in shoots of Scots pine: field measurements of net flux of sulphur in relation to stomatal conductance. *Plant, Cell and Environment*, 5, 75–83.

Johansson, C., and Granat, L. (1984). Emission of nitric oxide from arable land. *Tellus*, 36B, 25–37.

Jones, H. G. (1983). *Plants and Microclimate*. Cambridge University Press: 323 pages.

Manabe, S., and Stouffer, R. J. (1980). Sensitivity of a global model to an increase of CO_2 concentrations in the atmosphere. *J. geophys. Res.* 85 (C10), 5529–5554.

Ogura, Y. (1985). Modeling studies of convection. In Saltzman (Ed.) *Issues in Atmospheric and Oceanic Modelling. Advances in Geophysics*, 28, pp. 387–421. Academic Press.

O'Neill, R. V. (1988). Hierarchy theory and global change. (Chapter 3, this volume).

Panofsky, H. A. (1985). The planetary boundary layer. In Saltzman (Ed.) *Issues in Atmospheric and Oceanic Modelling. Advances in Geophysics*, 28, pp. 359–385. Academic Press.

Sellers, P. J., Mintz, Y., and Sud, Y. C. (1985). *A simple biospheric model (SiB) for use within general circulation models*. Report Laboratory for Atmospheres NASA/ Goddard Space Flight Center, Greenbelt, MD 2077.

Shukla, J., and Mintz, Y. C. (1982). Influence of land-surface evapotranspiration on the earth's climate. *Science*, **215**, 1498–1501.

Solomon, A. M., Tharp, M. L., West, D. C., Taylot, G. E., Webb, J. W., and Trimble, J. L. (1984). *Response of Unmanaged Forests to CO_2 Induced Climatic Changes: Available Information, Initial Tests and Data Requirements.* DOE/NBB–0053. Nat. Techn. Information Serv. U.S. Dept. Comm., Springfield, Virginia.

Verstraete, M. M. (1985). *A Soil–Vegetation–Atmosphere Model for Microclimatological Research in Arid Regions.* National Center for Atmospheric Research. Cooperative thesis No 88.

Scales and Global Change
Edited by T. Rosswall, R. G. Woodmansee and P. G. Risser
© 1988 Scientific Committee on Problems of the Environment (SCOPE)
Published by John Wiley & Sons Ltd.

CHAPTER 8

Simulation Models of Forest Succession[1]

HERMAN H. SHUGART[2], PATRICK J. MICHAELS[2], THOMAS M. SMITH[3],
DAVID A. WEINSTEIN[4] AND EDWARD B. RASTETTER[5]

INTRODUCTION

The current discussions of biospheric dynamics and global ecology (Risser, 1986) come at a time when there is a renewed interest in time- and space-scales in ecological systems. An appreciation of scales is a clear prerequisite to unifying the dynamics of atmospheric and oceanographic process with the dynamics of ecosystems on the terrestrial surface. Of particular importance is a knowledge of the patterns of dominance (in the sense of controlling pattern) of particular causal factors at particular scales. The categorization of controlling factors important at different space and time scales in particular ecosystems has been the topic of several reviews (Delcourt *et al.*, 1983; Pickett and White, 1985). In 'hierarchy theory' (Allen and Starr, 1982; Allen and Hoekstra, 1984; O'Neill *et al.*, 1986; Urban *et al.*, 1987), one sees a focus on expressing relevant mathematical developments in a manner that can provide insight into the ways ecosystems are structured at different scales.

Historically, the complex unravelling of ecological interactions was evident in A. S. Watt's (1925) early work on beech forests and elaborated in his now classic paper on pattern and process in plant communities (Watt, 1947). When one inspects Tansley's (1935) original definition of the ecosystem, one finds the same concepts that one sees in hierarchy theory, today.

Of course, the Watt/Tansley ecosystem paradigm has been reintroduced as a major ecosystem construct in ecological studies. One conspicuous

---

[1] Research supported by the National Science Foundation's Ecosystem Studies Program (Grant Number BSR−85−10099) to the University of Virginia.
[2] Department of Environmental Sciences, The University of Virginia, Charlottesville, Virginia 22903 USA.
[3] Department of Environmental Biology, Research School of Biological Sciences, The Australian National University, Canberra, A.C.T. 2001.
[4] Ecosystems Research Center, Corson Hall, Cornell University, Ithaca, New York 14850.
[5] The Ecosystems Center, Marine Biological Laboratory, Woods Hole, Massachusetts 02543.

re-introduction of these concepts was Whittaker's (1953) review which used the Watt pattern-and-process paradigm to redefine the 'climax concept'. These same ideas are also found in ecosystem concepts developed by Bormann and Likens (1979a, b) in their 'shifting-mosaic steady-state concept of the ecosystem' as well as in what Shugart (1984) called 'quasi-equilibrium landscape'.

The essential idea behind the Watt (1925, 1947) paradigm is that in ecosystems controlled by sessile organisms, the temporal dynamics at the scale of the individual organism are almost by necessity non-equilibrium dynamics. This is most apparent in forest systems where the spatial scale of the individual organisms (the canopy trees) is relatively large. Considering forests as an example, the space below a canopy tree has reduced light levels and a considerably altered microclimate due to the influence of the tree. These conditions determine the species of trees that can survive beneath the canopy tree. Upon the death of the canopy tree, the shading is eliminated and the environment is changed. In cases in which the canopy tree dies violently (e.g. broken by strong winds), the changes in the microenvironment are extremely abrupt. The death of the canopy tree initiates a scramble for dominance among the smaller trees that were persisting in the environment created by the canopy tree and seedlings that establish themselves in the high-light environment. Eventually, one of the trees becomes the canopy dominant. The establishment of a new canopy dominant represents the closure of the death/birth/death cycle that can be thought of as the typical small-scale behaviour of a forest.

In ecosystems other than forests but still dominated by sessile organisms, one would expect the same sorts of dynamics. This nonequilibrium behaviour at fine spatial scales has been noted in a diverse array of ecosystems including coral reefs (Connell, 1978; Huston, 1979; Pearson, 1981; Colgan, 1983), fouling communities (Karlson, 1978; Kay, 1980), rocky inter-tidal communities (Sousa, 1979; Paine and Levin, 1981; Taylor and Littler, 1982; Dethier, 1984), and a wide range of heathlands (Christensen, 1985).

The ecosystems that are both historically and currently the most studied in this regard are forests. For this reason, it is worthwhile to elaborate the details of the death/birth/death process in forests. In forests, the non-equilibrium dynamics are quasi-periodic with the period a function of the growth rates and potential longevities of the individual organisms (Shugart, 1984). This 'cycle' can be modified by a variety of factors. One important consideration is the manner of death of the dominant tree.

Some trees typically die violently catastrophically and the attendant alterations of environmental conditions at the forest floor (and thus the effect on the regeneration of potential replacements) are very abrupt. Typically these abrupt changes are associated with exogenous disturbances but there are some species of trees that are 'suicidal' in that mature trees flower but once and die to release canopy space to their progeny (Foster, 1977). Some trees tend to 'waste-away' after they die so that the changes in the microenvironment that they control are more continuous. Some trees tend to snap at the crown when torn down by winds; others are heaved over at the roots exposing mineral soil. All of these modes of death (and others) influence the stochastic regeneration success of the trees that form the next generation.

It is an open question as to whether mode of death or mode of regeneration is the strongest determinate of pattern of diversity in forests. Both are attributes of the various tree species and may be strongly interrelated. One aspect of the mortality of canopy trees and the associated opening in the forest canopy ('gap formation') is the size of the gap that is created. Several authors (van der Pijl, 1972; Whitmore, 1975; Grubb, 1977; Bazzaz and Pickett, 1980) have discussed species attributes that are important in differentiating the gap-size-related regeneration success of various trees. The complexity of the regeneration process in trees and its stochastic nature makes it nearly impossible to predict the success of an individual tree seedling. Most current reviewers recognize this and tend to discuss regeneration in trees from a pragmatic view that the factors influencing the establishment of seedlings can be usefully grouped in broad classes (Kozlowski, 1971a, b; van der Pijl, 1972; Grubb, 1977; Denslow, 1980).

In models of forests and other sessile-organism dominated ecosystems, the death and replacement of large individual organisms creates a cyclical, non-equilibrium response. There are good reasons to expect such dynamics to prevail in natural ecosystems as well (Whittaker and Levin, 1977).

In the next section we will review the functioning of several of the extant computer models of ecological section. These models all operate on the assumption that the landscape can be considered a dynamic mosaic of element at the scale of a single dominant plant—an approach that we have called the Watt paradigm and whose historical antecedents and importance in ecology we have already discussed. Following this review and some examples of ecological models at different resolutions, we will discuss the atmospheric influences on ecological dynamics. The focal ecosystems in these discussions will be forests.

ECOLOGICAL SUCCESSION MODELS

Over the past two decades there has been a remarkable development of computer models designed to simulate the dynamics of ecological succession. Many of the basic concepts used in these models originate in the works of Clements (an emphasis on dynamic interactions, 1916), Tansley (the ecosystem concept, 1935), Gleason (the importance of species attributes in dynamic systems, 1939), and Watt (the relationship between internal dynamics and spatial patterns, 1947). The theories that were developed by these early ecologists proved difficult to apply in a formal mathematical fashion to the complex natural systems (with a multiplicity of potentially important temporal dynamics) for which they were intended. The eventual development of mathematical models based on these concepts was clearly catalysed by the increased availability of computers.

Not only have computers become widely available to ecologists over the past several years, but there has also been a great reduction in the cost per computational unit operation. At present, there are several hundred succession models of forest ecosystems alone (Shugart, 1984). This set of succession models includes examples that have proven capable of predictions that can be used for purposes of application. Other models have inspired a continued theoretical development of ecological concepts. One of the conceptually important aspects of the use of succession models in both theoretical and applied contexts has been the emphasis that the developers of the models have placed upon 'scaling-up' the consequences of natural history of plants, physiology and demography.

Succession models take on a wide range of mathematical forms. Their richness in formulation originates to an extent in the diverse objectives and training of the model designers. These differences may also originate from different theoretical constructs as to what is important in the functioning of a given ecosystem. In this sense, the models represent a complex set of *a priori* hypotheses about the function and behaviour of ecosystems. The limitations, failures, and successes of these models potentially reflect functional patterns in the real systems that they represent.

We will review several of the more successful modelling paradigms used in succession models and to describe the underlying assumptions and thus the theoretical implications of these models. We will also discuss what we feel is a logical next area of development in succession models as tools for understanding global change. This is a more explicit consideration of the dynamics of the abiotic systems that drive ecosystem succession.

Most succession models have been developed for forested systems (Munro, 1974; Shugart, 1984). This is in part due to a simultaneous (and relatively independent) interest in modelling forest dynamics that developed in forestry and ecology in the early 1970s. In ecology, this was a period of great interest in systems approaches from engineering sciences and was also a period when several large coordinated research programmes used computer modelling as a way to organize a diverse array of studies of ecosystem processes. The most conspicuous of such programmes were associated with the International Biological Programme (IBP).

Many of the ecological succession models in use orginated from models developed in the early 1970s. The development of these models occurred at a time when research institutions were obtaining large fast computers, when there was strong motivation to increase the quantitative nature of ecology and when there was a considerable interest about the temporal dynamics of ecological systems.

At about the same time, there was also considerable interest in modelling, 'computerization', and quantitative methods in forestry (Munro, 1974). There were several new forestry practices being considered, such as the fertilization of forests and the development of genetically improved trees for plantations. These new and relatively untried practices were potentially beyond the prediction range of the forest yield tables that formed the backbone of predictive forestry. This was because yield tables had been calibrated with data based on more conventional forestry practices. Questions such as, 'How many trees per unit area should one plant to produce a maximum yield of pulp wood, if these trees grow 20% faster than the present stock?' provided an incentive for the development of models that could provide stand yield predictions based on tree growth.

Many of the forestry models were (and are) designed for the projection of tree size, tree density, timber yield, and other similar state variables for forests in which the regeneration of trees is controlled. For this reason, most of these models are not able to simulate the dynamics of a forest beyond one tree generation and are not succession models in the sense that most ecologists think about succession. However, the models are important in their role in the historical development of succession simulators and because they often contain a level of realism in the representation of the growth and competition processes that is rarely found in succession models.

A REVIEW OF SUCCESSION MODELS

There are several hundred computer models of ecological succession and a rich array of approaches. Reviews of the forestry literature are found in Munro (1974) and a compilation of examples from forestry in Fries (1974). Shugart (1984) reviews several ecological models of forest dynamics and explores the theoretical implications of gap models in particular. In the present discussion, we will review four different modelling approaches that are arranged along a spectrum according to the dimensionality of the interactions among the simulated plants. In this spectrum, Markov models simulate the change in state of a simulated area in time; gap models simulate plant-to-plant interactions in the vertical dimension; transect models simulate pattern in one horizontal dimension; spatial models simulate in two and, in some cases, three spatial dimensions.

Markov models

Markov models of succession are mathematically and conceptually the most straight-forward of the succession models that are presently in use. The models share obvious relationships to other quantitative approaches used in plant ecology. The models can be solved by hand (or on a small computer). Markov models are constructed by determining the probability that the vegetation on a prescribed (usually relatively small area) will be in some other vegetation type after a given time interval. It is an essential requirement of these models to have a scheme for classifying the vegetation into identifiable categories.

The manner in which the vegetation states are classified has varied across applications of Markov models. Horn (1975a, b; 1976) used the species of a canopy tree as the states of a well-known Markov model developed for a forest near the Institute of Advanced Studies in Princeton, New Jersey. The time interval of this particular model was the generation time of canopy trees. Waggoner and Stephens (1971) categorized the forest types according to the most abundant species (in terms of individual trees over 12 cm Diameter at breast height—DBH) on 0.01 ha plots located on the Connecticut Agricultural Experiment Station and applied the model over uniform time intervals. These two approaches for identifying the states of the forest (categorization by attributes of a dominant individual as in Horn's (1975a, b; 1976) model or by attributes of an aggregate of an individuals as in Waggoner and Stephens' (1971) model) represent most of the applications in ecology although a variety of other schemes could possible be used. For example, one could categorize a small plot of land by both the species of the largest individual and by the number of individuals (e.g. highly-stocked White Oak dominated type, understocked Loblolly Pine Stands, etc.). Hool (1966) used this approach in developing a Markov model of stand change over a large area.

In a Markov model, the number of model parameters is a function of the square of the number of states (or categories in the model). Thus, in the development of a Markov model, one is forced to trade-off between the increased resolution in being able to enumerate many different system states and the parameter estimation problems that attend this greater resolution. In Waggoner and Stephens' (1971) simulation, a 40-year long record of 327 regularly remeasured sample plots was used to compute the model parameters. This means that the change in state of about 16 plots on average was used to estimate each of the transition probabilities. One feature of Markov models is that the relatively uncommon transitions from one state to another need to be estimated with equivalent precision to that of the other more common transitions. This feature creates a need to observe the frequency of occurrence of rare transitions between states and causes an emphasis on large remeasurement data sets as necessary to parameterize a Markov model that has very many states.

An alternative to direct measurement to determine the parameters of a Markov model is to develop theoretical constructs that allow the estimation of the model parameters on some other basis. For example, Horn (1975a, b; 1976) assume that the proportion of trees of a given species found growing below a canopy tree indicated the transition probabilities. Noble and Slatyer (1978, 1980) have developed a concept called the 'vital attributes' concept which uses regeneration, response to disturbance, and longevity of plants to determine the parameters of a Markov model. They currently have an 'expert-system' under development on a small 'personal computer', a computer programme that queries the user as to the ecological attributes of the species in a successional system and then develops and implements a Markov model of the system. The development of these theoretical methods of estimating the transition probabilities creates the possibility of developing larger Markov models in which parameter estimation from data would normally be proscribed due to logistic difficulties (Cattelino *et al.*, 1979; Kessell, 1976; 1979a, b; Potter *et al.*, 1979; Kessell and Potter, 1980). The theories are based on biological attributes of the species and this approach is also found in the more complex models discussed below.

Gap models

Gap models are a subset of a class of forest succession models called individual-tree models (Munro, 1974) because the models follow the growth and fate of individual trees. The first model of this genre was the JABOWA model developed by Botkin *et al.* (1972); a similar modelling approach has been applied to several forests in different parts of the world (see Chapter 4 of Shugart, 1984, for a review of several of these applications, also see Kercher and Axelrod, 1984).

Gap models simulate succession by calculating the year to year changes in diameter of each tree on small plots. The plot size is determined by the size of the canopy of a single large individual. Forest succession dynamics are estimated by the average behaviour of 50 to 100 of these plots. The growth of each tree is determined by the average competitive influence of the neighbouring trees on a plot. Due to the small size of plots, gap formation events (the removal of canopy trees through mortality) strongly affect the resource availability on a plot which in turn affects tree growth.

The exact location of each tree is not used to compute competition in these models. Tree diameters are used to determine tree height, and then simulated leaf area profiles are computed to devise competition relationships due to shading. These models are spatial in that competition is computed in the vertical dimension. There is an implicit assumption that within a plot of a certain size the horizontal spatial patterns of the individual plants do not affect the degree of competitive stress acting on an individual to significant degree beyond that accounted for by the plants height (i.e. tree biomass and leaf area are considered to be homogeneously distributed across the horizontal dimension of the simulated plot).

The regeneration of seedlings on a plot and their subsequent growth is based on the silvicultural characteristics of each species, including site requirements for germination, sprouting potential, shade tolerance, growth potential, longevity, and sensitivity to environmental factors (water and nutrients). Under optimal growth conditions, the growth of a tree is assumed to occur at a rate that will produce an individual of maximum recorded age and diameter. This curvilinear function grows a tree to two-thirds of its maximum diameter at one half its age under optimal conditions. Modifications reducing this optimal growth are imposed on each tree based on the availability of light and, depending on the specific model, other resources. In most gap models, tree growth slows as the simulated plot biomass approaches some maximum potential biomass observed for stands of the given forest type. Growth is further reduced as climate stochastically varies. Death of individual trees is a stochastic process. The probability of an individual tree's death in a given year is inversely related to the individual's growth and the longevity of its species.

Gap model dynamics are based on information concerning the demography and growth of trees during the lifespan of species. The models have a capability to predict the sequence of replacement of species through time and other dynamics on the scale of the average tree generation time (Figure 8.1). At this scale, the success of a tree at growing into the canopy is more related to the opportunity for inseeding into a plot and the relative growth rate compared to other seedlings than it is related to the distribution of distances from other competing individuals.

The relationship of the height of the individual to the distribution of heights of competitors is assumed to be sufficient to determine the level of competitive stress experienced by an individual in relation to other trees on the plot. This implies that the distance of a tree to its competitors has no significant influence on the amount of light and other resources available to a given tree. In terms of implementing these models, these assumptions lead to a requirement that the dynamics of a large number of plots be averaged to better estimate the mean rate of success of canopy invasion of each species.

Because regeneration, growth, and death are modelled on a per-tree basis and the silvics of individuals vary among species, gap models are particularly useful tools for exploring the dynamics of mixed-aged and mixed-species forests. The models have been tested and validated against independent data (Shugart, 1984; Chapter 4). For these reasons, gap models can also be used to explore theories about patterns in forest dynamics at time scales that are sufficiently long to prohibit direct data collection. Such applications have been instrumental in developing a theoretical basis for understanding the coupled effects of tree death and regeneration in forest systems (Shugart, 1984).

One gap model that has been used in a large number of applications in complex, mixed-species, mixed-aged forests is the FORET model, a derivative of the JABOWA model (Botkin *et al.*, 1972). The JABOWA/FORET modelling approach has been the central topic of a pair of books on the dynamics of natural forests (Bormann and Likens, 1979a; Shugart, 1984). The FORET model and other analagous models have been modified and applied to

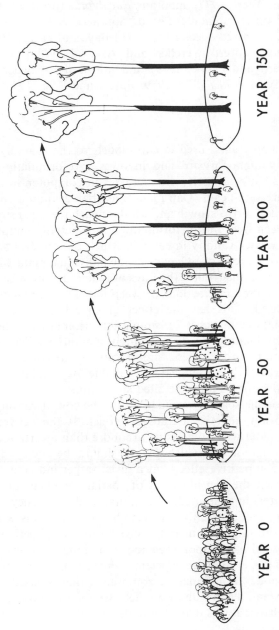

YEAR 0   YEAR 50   YEAR 100   YEAR 150

Figure 8.1 Example output from a gap model (The BRIND model of succession in Eucalyptus forests in the Brindabella Range, Australian Capital Territory, Shugart and Noble, 1981). The plots are drawn to scale (distance across each plot is 32 m) from model output for a single 1/12 ha. quadrat. The simulation time step of the model is one year but the output is displayed here on 50-year intervals. Trees shown are various species of Eucalyptus particularly *E. delegatensis* (the trees with half white and half black trunks) and *E. dalrympleana* (the trees drawn with white trunks)

simulate the dynamics of a wide range of forests: mixed hardwood forests of Tennessee (Shugart and West, 1977), montane *Eucalyptus* forests of Australia (Shugart and Noble, 1981), upland forest of Southern Arkansas (Shugart, 1984), eastern Canadian mixed species forest (El-Bayoumi *et al.*, 1984), the arid western coniferous forest (Kercher and Axelrod, 1984), a western coniferous forest (Reed and Clark, 1979) and northern hardwood forests (Botkin *et al.*, 1972; Aber *et al.*, 1978, 1979; Pastor and Post, 1985).

Transect models

The extension of the approaches used in gap models to the spatially explicit case is conceptually straight-forward and involves a reformulation of the competition function. Unfortunately, the increases in computer storage and computational time are significant. Gap models are based upon a computer-driven compilation of the birth, growth and death of each tree on a small plot. The interactions between trees in this formulation are spatially lumped. This lumped representation loses validity in cases in which the quadrat size is no longer small relative to the zone of influence of the individual plants. Thus, to extend model formulations that follow the fates of individual organisms to the spatially explicit case, paired interactions between individual simulated plant trees must be calculated. This has the effect of squaring the number of calculations; often making the cost of simulation prohibitive. There are, however, several approximations to the absolutely spatially explicit case that can reduce computation costs. These simplifications are often justified, independent of cost savings, because the data available for validation is not of sufficient resolution to justify a more detailed simulation.

One simplification of the spatially explicit case is the consideration of only one horizontal dimension, that is, a transect model. If the between-plant interactions can be ignored when the plants are more than a certain distance apart and the determination of which plants should be included in the determination of the competitive effects on a given target individual can be determined rapidly, then the consideration of spatially explicit interactions becomes less computer-time limited. Computational efficiency can be improved in a transect model because the search for individuals within the zone of influence of the target plant only need be in one direction. If the individuals are catalogued based upon their location along the transect, then this search can be made even more efficient. Shugart and Rastetter have developed a transect model of a mangrove community that is based on a gap model of the same system. The resultant model uses the same parameters as the gap model from whence it was derived and differs only in the formulation of the competition equations.

Transect models are most applicable to situations where the community is strongly zoned along some environmental gradient. Ecosystem/environment interactions are frequently conceptualized as responses of transects to gradients (see, for example, the illustrations in Watt's classic 1947 paper). It is surprising that transect models are not more in evidence given the interest that is manifested in transect representations of ecosystem processs.

Many dynamic physical processes that contribute to the pattern and responses of ecosystems over time can be simulated by models that are considerably simpler in a single horizontal dimension. Shugart *et al.* (1987) have developed a transect model that maintains the computational efficiency of a Markov model but also incorporates both a mechanistic formulation of the important population processes and the realism of spatial heterogeneity. The model is based upon a Markov chain representation of the life stage development of each plant species at intervals along a transect. If a species has 'n' life stages that are ecologically important, then that species is represented at each interval by an n-bit word, each bit signifying the presence or absence of a respective life stage and there are, therefore, 2^n possible simulation states for the species. In its current implementation on the computer, the number of life stages represented for the various species and the spatial resolution (interval width) along the transect can be defined by the model user. This allows the adjustment of the resolution of spatial patterns and life history detail to optimize detail and computational efficiency.

The state transition probabilities in the Shugart *et al.* (1987) model are calculated based on seed availability and environmental factors affecting sprouting, growth and mortality. There are 2^n (n = number of life stages) possible state transitions for each species at each location during any particular time step. For large n ($n > 3$), the dimensionality of the problem can be reduced by considering each life stage individually. This also facilitates a more mechanistic formulation of the transition probabilities incorporating the growth characteristics of the species.

There are four possible transitions for the individuals of a particular life stage at a particular location:

(1) they can all die

(2) they can all remain unchanged

(3) some can mature to the next life stage and some remain the same

(4) they can all mature to the next life stage.

Three other possibilities involving some plants dying, and some either remaining the same and/or maturing are indistinguishable from possibilities 2, 3, and 4 because only the presence or absence of individuals in each life stage is followed in the model. Since these four transitions represent all possibilities, the probabilities associated with them must sum to one. It is therefore only necessary to calculate three of the probabilities, the fourth can be calculated by

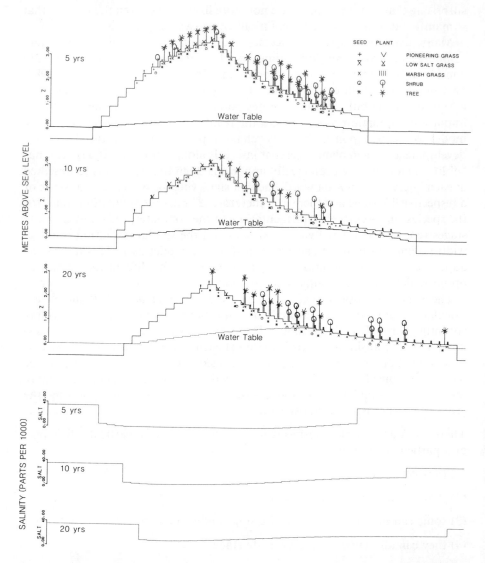

CROSS SECTION OF ISLAND AT 5, 10, and 20 YEARS

Figure 8.2 Example output from a computer model (Shugart *et al.*, 1987) simulating the pattern and dynamics of vegetation along a transect through a barrier island such as might be found along the Atlantic coast of North America. The open ocean is to the right of the figure. Over the time of the simulation the island moves landward and the water table on the right side of the island forms a brackish marsh with scattered shrubs. Trees and shrubs occupy a position behind the sheltering dune and the dune flattens

difference. Possibilities 3 and 4, however, do not exist for the oldest life stage, consequently only two probabilities must be calculated for this last life stage and one of these can be calculated by difference.

In addition, a recruitment probability must be calculated. Thus, a total of $3(n - 1) + 2 = 3n - 1$ probabilities must be calculated at each time step, for each species, at each location. The 2^n state transition probabilities can be calculated from these life stage transition probabilities by cross-multiplying the probabilities associated with each life stage with those of each of the other life stages and summing the probabilities of all redundant outcomes.

The model has been implemented to simulate the vegetation dynamics of coastal dune ecosystems (Shugart *et al.*, 1987). Because of the transect formulation of the model, several important physical variables can be simulated dynamically—notably the height of the water table, the height of the sand mass at any point and the salinity of the water table at each point. The model is driven by the position of the 0 height beach front. The successional dynamics of an example simulation (Figure 8.2) features the development of a horizontal gradient of vegetative pattern, a reduction of the height of the sand dune due to aeolian erosion, a landward displacement of the dune system (a consequence of vegetation-mediated aeolian transport of sand) and the eventual development of a back-dune marsh as the water table moves to the surface behind the dune. This particular model is presently in a prototype form but the example (Figure 8.2) is indicative of the richness of behaviour that can be developed from transect models.

Spatial models

Like gap models, spatial forest models simulate forest dynamics by modelling the establishment, growth, and mortality of individual trees within a defined area. Spatial models of other ecosystems tend to be developed as spatially extended Markov models of a form like the transect models discussed above (van Tongeren and Prentice, 1986). The spatial forest models differ from gap models in their explicit consideration of tree position in the horizontal plane. An inherent difference between gap models and spatial forest models is in the form of the competition functions. Because of the explicit consideration of horizontal position, spatial forest models generally use a measure of competition that is a direct function of the proximity and size of neighbouring individuals.

Although the competition indices used in spatial models vary greatly in their design, they can be classified into three major categories:

(1) Distance-based ratios

(2) influence-zone overlap indices

(3) growing-space polygons.

Distance-weighted size ratios (Hegyi 1974; Daniels, 1976) define the degree of competition between a given tree and a neighbouring individual as a function of the ratio of the sizes of the two trees (competitor/subject tree) multiplied by the inverse of the distance between the two individuals.

The influence-zone indices (Gerrard, 1969; Bella, 1971) are based on the assumption of a circular zone of influence around every tree, wherein direct competition occurs (Staebler, 1951). The extent to which this area overlaps the influence zone of neighbouring trees represents a measure of encroachment and crowding of a tree's optimal functional environment. These indices vary with regard to the type of overlap expressions used (i.e. linear, angular, areal).

Growing-space polygons (Brown, 1965; Moore *et al.*, 1973; Alard, 1974; Pelz, 1978; Doyle, 1983) represent geometrical designs to calculate non-overlapping crown area of a tree as limited by the proximity and size of neighbouring individuals.

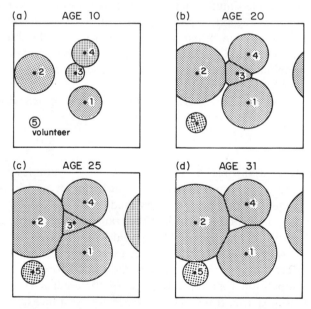

Figure 8.3   Simulated growth of the crowns of five trees from Mitchell's (1975) model of Douglas-fir (*Pseudotsuga menziesii*). Tree 3 is subjected to active competition and eventually dies in year 25. The crown of tree number 2 has grown outside of the simulated plot by year 20 and is 'wrapped' to re-enter the plot on the right side. This is a typical device used in spatially explicit models to eliminate the effects of plot boundaries. The Mitchell model simulates the interaction of the trees by branch pruning as the crowns overlap

Although each of these indices is based on the relative horizontal position of individuals on the plot, they vary in the methods that are used to determine which neighbouring individuals to consider as potential competitors, in defining the size of the zone of influence for a given individual, in consideration of size of competing individuals relative to the target tree, and in their consideration of potential differences in competitive ability among species.

With the exception of Doyle (1983), the above-mentioned competition indices are based on statistical models and the calculated values of the competition indices are regressed against observed growth rate of individuals to determine the functional relationship between competition and growth. As a result, the functional form of the relationship between the competition index and the growth rate is site specific (i.e. related to site factors such as nutrient and moisture availability) and data intensive. For these reasons, most spatial models have been developed for managed forests (e.g. Mitchell, 1975 and Figure 8.3) and, with the exception of the Ek and Monserud (1974) FOREST model, simulate monospecific stands.

FUTURE APPLICATIONS AND CHALLENGES

The development of simulation models initially resulted from an interest in forecasting the patterns of change in plant communities. Over time this interest in forecasting has matured into a more theoretical focus of understanding the scaled-up consequences of various physiological and ecological processes. This shift in interest from forecasting to theory testing is a consequence of several factors:

(1) The actual data base for our understanding of succession is extremely sparse and the goal of forecasting in such situations is often unrealistic

(2) Successional studies are frequently based on inferences used to piece together data sets that relate important features of ecosystems into a rational pattern. One frequent method to organize successional data is to order systems sampled at the same time but over a large space into an inferred chronosequence

(3) The rate of addition of new data-rich studies of succession to the ecological literature has decreased over the past several decades

(4) An increased interest in the non-equilibrium nature of ecosystems requires a data base that is beyond the spatial and temporal scale of most data sets that are available and has created an arena for theorization about ecosystem dynamics.

A potentially important class of theoretical investigations to which models can be usefully employed regards the response of ecosystems to the dynamics in the systems that drive them. The problem is one of understanding which scales of dynamics in the driving system interact with the ecosystem to produce responses. There is also a related problem of understanding how changes in ecosystems in response to the environment are fedback to alter the environment. In some cases these problems are global in their scale.

Cowles (1899) once characterized succession as 'a variable approaching a variable' and thus conceptualized ecosystem dynamics as a system perturbed away from—but moving toward—an equilibrium that was itself changing. The present problem is to understand both of these 'variables' and the interaction between them. Successional models are the tools to investigate the responses of systems such as those conceived by Cowles. In the following discussion we will focus on atmosphere/ecosystem interactions to provide what we feel is an important example of this class of problem. The emphasis on these interactions stems from the implications of terrestrial ecosystem dynamics to the global scale, from the important and unique role of models in understanding these interactions, and to the degree to which including these interactions challenges our understanding of ecosystem function.

The addition of realistic climatic factors to forest succession models

It is interesting to note that most of the classical static climatic classifications, such of those of Koeppen (summarized in Oliver and Hidore, 1984) were originally subdivided by biotic unit. Their use in climate/biota research therefore ensnares the investigator in a circular trap of guaranteed results.

Only rather recently has climatology evolved, from such endless series of tables and classifications, into a much more dynamic science based upon physics and theory. Dynamic climatology is the spatial and temporal integration of the dynamic meteorology evident on the daily surface and upper atmospheric maps, where fronts, airmasses, and cyclones are modulated by complicated dynamic processes. Like ecosystems, these phenomena exhibit discontinuous and preferred scales of motion. Many of these are characterized as aggregations of physical variables that have quantifiable impact on tree mortality and consequent community distribution.

Thompson Webb III, in his doctoral dissertation (1971), used multivariate statistical techniques to relate both standard climatic variables and more integrative airmass durations and frequencies (i.e. a rudimentary dynamic climatology) to arboreal pollen assemblages over North America. The result was a partial specification of holocene atmospheric circulation patterns related to and predicted from community migration. The implication is that one can indeed quantitatively relate atmospheric circulation systems and community distribution. Thus the prospect of defining functional relationships between the dynamic climate and changing ecosystems seems attainable.

Yet, atmospheric perturbations—weather, climate, and their moments of variability—have traditionally been parameterized in simplistic fashion in most successional models. Further, these variables are assumed to exert primarily first-order effects rather than acting through intermediaries such as defoliators or herbivores. Concurrent with the development of successional modelling has been a substantial increase of our understanding of climate and its variability. It remains an important research focus in understanding global change to interface these two developing fields.

Barry and Perry (1973) have written what remains the standard text on comprehensive analyses of climatic data; it describes a number of method-ological and theoretical considerations that should be considered in the interface of climate and successional models. Fujita (1981) has written an interesting contextual background, which details the scales of atmospheric motion. However, neither of these works are directed to an ecological audience.

The heart of the research problem is the discrimination between weather and climate events that have significant successional effects and those that do not. This is one of the important scale questions that was mentioned in the introduction. Verified successional models are appropriate tools to determine the dimensions of those events. This is true in particular because the historical (non-proxy) record of climatic variability is insufficiently long to relate directly to successional impact. However, the more sophisticated classifications of the nature and parameters of climatic variability can be used with successional models to isolate the successionally important climatic events.

For an atmospheric process to exert a significant successional effect, it must impact the ecosystem in a manner that directly allows for the alteration of local community composition. Some processes—including gap openers such as windfalls, lightning strikes, and ice-induced crown damage—appear as prom-inent and persistent. Others, such as differential mortality resulting from unusual temperatures, are much more subtle.

Fujita (1981) provides strong evidence for five relatively discrete scales of atmospheric motion (Figure 8.4), some of which can influence succession. The dynamic nature of successional models can be used to quantify that influence. Fujita's five scales are each separated by approximately two orders of magnitude in spatial extent, with the largest, or 'A' scale, corresponding to the planetary and synoptic circulations. The planetary scale circulations do not directly impinge upon forest community processes. Rather, it is generally the 'E' ('mesoscale') circulations that are more ecologically active.

The planetary scales of motion dissipate a substantial portion of their energy in the form of the more familiar secondary scales—mid latitude cyclones, their associated fronts, and the anticyclones that comprise the airmasses that are separated by frontal discontinuities. The secondary scales have embedded

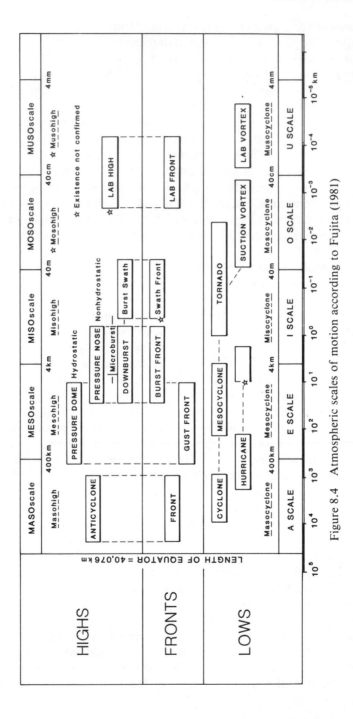

Figure 8.4   Atmospheric scales of motion according to Fujita (1981)

within a variety of mesoscale phenomena whose effects, temporal, and spatial characteristic should be input to successional models.

One exception to this general scheme is the planetary-scale intertropical convergence zone (ITCZ), a region of relative upward motion that roughly corresponds to the earth's thermal equator. The ITCZ produces only ill-defined secondary circulations that tend to be confined to the Asian Subcontinent. Instead, the primary mesoscale circulations of ecological import are directly produced by the ITCZ and are in the form of strong thunderstorms with attendant downburst winds and lightning strikes.

It is noteworthy that the ITCZ takes on several forms, some of which bear no strong relation to seasonality. In fact, its position and conformation from day-to-day is surprisingly unpredictable (Chang, 1972), negating the common misconception that its movement is one of the primary predictable forcing functions that stabilizes tropical forest systems. It therefore seems appropriate that tropical successional models (e.g. Doyle, 1981; Shugart *et al.*, 1981) be modified to investigate the ecological implications of this phenomena.

The occurrence or non-occurrence of thunderstorms should be treated as a random variable within broad seasonal limits. However, the ecologically significant effects—immediate gap opening resulting from downbursts, and slowly opening gaps resulting from lightning-induced mortality—should be given exponentially increasing likelihoods related to a relative relief and topography.

As noted above, the primary successional disturbances are generated by mesoscale meteorological phenomena embedded in the macroscale. Those which may have considerable successional significance include tropical cyclones, mesoscale convective complexes ('MCCs'), severe thunderstorms, and frozen precipitation. Temperature effects are dealt with separately below.

1. Tropical cyclones

Tropical cyclones are warm-core eddies that develop primarily as embedded disturbances within the trade wind regime, which itself is the return flow that maintains the mass continuity of the ITCZ. Winds often exceed 100 km/hr, and one-minute velocities at landfall as high as 300 km/hr have been noted. While their most obvious damage is done primarily by storm surge waves impinging upon pericoastal regions, they also produce broad areas of up-rooted trees and mid-crown damage within 200 km of landfall.

Ecologists interested in exogenous disturbances and successional response, particularly in North America, should be aware of the historical climatology of tropical cyclone tracks produced by Neumann *et al.* (1981) for the previous 110 years. Also, attention should be paid to a paper by Shapiro (1982) that details secular change in the track regime through the 20th century. The combined analyses suggest a significant nonstationarity in the time series over the fire-stabilized extensive Oak-Pine forests of the southeastern United States. This indicates that using simple mean and variance estimates for climatic input into succession models may be an inappropriate simplification if a high level of realism is the modelling goal.

Tropical cyclone movement is primarily related to the strength of upper-tropospheric winds which are subject to secular changes associated with hemispheric temperature fluctuations. For example, it is quite plausible that the relative cooling of the late 1950s and 1960s, compared to the earlier three decades, as noted by Diaz and Quayle (1980) would have resulted in stronger mid-atmospheric westerlies and a concomitant increasing northward deflection of tropical systems up the east coast of North America observed during that period. Attempts to include prospective climatic change in succession models should include such phenomena.

2. Mesoscale convective complexes and severe thunderstorms

Mesoscale Convective Complexes ('MCCs') and severe thunderstorms are two notable E-scale phenomena that tend to occur in preferred regions of A-scale fronts and cyclones. MCCs are concentrated areas of intensive convection, approximately 10 000 km^2, often accompanied by intense cloud-to-ground lightning. They are only sometimes accompanied by heavy precipitation. The synoptic climatology of MCCs, including tracks and background conditions, have been documented by Maddox (1983). Approximately 30 occur per year over North America east of the Rocky Mountain system, and they tend to appear ahead of northward moving thickness patterns ('warm fronts').

The synoptic climatology of MCCs, which were initially detected from weather satellites, has only been documented for recent years (see Maddox, 1983). Further, a preferred track analysis, similar to that for tropical cyclones in Shapiro (1982) or for temperate cyclones (Hayden, 1981) has yet to be generated. But it seems reasonable to conclude that their association with frontal situations would also suggest a nonstationary time series forced by hemispheric climatic change.

Severe thunderstorms, which produce frequent lightning and are accompanied sometimes either by strong downburst winds exceeding 150 km/hr (Fujita, 1981), or by tornadoes, are primarily a factor during the transition from winter to summer, over most regions, although they have been noted in all seasons over eastern North America. As yet, no objective synoptic climatology of either severe thunderstorms or tornadoes has been generated, so it is impossible to state the broad parameters of temporal and spatial variation. However, as is the case for all lightning and wind generators, it is likely that the most significant effects will tend to be concentrated over the high elevations in areas of relatively high relief. Thus, similar to the case for ITCZ thunderstorms, it would seem appropriate to apply some exponentially increasing risk function for gap opening in such areas. Canham and Loucks (1984) recently detailed the community alteration resulting from a strong downburst in a severe thunderstorm.

Tornadoes are sufficiently uncommon over most of the forested areas of North America as to exert only a minimal effect. However, this is not true for some of the forested regions extending from Alabama to East Texas, where calculations based upon average track length of 2 km^2 suggests the a return period of about 2000 years at a given point, based upon data from the National Climatic Data Center (1979).

3. Frozen precipitation

The synoptic climatology of frozen precipitation, primarily in the forms of snow or freezing rain, may have particular bearing upon successional transformations. For example, it is generally noted that boreal coniferous species shed snow more efficiently than their deciduous counterparts. Thus Bryson's (1966) conclusion that the mean southward (winter) and northward (summer) position of the polar front circumscribes the boreal zone seems logical, as it is between these areas that substantial precipitation is likely to fall as snow.

Freezing rain can induce substantial and extensive crown damage. It can occur over extensive regions where shallow layers (less than 2 km) of cold air are trapped either because of dynamic or topographic processes. It is thus a phenomenon confined primarily to the excursionary regions of the wintertime polar front, and accentuated by topography. Regions such as the eastern slope of the Appalachian mountains and nearby piedmont, from approximately 30° to 40° North, are unusually subject to this phenomenon. It is also likely, but undocumented in the western literature, that the same problem occurs along the eastern slope and piedmont of the Ural Mountains. As a mix of coniferous and deciduous vegetation dominates these regions it would appear that the coniferous species, with their more extensive surface area during the cold portion of the year, would be more substantially impacted.

4. Temperature

The previous discussion ignores the effects of changing or unusual temperatures on successional phenomena. Successionally important temperature fluctuations are primarily controlled at the 'A' level in Fujita's scheme, denoted by either warm anticyclonic conditions at the subtropical level or by cold anticyclonic weather associated with the export of shallow airmasses from arctic regions.

Overlaying the entire temperature picture are the scenarios for increasing trace gas-related warming, noted with perspective by the US National Research Council (1983) Report. The caution with which that report was tendered cannot be understated; however, it is clear that some hemispheric warming should be expected over the coming decades. Predictive successional models should therefore be conditioned with temperature data suggestive of both increased variance and increased mean temperature values, if both the findings of Karl *et al.* (1984) and the National Research Council Report are accepted.

CONCLUSION

As noted above, resolution of the relationships between tree mortality, distribution, and the dynamic atmosphere can be approached at a number of levels. Signal progress will be achieved when the atmospheric circulation system regime, rather than its derived physical variables, serves as the climatic input to dynamic models of the forestscape.

Given the diversity of climatic phenomena that can affect succession and the current debate with regard to future climate, an important future challenge is to make succession models become more responsive to changes in their base atmospheric input. This includes the design of models with appropriate local climatology. The greatest need appears to be realistic input of the synoptic climatological variables that are probably associated with successional disturbance.

It is important to determine what degrees of climatic fluctuation are necessary to affect successional change. This can be accomplished by examining the responses of ecological models for significant differences in the simulated communities that result when differing climatic data are used as input in realistic ecological models. We have only provided some general guidelines here concerning the important phenomena. To properly answer this question will take a long-term interdisciplinary effort in ecology and climatology.

In developing this review we initially discussed scales of controlling external phenomena and scales of ecosystem response as both a present issue in ecological studies and as a classic historical topic in ecology's history. The understanding of resolution of scale problems in ecological systems is clearly a central problem in understanding global change. In this review one sees that there exist a considerable number of modelling tools for projecting the long-term dynamics of forest systems and these tools have analogues for other ecological systems. However, when one considers the scales at which atmospheric phenomena can affect the dynamics of forests, for example, one quickly realizes that the effects can occur at a wide range of time and space scales.

What this all implies is that the use of models to project ecosystem response to climatic change will be considerably more complex than simply taking a model 'off the shelf'. The models will need to be honed to the task of interfacing with projections from atmospheric models to a considerable degree. The present generation of ecosystem models has been developed to meet a diverse set of problems and objectives. As a result, even a well tested and seemingly reliable model will not necessarily contain the responses to critical driving variables that might be effected by a change in atmospheric systems. The development and testing of models either created or modified to project forest (or other ecosystem) response to change in the atmospheric systems is not a solved problem by any means. The richness of scales of the abiotic driving variables may proscribe use of a single model for these purposes. One approach is the application of several models that are internally consistent in their basic assumptions but designed to operate at different time and space scales to problems of interfacing ecological and atmospheric systems.

REFERENCES

Aber, J. D., Botkin, D. B., and Melillo, J. M. (1978). Predicting the effects of differing harvest regimes on forest floor dynamics in northern hardwoods. *Can. J. For. Res.*, **8**, 306–315.

Aber, J. D., Botkin, D. B., and Mellilo, J. M. (1979). Predicting the effects of differing harvest regimes on productivity and yield in northern hardwoods. *Can. J. For. Res.*, **9**, 10–14.

Aber, J. D., Hendry, G. R., Francis, A. J., Botkin, D. B., and Melillo, J. M. (1982). Potential effects of acid precipitation on soil nitrogen and productivity of forest ecosystems. In Ditri, F. M. (Ed.) *Acid Precipitation: Effects on Acid Precipitation*, pp. 411–434. Ann Arbor Science, Ann Arbor, Michigan.

Alard, P. G. (1974). Development of an empirical competition model for individual trees within a stand. In Fries, J. (Ed.) *Growth Models for Tree and Stand Simulation*, pp. 22–37. Royal College of Forestry, Stockholm, Sweden.

Allen, T. F. H., and Hoekstra, T. W. (1984). Nested and non-nested hierarchies: a significant distinction for ecological systems. In Smith, A. W. (Ed.) *Proceedings of the Society for General Systems Research. I. Systems Methodologies and Isomorphies*, pp. 175–180. Intersystems Publ., Coutts Lib. Serv., Lewiston, N.Y.

Allen, T. F. H., and Starr, T. B. (1982). *Hierarchy: Perspectives for Ecological Complexity*. Univ. of Chicago Press, Chicago, Illinois.

Barry, R. G., and Perry, A. H. (1973). *Synoptic Climatology Methods and Applications*. Methuen, London: 555 pages.

Bazzaz, F. A., and Pickett, S. T. A. (1980). Physiological Ecology of tropical succession. *Annu. Rev. Ecol. Syst.*, **11**, 287–310.

Bella, I. E. (1971). A new competition model for individual trees. *Forest Science*, **17**, 364–372.

Bormann, F. H., and Likens, G. E. (1979a). *Patterns and Process in a Forested Ecosystem*. Springer-Verlag, New York.

Bormann, F. H., and Likens, G. E. (1979b). Catastrophic disturbance and the steady state in northern hardwood forests. *Amer. Scientist.*, **67**, 660–669.

Botkin, D. B., Janak, J. F., and Wallis, J. R. (1972). Some ecological consequences of a computer model of forest growth. *J. of Ecol.*, **60**, 849–872.

Brown, G. S. (1965). *Point Density in Stems per Acre.* Forest Research Institute. New Zealand Forest Service, Forest Research Notes No. 38: 11 pages.

Bryson, R. A. (1966). Air masses, streamlines, and the boreal forest. *Geogr. Bull.*, **8**, 228–269.

Canham, C. D., and Loucks, O. L. (1984). Catastrophic windthrow in the presettlement forest of Wisconsin. *Ecology*, **65**, 803–809.

Cattelino, P. J., Noble, I. R., Slatyer, R. O., and Kessell, S. R. (1979). Predicting the multiple pathways of plant succession. *Environ. Manage.*, **3**, 41–50.

Chang, J.-H. (1972). *Atmospheric Circulation Systems and Climates.* Oriental, Honolulu: 360 pages.

Christensen, N. L. (1985). Shrubland fire regimes and their evolutionary consequences. In Pickett, S. T. A., and White, P. S. (Eds.) *The Ecology of Natural Disturbance and Patch Dynamics*, pp. 86–100. Academic Press, New York.

Clements, F. E. (1916). *Plant Succession: An Analysis of the Development of Vegetation.* Carnegie Inst. Pub. 242. Washington, DC: 512 pages.

Colgan, M. W. (1983). Succession and recovery of a coral reef after predation by *Acanthaster planci* (L.) *Proc. Int. Coral Reef Symp., 4th. Manilla, 1981*, **2**, 333–338.

Connell, J. H. (1978). Diversity in rain forests and coral reefs. *Science*, **199**, 1302–1310.

Cowles, H. C. (1899). The ecological relations of the vegetation on the sand dunes of Lake Michigan. *Bot. Gaz.*, **27**, 95–117, 176–202, 281–308, 361–369.

Daniels, R. F. (1976). Simple competition indices and their correlation with annual loblolly pine growth. *Forest Science*, **22**, 454–456.

Delcourt, H. R., Delcourt, P. A., and Webb, T. (1983). Dynamic plant ecology: The spectrum of vegetational change in space and time. *Quat. Sci. Rev.*, **1**, 153–175.

Denslow, J. S. (1980). Gap partitioning among tropical rain forest trees. *Biotropica*, **12**, (Suppl.) 47–55.

Dethier, M. N. (1984). Disturbance and recovery in intertidal pools: maintenance of mosaic pattern. *Ecol. Monogr.*, **54**, 99–118.

Diaz, H. F., and Quayle, R. G. (1980). The climate of the United States since 1895: Spatial and temporal changes. *Mon. Wea. Rev.*, **108**, 249–266.

Doyle, T. W. (1981). The role of disturbance in the gap dynamics of a montane rain forest: An application of a tropical forest succession model. In West, D. C., Shugart, H. H., and Botkin, D. B. (Eds.) *Forest Succession: Concepts and Application.* New York, Springer-Verlag.

Doyle, T. W. (1983). *Competition and growth relationships in a mixed-age, mixed-species forest community.* Ph.D. dissertation, University of Tennessee, Knoxville: 86 pages.

Ek, A. R., and Monserud, R. A. (1974). *FOREST: Computer Model for the Growth and Reproduction Simulation for Mixed Species Forest stands.* Research Report A2635. College of Agricultural and Life Sciences, University of Wisconsin, Madison: 90 pages.

El-Bayoumi, M. A., Shugart, H. H., and Wein, R. W. (1984). Modelling succession of eastern Canadian Mixedwood Forest. *Ecological Modelling*, **21**, 175–198.

Foster, R. B. (1977). *Tachigalia versicolor* is a suicidal neotropical tree. *Science*, **268**, 624–626.

Fries, J. (Ed.) (1974). *Growth Models for Tree and Stand Simulation.* Royal College of Forestry, Stockholm. Res. Notes 30.

Fujita, T. T. (1981). Tornadoes and downbursts in the context of generalized planetary scales. *J. Atm. Sci.*, **38**, 1511–1534.

Gerrard, D. J. (1969). Competition quotient: *A New Measure of the Competition Affecting Individual Forest Trees.* Michigan State University Agricultural Experiment Station Research Bulletin No. 20. 32 pages.

Gleason, H. A. (1939). The individualistic concept of the plant association. *Am. Midl. Nat.*, **21**, 92–110.

Grubb, P. J. (1977). The maintenance of species-richness in plant communities: The importance of the regeneration niche. *Biol. Rev.*, **52**, 107–145.

Hayden, B. P. (1981). Secular variation in Atlantic Coast extratropical cyclones. *Mon. Wea. Rev.*, **109**, 159–167.

Hegyi, F. (1974). A simulation model for managing jack-pine stands. In Fries, J. (Ed.) *Growth Models for Tree and Stand Simulation*, pp. 74–90. Royal College of Forestry, Stockholm, Sweden.

Hool, J. N. (1966). A dynamic programming-Markov chain approach to forest production control. *For. Sci. Monogr.*, **12**, 1–26.

Horn, H. S. (1975a). Forest Succession. *Sci. Am.*, **232**, 90–98.

Horn, H. S. (1975b). Markovian properties of forest succession. In Cody, M. L., and Diamond, J. M. (Eds.) *Ecology and Evolution in Communities*, pp. 196–211. Harvard University Press, Cambridge.

Horn, H. S. (1976). Succession. In May, R. M. (ed.) *Theoretical Ecology*, pp. 187–204. Blackwell Scientific Publications, Oxford.

Huston, M. (1979). A general hypothesis of species diversity. *Am. Nat.*, **113**, 81–101.

Karl, T. R., Livezey, R. E., and Epstein, E. S. (1984). Recent unusual mean winter temperatures across the contiguous United States. *Bull. Amer. Met. Soc.*, **65**, 1302–1309.

Karlson, R. H. (1978). Predation and space utilization patterns in a periodically disturbed habitat. *Bull. Mar. Sci.*, **30**, 894–900.

Kay, A. M. (1980). *The Organization of Sessile Guilds on Pier Pilings.* Ph.D. thesis. University of Adelaide, South Australia.

Kercher, J. R., and Axelrod, M. C. (1984). A process model of fire ecology and succession in a mixed-conifer forest. *Ecology*, **65**, 1725–1742.

Kessell, S. R. (1976). Gradient Modeling: A New Approach to Fire Modeling and Wilderness Resource Management. *Environ. Manage.*, **1**, 39–48.

Kessell, S. R. (1979a). *Gradient Modeling: Resource and Fire Management.* Springer-Verlag, New York.

Kessell, S. R. (1979b). Phytosociological inference and resource management. *Environ, Manage.*, **3**, 29–40.

Kessell, S. R., and M. W. Potter, 1980. A quantitative succession model for nine Montana forest communities. *Enviro. Manage.*, **4**, 227–240.

Kozlowski, T. T. (1971a). *Growth and Development of Trees: Vol I. Seed Germination, Ontogeny and Shoot Growth.* Academic Press.

Kozlowski, T. T. (1971b). *Growth and Development of Trees: Vol II. Cambial Growth, Root Growth and Reproductive Growth.* Academic Press.

Maddox, R. A. (1983). Large scale meteorological conditions associated with mid-latitude mesoscale convective clusters. *Mon. Wea. Rev.*, **111**, 2123–2128.

Mitchell, K. J. (1975). Dynamics and simulated yield of Douglas-fir. *Forest Science Monogr.*, **17**, 1–39.

Moore, J. A., Budelsky, C. A., and Schlesinger, R. C. (1973). A new index representing individual tree competitive status. *Can. J. For. Res.*, **3**, 495–500.

Munro, D. D. (1974). Forest growth models: A prognosis. In Fries, J. (Ed.) *Growth Models for Tree and Stand Simulation*, pp. 7–21. Res. Note 30, Royal College of Forestry, Stockholm.

National Climatic Data Center. (1979). *Climatological Data, National Summary*. NOAA, Asheville, NC: 113 pages.

National Research Council. (1983). *Changing Climate. Report of the Carbon Dioxide Assessment Committee*. National Academy Press. Washington, DC: 496 pages.

Neumann, C. J., Cry, G. W., Caso, E. L., and Jarvinen, B. R. (1981). *Tropical Cyclones of the North Atlantic Ocean*, 1871–1980. NOAA, Asheville, NC: 170 pages.

Noble, I. R., and Slatyer, R. O. (1978). The effect of disturbances on plant succession. *Proc. Ecol. Soc. Ausat.*, **10**, 135–145.

Noble, I. R., and Slatyer, R. O. (1980). The use of vital attributes to predict successional changes in plant communities subject to recurrent disturbances. *Vegetatio*, **43**, 5–21.

Oliver, J. E., and Hidore, J. J. (1984). *Climatology*. Merrill, Columbus, Ohio.

O'Neill, R. V., DeAngelis, D. L., Waide, J. B., and Allen, T. F. H. (1986). *A Hierarchical Concept of the Ecosystem*. Princeton University Press, Princeton, New Jersey.

Paine, R. T., and Levin, S. A. (1981). Intertidal landscapes: disturbance and the dynamics of pattern. *Ecol. Monogr.*, **51** 145–178.

Pastor, J., and Post, W. M. (1985). *Development of a linked forest productivity-soil process model*. ORNL/TM–9519. Oak Ridge National Laboratory, Oak Ridge, TN.

Pearson, R. G. (1981). Recovery and recolonization of coral reefs. *Mar. Ecol.: Prog. Ser.*, **4**, 105–122.

Pelz, D. R. (1978). Estimating tree growth with tree polygons. In Fries, J., Burkhart, H. E., and Max, T. A. (Eds.) *Growth Models for Forecasting of Timber Yields*, pp. 172–178. School of Forestry and Wildlife Resources, Va. Polytech. Inst. and State Univ. FWS–1–78.

Pickett, S. T. A., and White, P. S. (Eds.) (1985). *The Ecology of Natural Disturbance and Patch Dynamics*. Academic Press, New York.

van der Pijl, L. (1972). *Principles of Dispersal in Higher Plants*. Springer-Verlag, Berlin.

Potter, M. W., S. R. Kessell and P. J. Cattelino. (1979). FORPLAN: A FORest Planning LANguage and simulator. *Environ. Manage.*, **3**, 59–72.

Risser, P. G. (1986). *Spatial and temporal variability of biospheric and geospheric processes: Research needed to determine interactions with global Environmental change*. International Council of Scientific Unions Press, Paris: 53 pages

Reed, K. L., and Clark, S. G. (1979). *SUCcession SIMulator: a coniferous forest simulator model description*. Coniferous Forest Biome, Ecosystem Analysis Studies, U.S./International Biological Program, Bull. 11, University of Washington, Seattle, WA: 96 pages.

Shapiro, L. J. (1982). Hurricane Climatic Fluctuations. Part 1: Patterns and Cycles. *Mon. Wea. Rev.*, **110**, 1007–1023.

Shugart, H. H. (1984). *A Theory of Forest Dynamics: The Ecological Implications of Forest Succession Models*, Springer-Verlag, New York: 278 pages.

Shugart, H. H., Bonan, G. B., and Rastetter, E. B. (1987). Niche theory and community organization. *Canadian J. Bot.* (in press).

Shugart, H. H. and I. R. Noble. (1981). A computer model of succession and fire response of the high-altitude Eucalyptus forest of the Brindabella Range, Australian Capital Territory. *Australian. J. Ecology*, **6**, 149–164.

Shugart, H. H., and West, D. C. (1977). Development of an Appalachian deciduous forest succession model and its application to assessment of the impact of the chestnut blight. *Journal of Environmental Management*, **5**, 161–170.

Shugart, H. H., and West, D. C. (1981). Long-term dynamics of forest ecosystems. *Am. Sci.,* **69**, 647–652.

Shugart, H. H., West, D. C., and Emanuel, W. R. (1981). Patterns and dynamics of forests: An application of simulation models. In West, D. C., Shugart, H. H., and Botkin, D. B. (Eds.). *Forest Succession: Concepts and Application*, pp. 74–79. Springer-Verlag, New York.

Sousa, W. P. (1979). Disturbance in marine intertidal boulder fields: the nonequilibrium maintenance of species diversity. *Ecology*, **60**, 1225–1239.

Staebler, G. R. (1951). *Growth and spacing in an even-aged stand of Douglas-fir.* M.F. Thesis, University of Michigan: 46 pages.

Tansley, A. G. (1935). The use and abuse of vegetational concepts and terms. *Ecology*, **16**, 284–307.

Taylor, P. R., and Littler, M. M. (1982). The roles of compensatory mortality, physical disturbance, and substrate retention in the development and organization of a sand-influenced, rocky-intertidal community. *Ecology*, **63**, 135–146.

van Tongeren, O., and Prentice, I. C. (9186). A spatial simulation model for vegetation dynamics. *Vegetatio*, **65**, 163–173.

Urban, D. L., O'Neill, R. V., and Shugart, H. H. (1987). Landscape Ecology. *Bioscience*, **37**, 119–127.

Waggoner, P. E., and Stephens, G. R. (1971). Transition probabilities for a forest. *Nature*, **225**, 93–114.

Watt, A. S. (1925). On the ecology of British beechwoods with special reference to their regeneration. Part 2, Sections II and III. The development of the beech communities on the Sussex Downs. *J. Ecol.,* **13**, 27–73.

Watt, A. S. (1947). Pattern and process in the plant community. *J. Ecol.,* **35**, 1–22.

Webb, T. (1947). *The late and postglacial sequence of climatic events in Wisconsin and eastcentral Minnesota: Quantitative estimates derived from fossil pollen spectra by multivariate statistical analyses.* PhD Diss. University of Wisconsin, Madison.

Whitmore, T. C. (1975). *Tropical Rain Forests of the Far East.* Clarendon Press, Oxford.

Whittaker, R. H. (1953). A consideration of climax theory: The climax as a population and a pattern. *Ecol. Monogr.,* **23**, 41–78.

Whittaker, R. H., and Levin, S. A. (1977). the role of mosaic phenomena in natural communities. *Theoretical Population Biology*, **12**, 117–139.

Scales and Global Change
Edited by T. Rosswall, R. G. Woodmansee and P. G. Risser
© 1988 Scientific Committee on Problems of the Environment (SCOPE)
Published by John Wiley & Sons Ltd.

CHAPTER 9

Processes in Soils—from Pedon to Landscape

R. G. KACHANOSKI

Dept. of Land Resource Science,
University of Guelph,
Guelph, Ontario, Canada

ABSTRACT

This chapter examines the influence of topography on the spatial distribution of soil properties and processes at different spatial scales. Three data sets are discussed to illustrate a variety of scaling problems. Topography is characterized by elevation, gradient, and vertical and horizontal curvature. Scale relationships between variables are examined using linear stochastic theory and the theory of regionalized variables. The soil pedon is described as the basic representative elementary volume of soil horizon properties. The data sets, however, indicate different properties have different spatial variance relationships. Even horizon thickness has different representative spatial scales depending on surface or subsurface influences. The nature of the spatial variance relationships are highly dependent on the spatial distribution of topographical parameters operating at specific scales which control the transfer of mass and energy in the landscape. Integration of small scale observation to large scale units should be based on the flow of mass and energy. Thus the drainage basin becomes an obvious level of integration. The development of methods for describing the joint distribution of soil and topography should continue to be a high priority for research.

INTRODUCTION

Soils are viewed as natural bodies formed on the land surface, occupying space and having unique morphology (Simonson, 1968). Soil as an isotropic or anisotropic body is also a continuum rather than particulate (Knox, 1965). Thus in the study, classification, and mapping of soils, there is an inherent dependence on scale of observation and scale linkages. Finkl (1982) has presented a collection of benchmark papers which present a historical development of the present day concept of soil. Included in these papers is the concept of a pedon (Soil Survey Staff, 1960) which is defined as the smallest three-dimensional spatial unit of the surface of the earth that is considered as a soil. A pedon has lateral dimensions such that it contains 'representative variations' of the soil being studied. The lateral dimensions are 1 m if 'ordered variation' in genetic horizons is present. If horizons are cyclical or intermittent and are repeated every 2 to 7 m, then the lateral dimensions are one-half the cycle. The concept of a pedon has been the traditional method of dealing with short range spatial variations. Mean values of soil properties for the pedon are used to place that individual spatial soil unit into a classification system.

The concept of a pedon is still a topic of considerable debate. For non-repetitive variation no formal definition of 'representative variation' is given and the term 'ordered variation' is confusing. However, a more formal treatment of the problem using continuum theory, the concept of a representative elementary volume (Bouma, 1984; Waganet, 1984), and analysis of spatial variance structure using geostatistics can lead to a similar definition for a basic soil unit.

Soil classification systems are concerned with organizing information and ideas about soils in a logical and useful fashion; it is not generally concerned with spatial ordering. The purpose of soil survey is to produce a soils map. It is a land mapping system which breaks the soil continuum into spatial units that have less variance for selected soil properties than the continuum (Wilding, 1984). Spatial mapping units are theoretically formed by grouping similar and contiguous pedons into polypedons and grouping polypedons into larger and larger ensembles depending on the scale of the map. In general because of the close relationship between topography and soils, spatial units correspond with landform features. The description of the mapping unit usually involves classification of the included soils, thereby tying the soil classification system to the mapping system at different scales. In the process of mapping, surveyors subjectively decide on the variation allowed in the mapping unit (Bridges, 1982), both in the variation allowed within soil classes, and the level of inclusion of unspecified soils within map unit areas. In many cases however, at least half of the variability of soil properties exists on a scale of $< 1 \text{ m}^2$ (Webster and Beckett, 1968) which is less than the dimensions of the pedon (as defined).

The condensed summary above is somewhat superficial, but it does define the basic concept of a soils map and how soil variability is considered. Soil surveyors have been concerned with soil variability and scale on a day to day basis, but have generally dealt with it by constructing conceptual models of systematic variability that have ignored short range variability characteristics of most soil landscapes. This variability is important because soils maps are curently being used to extrapolate point information about behaviour of soils to larger scales (Bridges and Davidson, 1982; Jarvis, 1982). Mean values of soil properties for arbitrarily selected, typical pedons are used as input to models for prediction over a given spatial unit. The predictions from each unit are further averaged (weighted according to actual area of each unit) over larger areas until the region of interest has been covered. In some cases estimates of the probability density functions (PDF) of the soil variables within the spatial unit are used for prediction at a given level of probability.

LANDFORM ANALYSIS

A number of quantitative methods have been developed for describing landscape form, but their application for studying the relationship between soils and topography is limited. Description of landscape form is the basis of geomorphology. Specific geomorphometry is the measurement and analysis of specific land surface features which are defined and separated from adjacent land areas according to clearly defined criteria (Evans, 1981).

With an elevation grid (matrix) landform shape parameters at every point in the grid can be calculated (Young and Evans, 1978). Landform shape can be characterized in part by five parameters: elevation, gradient, aspect, vertical curvature, and horizontal curvature. Gradient and aspect describe a plane tangential to the surface at the specified point and together define the concept of slope. Gradient is the maximum rate of change of elevation while aspect is the direction of the gradient usually given in degree units using North as a reference. Vertical curvature is the rate of change of gradient or the second vertical derivative of elevation. By convention, concave and convex curvature have negative or positive values, respectively. Horizontal curvature is the rate of change of aspect (i.e. direction of flow) along a contour line with negative and positive values indicating convergence or divergence of flow lines respectively. The parameters are estimated by fitting a least square quadratic surface to a subgrid of elevation readings around each point of interest, and calculating the derivatives of the surface.

For basins, after a value for aspect (surface flow direction) has been obtained for each grid point, a sixth parameter, catchment area index (CAI) can be calculated. CAI is the total number of unit cell areas (area associated with one grid point) that contributes to flow at a given grid point. The CAI is calculated as a running summation of all flow lines converging at a specific location.

The mathematical definition of slope and curvature parameters is unambiguous and, ideally, an instantaneous point measurement. The measurements are, however, made over a specific area and equations using least squares calculation will give a smoothed surface form. A discussion of scale effects on estimation of topographic parameters has been given by Evans (1972). He defines the estimates on the smallest scale as 'local' values whose scale is pre-determined by the sampling interval. Estimates based on increasing scales are regional values, and are based on statistical moments of elevation rather than derivatives. Any area can be characterized by a probability distribution function of elevation, which can in turn be summarized by its statistical moments. For example, the third moment of elevation is skewness which is an inverse measure of the region curvature. Positive skew indicates proportionally more higher (than the mean) than lower elevations which is an indication of concave surfaces (negative curvature). Thus regional curvature is measured by the third moment of the original elevation while local curvature is measured by the first moment of the second derivative of elevation.

LANDFORM AND SOIL

Although relationships between topography and soils are well documented (Gerrard, 1981; Huggett, 1982), limited success in quantitatively or statistically describing the relationship has been achieved. With a significant proportion of soil variability occurring over very short distances, there appears to be as much variability within slope class designations as between classes (Joel, 1933; Ball and Williams, 1968; King *et al.*, 1983). In addition there are fundamental differences in soil relationships on convex, concave and straight slopes (Gerrard, 1981).

In a detailed examination of soil-landscape relationships in Saskatchewan, Canada, from a soil survey perspective, King *et al.* (1983) concluded that the only realistic division for map units is by convex upper slopes (shallow soils), concave lower slopes (deep soils) and depressional areas (gleyed soils). Short range variability was so great that it was not possible to create meaningful map-units on a smaller scale or by using more refined soil groupings such as soil series. As the authors stated, the divisions suggested by the study are significant in that they correspond with landscape elements recognized as meaningful in terms of crop and vegetative productivity.

Even though considerable variability of landforms and soils exist, it is generally conceded that the variance is not necessarily random. That is, measurements of a property taken close together are more alike than values taken farther apart. This connectiveness is called spatial variance structure. Methods for describing the spatial variance structure are usually derived from the theory of regionalized variables using the semi-variogram or covariogram (Matheron, 1971) or from linear stochastic theory using the autocorrelogram, cross-correlogram or spectral analysis (Jenkins and Watts, 1968). Spectral analysis is a one-way analysis of variance as a function of frequency or scale. Inter-relationships between variables can also be examined in the frequency domain. All of the methods are part of space-time systems analysis (e.g. Bennett, 1979). Recent reviews of spatial statistical analysis of soil properties have been given by Peck (1983), Warrick *et al.* (1986) and Nielsen and Bouma (1984). A historical review or spectral analysis of landforms has been given by Pike and Rozema (1975). Problems associated with the scale of observation in the application of autoregressive theory to topographic data have been discussed by Thornes (1973).

The purpose of this paper is to present results from studies where the spatial distribution of soil properties and the influence of topography at different scales are the focal point. Three data sets will be given which illustrate a variety of scaling problems. The relationship of the soil properties to soil processes occurring over long and short time scales will be discussed.

Description of data sets

The Delhi data set will be used to illustrate the problem of soil variability at different scales. The Weyburn data set will be used to examine the relationship between topographic parameters and soil variability on a scale of < 50 m. With the Floral data we examine the relationship between soil electrical conductivity (a function of the water and salt content of soil and soil texture) and landform across a small agricultural drainage basin. Spatial autocorrelations were calculated using equations given by Davis (1973). Power spectra, co-spectra, and coherency estimates were calculated using smoothed Fourier methods (Brillinger, 1981; Otnes and Enochson, 1978). A summary of the equations used has been given by Kachanoski *et al.* (1985a).

Delhi data set

The site is located on the Delhi Agricultural Research Station in Southern Ontario, Canada. The samples have been collected, analysed and tabulated by Protz *et al.* (1987) as part of an ongoing study on soil variability. Three transects, 1000 m, 100 m, and 10 m long, were sampled in both virgin forest and cultivated fields. The site was selected for its general uniformity in topography over the study area (< 1% slope). Sampling intervals were 10 m, 1 m, and 0.1 m in the 1000 m, 100 m and 10 m long transects respectively. There were 100 points, termed profiles, where soil horizons were measured from the soil surface to a depth of about 2 m. For the 1000 m and 100 m long transects, soil pits were excavated at each location and the soil horizon thicknesses recorded and soil cores taken. The 10 m transect data were obtained from a 10 m trench, sampling the face of the trench every 0.1 m. Approximately 3600 soil horizon samples were taken.

Horizon thicknesses for profiles along the forest transects are illustrated in Figure 9.1. The statistical moments of the major soil horizons (Table 9.1) indicate that a significant amount of variability exists on a scale of < 10 m. However, the 10 m scale also shows a certain amount of 'smoothing out'

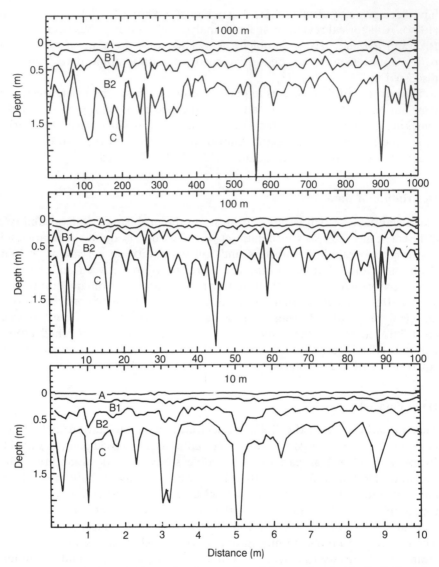

Figure 9.1   Horizon thickness for three scales of observation for the Delhi forest transects

compared with the other transects (Figure 9.1). Similar variations were present in the cultivated transects. The 1000 m transect has a significantly greater horizon and solum thickness than the other transects, which is due to the first 38 profiles. The last 62 profiles of the 1000 m transect have the same moments as those of the 10 m and 100 m transect.

Table 9.1 Means and coefficients of variation (%) for the thickness of soil horizons in the Delhi data set

Soil Horizon	Transect		
	10 m	100 m	1000 m
LFH	0.026(46)[*]	0.030(45)	0.034(39)
Ah	0.119(19)	0.118(30)	0.128(19)
B	0.754(50)	0.720(52)	0.840(43)
SOLUM	0.899(42)	0.868(44)	0.100(35)

[*]mean (% coeff. of variation)

The solum thickness is the depth from the surface to the C_K horizon boundary and is an indication of long term leaching and weathering intensity where soils are freely drained. The autocorrelation values of solum thickness of the native forest (Figure 9.2) for the 10 m and 1000 m transects do not approach 1.0 as the lag approaches zero. As indicated by the autocorrelation functions of the 100 m and 1000 m transects, a discontinuity at the origin exists which is called the nugget effect. The nugget effect can be attributed to either sampling error or variance microstructure (Rendu, 1978). Because the sampling error for horizon thickness would be small, the nugget must be due to the small scale spatial dependence which is illustrated by the autocorrelogram of the 10 m transect. The 1 m and 10 m sampling intervals are too wide to pick up the short range (<2 m) variance structure indicated by the 0.1 m sampling interval. The strong autocorrelation for the 10 m transect at small lags (<0.5 m) and the convergence of the autocorrelation to 1.0 as the lag approaches zero (Figure 9.2) indicate the 0.1 m sampling interval is sufficiently small to characterize the spatial structure of the horizon continuum. In addition, the strong autocorrelation peak at the 2 m lag indicates that sudden increases in solum thickness occur, on average, every 2 m. Other smaller fluctuations in the autocorrelogram are present and would not normally be considered significant. However, the solum thickness autocorrelograms (Figure 9.3) for the 10 m transect from both the forested and cultivated site (the transects were about 1 km apart) are remarkably similar.

The stability of the spatial variance structure over space (in this case over 1 kg) is encouraging since it suggests a regularity or stationarity of the soil forming processes which can be used for spatial prediction using linear stochastic models (e.g. kriging or autoregressive equations). Examples of using spatial variance structures of soil properties for constructing linear prediction models such as Kriging and autoregressive equations have been given by Webster and Burgess (1980, 1980b), Sisson and Wierenga (1981), Vieira *et al.*, (1981), Vauclin *et al.*, (1983), and others. However the non-stationarity of the

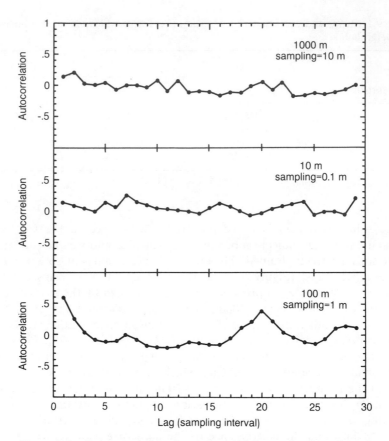

Figure 9.2   Autocorrelation for solum thickness for 3 scales of observation for the Delhi forest data

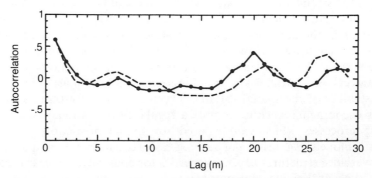

Figure 9.3   Comparison of solum thickness autocorrelation for the cultivated and forest 10 m transects (Delhi data, dotted line is the cultivated site)

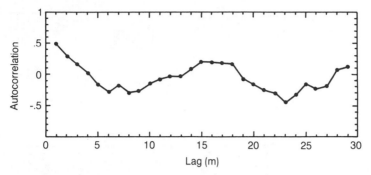

Figure 9.4   Autocorrelation of the LFH horizon thickness for the Delhi 10 m forest transect

mean such as in profiles 1 to 38 of the 1000 m transect, (Figure 9.1) may limit the transferability of these variance structures.

The average 2 m repetitive nature of the solum thickness is consistent with the concept of a pedon. In this case the pedon would have average lateral dimensions of 1 m (half the average cycling frequency). The 30–60 m section of the 100 m transect (Figure 9.1) indicates that the A horizon increases in thickness over a short distance (44–46 m) while the lower horizons are affected at successively wider and wider distances with soil depth. This is consistent with a point source of water at the surface and subsequent radial flow affecting the weathering intensity over wider and wider areas with depth. The above scenario is not entirely consistent with the concept of a pedon since the scale of observation for horizons is changing with depth. Solum thickness is the cumulative effect of soil forming conditions over thousands of years. The surface litter layer (LFH horizon) is a more dynamic horizon and is an expression of more recent soil forming conditions. The autocorrelation of the LFH thickness (Figure 9.4) indicates a spatial pattern which is different from the mineral horizon.

Weyburn data set

The study area is situated approximately 30 km east of Weyburn, Saskatchewan, Canada. The parent material of the soil is glacial till underlain by a second till formation of earlier age. The uppermost till is approximately 1–2 m thick and was deposited about 20 000 years ago B.P. The second till lies below the first till, has significantly different physical and chemical characteristics than the overlying till, and was deposited approximately 38 000 years ago B.P. The stratigraphic surface between the two tills were characterized by a sand-gravel layer of variable thickness.

Soil cores (76 mm diameter, 2 m long) were taken every 1 m in a number of transects in both a native (never cultivated) grassland and a portion of the field

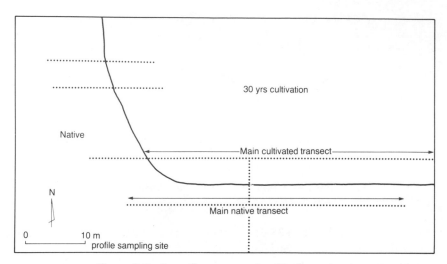

Figure 9.5   Sampling pattern for Weyburn data

which had been cultivated for 30 yrs (Figure 9.5). The soil cores were separated
into horizons, horizon thickness recorded, and a variety of soil properties
measured including bulk density, soil water, and organic carbon. Around each
sampling point, elevation readings were taken every 1 m in a 3 m × 3 m grid.
Microtopographic parameters including gradient and vertical curvature were
calculated at each point using the methods of Young and Evans (1978). The
study area is situated on a relatively smooth (overall slope < 0.5%) upper slope
portion. Thus, the soil variability would be classified as 'within slope
variability' according to soil survey. The soil type is similar as that studied by
King *et al.* (1983) who concluded that within slope variability was so great that
soil survey could not practicably delineate meaningful units other than upper
(shallow soils) and lower (deep soils) slope areas. A more detailed analysis and
description of the Weyburn data set is contained in recent papers (Kachanoski
et al., 1985a, 195b, 1985c; Singh *et al.*, 1985; Selles *et al.*, 1985).

Statistical moments (Table 9.2) and the correlation matrix (Table 9.3) for
selected soil properties of the native grassland are given. Although the A
horizon thickness and mass were significantly correlated to vertical surface
curvature, only 14% of the variability could be explained. Thickness and mass
of the B horizon were not significantly correlated (5% level) to any microtopographic parameter. Multiple correlation between microtopographic parameters and horizon thickness accounted for 25% and 9% of the variability of
the A horizon thickness and B horizon thickness respectively. In contrast, the
depth to the intertill sand–gravel layer was significantly correlated (1%
probability level) to thickness and mass of the B horizon, but not the A
horizon.

Table 9.2   Means and coefficients of variation (%) for the Weyburn data set

Soil Property	A horizon	B horizon	Solum (A + B)
Thickness (m)	0.172(28)[*]	0.215(33)	0.384(34)
Bulk density (Mg m^{-3})	1.14(8)	1.47(6)	1.32(5)
Mass (Mg m^{-2})	0.197(29)	0.315(34)	0.512(22)
Total Carbon (Kg m^{-2})	7.2(24)	4.0(41)	11.2(22)

[*]mean (% coeff. of variation)

The power spectra of curvature and A horizon mass (native grassland) are remarkably similar (Figure 9.6) with a strong increase in variance power centered around 0.14 cycles m^{-1} (7 m period) and secondary increases around 0.32 cycles m^{-1} (3 m period). The spectrum of A horizon mass does not show a significant increase in variance at low frequencies (large scale) which indicates that the variability is mainly on a scale of $\leqslant 10$ m. Coherency estimates indicated significant correlation (5% level) between curvature and A horizon mass at the 7 m cycling frequency.

The power spectra of B horizon thickness and depth to the intertill sand lens are given in Figure 9.7. Both spectra have a significant peak at 0.23 cycles m^{-1} (4.5 m period) and significant coherency (correlation) at that frequency. The concentration of variance at 0.23 cycles m^{-1} is not found in the A horizon spectrum.

Table 9.3 Correlation between topography, depth to sand layer and soil properties for the Weyburn data set

		Microtopography Properties			Sand lens depth
Soil Property		Elevation	Gradient	Curvature	
A horizon	Thickness	0.14	0.17	0.38[**]	0.15
	Bulk density	−0.18	0.34[*]	0.12	−0.22
	Mass	0.07	0.27	0.37[**]	0.08
	Total Carbon	0.01	0.08	0.14	0.12
B horizon	Thickness	−0.08	−0.03	−0.27	−0.41[**]
	Bulk density	−0.41[**]	0.46[**]	−0.01	−0.03
	Mass	−0.16	0.06	−0.28	−0.41
	Total carbon	−0.14	−0.06	−0.10	−0.38
Solum	Thickness	0.01	0.08	−0.03	−0.30[*]
	Bulk density	−0.50[**]	0.50[**]	−0.17	−0.35[*]
	Mass	−0.12	0.19	−0.07	−0.35[*]
	Total Carbon	−0.10	0.02	−0.03	−0.17

[*]   sign. at 0.05 probability
[**] sign. at 0.01 probability

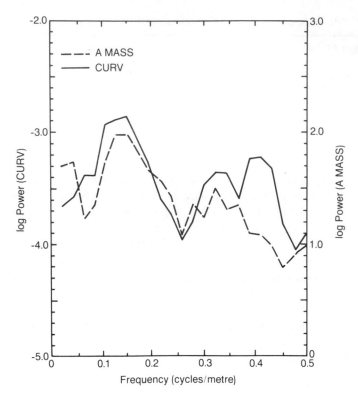

Figure 9.6   Comparison of the A horizon mass and surface curvature spectra, native Weyburn site

The data suggest that processes controlling A horizon formation do not have the same spatial variance relationships as those controlling the formation of B horizons. The A horizon variability is affected by the variability of surface curvature which would influence the redistribution of rainfall and the moisture in the rootzone of the vegetation and, therefore, both biomass production and leaching potential. At greater depths, the factor affecting the redistribution of soil water appears to be depth to the intertill sand layer. For unsaturated flow, which is the normal condition at this site, the sand interlayer would act as an impedance resulting in higher moisture conditions and thus increased weathering potential in the layers above it. The net result is the significant negative correlation between depth to the stratigraphic sand layer and B horizon thickness and mass.

The amount of accumulated organic carbon and the formation of A and B horizons in prairie soils (Mollisols) are intimately related. In native conditions, for a given slope position, vegetative growth will ultimately determine

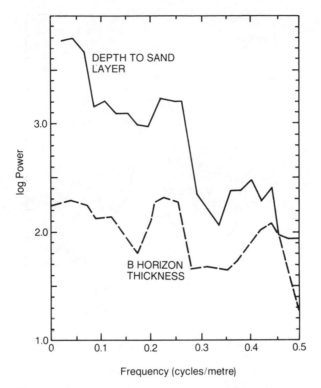

Figure 9.7 Comparison of the spectra for B horizon
mass, and depth to sand layer spectra (Weyburn data)

soil organic carbon levels because it will control the type and amount of
organic matter added to the soil and will vary mainly with moisture conditions.
The similarity of the spectra for total solum carbon and depth to intertill sand
layer (Figure 9.8) confirms the influence this layer is having on the long term
moisture regime.

The interrelationships among horizon mass, curvature, and depth to the
sand layer produce an interesting relationship between surface curvature and
total solum carbon. The coherency and cospectra for these variables (Figure
9.9) indicate significant (1–5% level) negative correlation for scales greater
than 4.0 m (frequency <0.25 cycles m^{-1}) and significant (1–5% level) positive
correlation on a scale less than 4 m (frequency >0.25 cycles m^{-1}). The
positive correlation at one scale cancels the negative correlation at the other
scale so the overall (standard) correlation is essentially zero (i.e. $r = 0.03$,
Table 9.3).

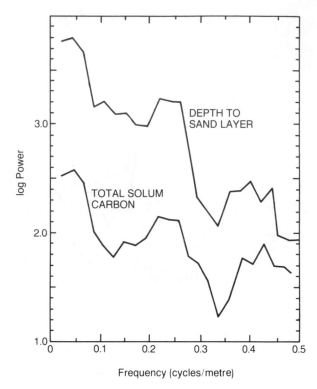

Figure 9.8   Comparison of the spectra of Total Solum
Carbon and depth to sand layer (Weyburn data)

Because the native grassland was sampled at the end of the growing season
approximately 25 days after the last rainfall, the measured soil water content
values would represent the permanent wilting point (PWP) of the profile. The
semi-variograms of soil water (Figure 9.10) indicate significantly different
spatial correlation ranges for the A horizon and B horizon PWP.

The relationships discussed so far have been for the native grassland. The
spectrum of total solum carbon for the cultivated (30 yrs) portion of the field
(Figure 9.11) also indicates increases in variance at the 7 m and 3 m periods
which are the same as those of the curvature spectrum in the native field. The
spectrum of surface curvature in the cultivated field showed no spectral peaks
at these scales due to smoothing and infilling from the tillage operations
(Kachanoski, 1985c). Coherency analysis (Figure 9.11) indicated the native
surface curvature was correlated to solum properties in the cultivated field,
while present day local curvature was not correlated.

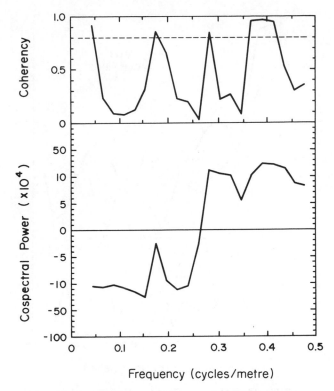

Figure 9.9 Coherency and Co-spectrum between total solum carbon and surface curvature. (Dashed line indicates 5% probability level.)

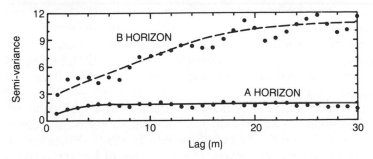

Figure 9.10 Semi variograms for A horizon and B horizon soil water (Weyburn data)

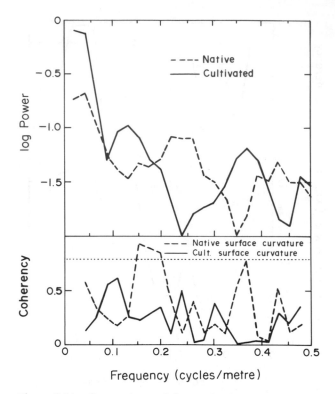

Figure 9.11   Comparison of the total solum carbon spectra from the native and cultivated fields including the coherency to curvature (Weyburn data, dashed line = 5% probability)

Floral data set

The previous data sets have dealt with soil properties that take medium to long periods of time to change perceptibly. In this data set, the spatial distribution of soil electrical conductivity (EC) is examined in relation to topographic parameters measured over a small drainage basin. The EC is influenced by soil moisture and dissolved salts, dynamic properties strongly affected by recent hydrological processes.

The study area is a 57 ha agricultural drainage basin approximately 30 kg east of Saskatoon, Saskatchewan, Canada. The soils are silt loam to sandy loam textured Mollisols. The basin was part of the Saskatchewan Research Council basin study (1960–1969). Contour lines (0.3 m interval) had been determined for the basin. The contour map of the basin was digitized and interpolated to produce an elevation grid (6 m inverval). Topographic parameters were calculated in a manner similar to the Weyburn data set. EC readings

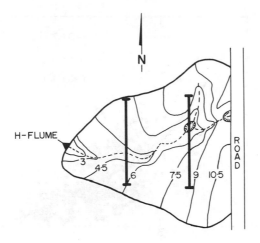

Floral basin (57 ha)

Figure 9.12   Floral drainage basin with location of first and last EC transects

(0–1.0 m depth) were taken on a 6 m grid in eight 540 m long transects which ran perpendicular to the main drainage pattern, using a non-contacting electromagnetic probe (Geonics EM-38). A diagram of the drainage basin showing the 1.5 m contour lines is given in Figure 9.12.

A summary of the average topographical parameters of the transects is given in Table 9.4. At the basin scale there is a significant correlation (<0.001 probability level) between EC and surface elevation (Figure 9.13). Multiple correlation indicated that the deviations from the regression line (Figure 9.13) were significantly correlated to local surface curvature (i.e. measured on an

Table 9.4   Mean elevation and soil conductivity for the Floral Basin Transects

Basin Transect	Relative Elevation (m)	Soil Conductivity (0–0.5 m, dSm^{-1} × 100)
1	23.0	14.1(46)[*]
2	19.3	14.2(51)
3	17.2	29.0(54)
4	16.9	28.4(46)
5	16.4	27.0(54)
6	15.1	48.4(47)
7	13.8	52.5(37)
8	11.3	62.1(27)

[*]Mean (% Coeff. of variability)

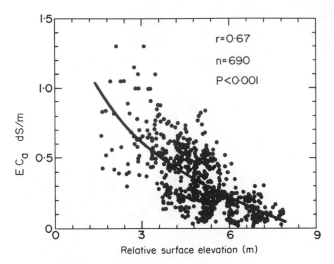

Figure 9.13   Regression relationship between EC and surface elevation, Flora drainage basin

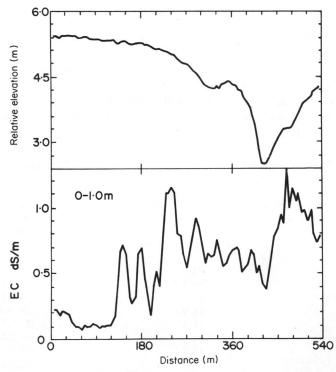

Figure 9.14   Relative elevation and EC measurements for the
seventh transect (Floral data)

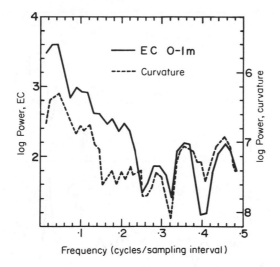

Figure 9.15   Comparison of the EC and surface
curvature spectra for the seventh transect (Floral
data)

areal scale = 0.014 ha). Elevation and curvature together accounted for
approximately 60% of the variability of measured EC values. The negative
relationship between EC and elevation is probably a reflection of the regional
(basin) groundwater flow system which would influence the overall soil
moisture and salt accumulation regimes. Spatial variations in local curvature
cause fluctuations away from this regional relationship. This can be seen in
transect seven (Figure 9.14) where there is a general increase in EC with
decrease in elevation, with local fluctuation superimposed. The spectra of local
curvature and EC for this transect (Figure 9.15) indicate that for fluctuations
on a scale <25 m (>0.24 cycles per 6 m) the variance decompositions are
almost identical.

The increase in EC variance at low frequencies (large scale) is due to the
general increase (trend) in EC over the length of the transect (Figure 9.14)
which, as mentioned earlier, can be related to the general decrease in elevation.
Coherency analysis indicated significant correlation between elevation and EC
at low frequencies, and curvature and EC at high frequencies. Strictly
speaking, a power spectrum is not defined if a trend is present, however, in
practice this is not a serious problem because the spectrum still isolates the
variance contribution at different scales and the presence of a trend will only
result in increased variance at low frequencies (Jenkins and Watts, 1968). The
spatial variance distribution of soil EC is clearly a combination of large scale
(low frequency) variations in elevation and smaller scale (high frequency)

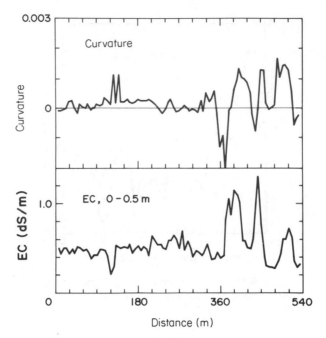

Figure 9.16   Comparison of EC and surface curvature
for the eighth transect (Floral data)

variations in surface curvature. In other transects, (e.g. transect eight in Figure
9.16), the relationship between local changes in surface curvature and EC are
so apparent that spectral analysis is not necessary.

SUMMARY

The purpose of any spatial model is to simplify, organize, and extrapolate
information about the soil system. Soil systems are particularly complex
because of the large number of interacting variables. The utility of a simplified
description of a space-time system will depend on how well the linkages
between spatial units are described (Bennett, 1979).

Two frequently occurring spatial linkages are (1) hierarchical processes and
(2) contiguity (lag) processes (Bennett, 1979). Anderson *et al.* (1983) merged
the concept of hierarchical land classification systems and the level of
integration concept of ecology (Rowe, 1961) and proposed a hierarchical
classification of agro-ecosystems for the basis of compiling and managing
data, and extrapolation into the future. Ecosystem levels include the pedon,
polypedon, soil catena, and higher levels. Contiguity (lag) processes are
governed according to lag, or the co-spatial dependency of adjacent spatial

units. The behaviour of larger regions is dependent on lagged diffusion terms between subregions. These lagged diffusion terms are frequently described by the partial differential equations of flow, flux, and potential in air and water movement (Bennet, 1978). Contiguity (lag) processes describe the law that everything is related but near things are related more than distant things. This is represented by the autocorrelograms, spectra, cospectra, etc. which have been presented for the data sets. The co-dependency of spatial processes due to spatial positioning and distribution has led to the suggestion that the erosional drainage basin, which accommodates the three dimensional characteristics of soil and soil related processes, should be adopted as the basic functional study unit (Vreeken, 1973; Huggett, 1973, 1982).

The data sets given in this chapter have been concerned with representing and describing the continuous distribution of soil properties by examining the spatial variance structure, that is the contiguity (lag) relationships. The nature of the contiguity (lag) relationships has been shown to be highly dependent on the spatial distribution of topographic parameters (at different scales) which control the transfer of mass and energy in the landscape. This is as it should be, because soil properties are dependent variables, determined by the major soil forming factors.

A complex system such as the soil will have a variety of linkage processes occurring. The hierarchical agro-ecosystem classification (Anderson *et al.*, 1983) requires that the integration levels (scales) be related to each other in a definable and quantitative sense. Integration levels are related in a definable way in that the objects at one level are the environment for those at the next lowest level. The levels, however, are defined in a descriptive manner and are thus morphological models (Dijkerman, 1974). More realistically, levels should be related in a functional and quantitative manner; the cascading system defined by the path followed by flows of energy and mass (Dijkerman, 1974) seems most appropriate. If the quantitative criteria are based on the physical laws determining the flow of mass and energy to higher and lower levels then the drainage basin becomes an obvious level of integration. Sub-basins, geomorphic elements within basins (valley sides, uplands, flood-plains), soil catenas, and convex (shallow soil) and concave (thick soil) slopes as described by King *et al.* (1983) are all successively lower levels of integration. Soil catenas rather than surface form or topography alone is an important element, in that the properties of the soil (texture, structure) become part of the system. Local changes in surface curvature can be used to determine even smaller levels of integration within a slope position as indicated by the relationship between EC and surface curvature on a scale <25 m (Figure 9.15, Floral data set) and surface curvature and A horizon thickness (Figure 9.7, Weyburn data set). In many cases this level will coincide with the scale of a pedon. The interpolation can continue to smaller scales using the spatial variance (contiguity) structure of representative elementary volumes

(REV) of porous media (Wagenet, 1984; Bouma, 1984). The REV represents the smallest sample (soil) in a practical sense since it determines the stability of the phenomenological relationships used to describe the flux of energy or mass (e.g. Fick's Darcy's and Fourier's law) which have transfer coefficients with area (L^{-2}) units.

The pedon is a concept for describing the REV of soil horizon properties. Unfortunately, as the data sets indicate, different properties have different spatial variance relationships. Even horizontal thickness has different spatial scales depending on surface or subsurface influences. In the Weyburn data the native A horizon was influenced by the local curvature of the present surface curvature while the B horizon was influenced by the surface which existed 20 000 B.P. (i.e. the depth to the sand layer which lies on the older till at the site). In addition because mass and energy transfer with depth is not uni-directional (vertical), surface properties influence what is happening both ahead and behind in the horizontal plane. These co-spatial interrelationships pose significant difficulties in interpreting and using mean values for soil horizon properties of a pedon for input into existing soil process models at a more general scale. The problems associated with sampling on a pedon basis have resulted in most spatial process models using point sampling techniques. However, similar problems exist at this sampling scale with respect to the size of the core used.

In many cases it will not be possible to identify the factors influencing the spatial variance structure of soil variables. For example, the local surface curvature which controlled the variance distribution of the native A horizon (Weyburn data) is no longer present because of the effects of cultivation. Aggregation to the next level of integration (convex upper, etc.) as done by King *et al.* (1983) on this soil type may be necessary. Prediction at smaller scales may require measurement of the microvariance structure which is time-consuming, but may have wider application if the variance structure remains stationary over larger areas as in the Delhi data (Figure 9.3). In some soil systems such as the Floral basin it may be possible to explain the local fluctuations of soil processes based solely on digital topographic analysis.

Almost all point soil process models will require water content as an input. On this basis alone, it is necessary that 'real' space-time modelling uses digital terrain analysis methods. In addition, because surface landform shape is a good indication of the spatial variance distribution of soil properties, it seems reasonable that finding methods of describing the joint distribution of soil and topography should continue to be a high priority.

REFERENCES

Anderson, D. W., Heil, R. D., Cole, C. V., and Deutsch, P. C. (1983). Identification and characterization of ecosystems at different integration levels. In *Nutrient*

Cyclying in Agric. Ecosystems, pp. 517–529. Univ. of Ga., College of Agric. Exp. Stations. Spec. Pub. 23.

Ball, D. F., and Williams, W. M. (1968). Variability of soil chemical properties in two uncultivated brown earths. *J. Soil Sci.*, **19**, 379–391.

Bennett, R. J., (1979). *Spatial Time Series*, pp. 404–457. Pion Limited Publishers, London.

Bouma. J. (1984). *Soil Variability and Soil Survey. Soil Spatial Variability*, pp. 130–150. Proc. Int. Soc. Soil. Sci. and Soil. Sci. Soc. Amer. workshop. Las Vegas, USA. PUDOC Publishers, Wageningen Netherlands.

Box, G. E. P., and Jenkins, G. M. (1970). *Time Series Analysis, Forcasting and Control*. Holden Day Inc., Publishers, San Francisco.

Bridges, E. M. (1982). Techniques of modern soil survey. In Bridges, E. M., and Davidson, D. A. (Eds.) *Principles and Applications of Soil Geography*, pp. 1–28. Longman Inc. Publishers, N.Y.

Bridges, E. M., and Davidson, D. A. (1982) (Eds.) *Principles and Applications of Soil Geography*. Longman Inc., Publishers. NY: 297 pages.

Brillinger, D. R. (1981). *Time Series, Data Analysis and Theory*. Holden-Day Inc., San Francisco, CA: 540 pages.

Burgess, T. M., and Webster, R. (1980b). Optimal interpolation and isarithmic mapping of soil properties II. Block kriging. *J. Soil Sci.*, **31**, 333–341.

Davis, J. C. (1973). *Statistics and Data Analysis in Geology*. John Wiley and Sons. Inc., New York: 550 pages.

Dijkerman, J. C. (1974). Pedology as a science. The role of data, models, and theories in the study of natural systems. *Geoderma*, **11**, 78–93.

Evans, I. S. (1972). General geomorphometry, derivatives of altitude and descriptive statistics. In Chorley, R. J. (Ed.) *Spatial Analysis in Geomorphology*, pp. 17–91 Methuen and Co. Ltd., London.

Evans, I. S. (1981). General geomorphometry. In Goudie, A. (Ed.) *Geomorphological Techniques*, pp. 31–36. George Allen and Unwin Publishers, London.

Finkl, C. W., Jr. (1982). *Soil Classification Benchmark Papers in Soil Science 1*. Hutchinson Ross publishers. USA: 391 pages.

Gerrard, A. J. (1981). *Soils and Landforms*. George Allen and Unwin Publishers. London, UK: 219 pages.

Huggett, R. J. (1973). Soil landscape systems: a model of soil genesis. *Geoderma*, **13**, 1–22.

Huggett, R. J. (1982). Models and spatial patterns of soils. In Bridges, E. M., and Davidson D. A. (Eds.) *Principles and Applications of Soil Geography*, pp. 132–171. Longman Inc. Publishers. NY.

Jarvis, M. G. (1982). Non agricultural uses of soil survey. In Bridges, E. M., and Davidson, D. A. (Eds.) *Principles and Applications of Soil Geography*, pp. 216–256. Longman Inc. Publishers.

Jenkins, G. M., and Watts, D. G. (1968). *Spectral Analysis and its Applications*. Holden Day Int. Publishers. San Francisco, CA: 525 pages.

Joel, A. H. (1933). The zonal sequence of soil profiles in Saskatchewan. *Soil Sci.,* **36**, 173–184.

Kachanoski, R. G., Rolston, D. E., and De Jong, E. (1985a). Spatial and spectral relationships of soil properties and microtopography: I. Density and thickness of A horizon. *Soil Sci. Soc. Amer. J.*, **491**, 804–812.

Kachanoski, R. G., De Jong, E., and Rolson, D. E. (1985b). Spatial and spectral relationships of soil properties and microtopography: II. Density and thickness of B horizon. *Soil Sci. Soc. Amer. J.*, **49**, 812–816.

Kachanoski, R. G., D. E. Rolston, and E. De Jong. (1985c). Spatial variability of a cultivated soil as affected by past and present topography. *Soil Sci. Soc. Amer. J.,* **49,** 1082–1087.

King, G. J., Acton, D. F., and St. Arnauld, R. J. (1983). Soil landscape analysis in relation to soil distribution and mapping at a site in the Weyburn Association. *Can. J. Soil Science.,* **63,** 657–670.

Knox, E. G. (1965). Soil individuals and soil classifications. *Soil Sci. Soc. Amer. Proc.,* 29, 79–84.

Matheron, G. (1971). *The Theory of Regionalized Variables and its Application.* Les cahiers du Centre de Morphologic Mathematique de Fontainbleau, Ecole Nationale Superieuve des Mines de paris.

Nielsen, D. R., and Bouma, J. (Eds.) (1984). *Soil Spatial Variability.* Proc. Int. Soc. Soil Sci. and Soil Sci. Soc. Amer. workshop. Las Vegas, USA. PUDOC publishers, Wageningen, Netherlands: 235 pages.

Otnes, R. K., and Enochson, L. (1978). *Applied Time Series Analysis Vol. 1: Basic techniques.* John Wiley and Sons publishers: 449 pages.

Peck, A. J. (1983). Field variability of soil physical properties. *Advances in Irrigation,* **2,** 189–221.

Pike, R. J., and Rozema, W. J. (1975). Spectral analysis of landforms. *Ann. Assoc. Amer. Geog.,* **65**(4), 499–516.

Protz, R., Fischer, J., Bolton, K., Shipitalo, S. E., and Lapalme, A. (1987). *Spatial Dependence of Selected Properties of the Fox Sandy Loam. I. Sampling Design and Initial Data.* Tech. memo. 87–2. Land Resource Science, Univ. of Guelph, Guelph, Ontario N1G 2W1.

Rendu, J. M. (1978). *An Introduction to Geostatistical Methods of Mineral Evaluation.* South African Inst. of Mining and Metallurgy: 83 pages.

Rowe, J. S. (1961). The level-of-integration concept and ecology. *Ecology,* **42,** 420–427.

Selles, F., Karamanos, R. E.., and Kachanoski, R. G. (1986). The spatial variability of nitrogen-15 and its relation to the variability of other soil properties, *Soil Sci. Soc. Amer. J.,* **50,** 105–110.

Simonson, R. W. (1968). Concept of soil. *Adv. Agron.,* **20,** 1–47.

Singh, J. P., Karamanos, R. E., and Kachanoski, R. G. (1985). Spatial variation of extractable micronutrients in a cultivated and native prairie soil. *Can. J. Soil Science.,* **65,** 149–157.

Sisson, J. B., and Wierenga, P. J. (1981). Spatial variability of steady state infiltration rates as a stochastic process. *Soils Sci. Soc. Amer. J.,* **45,** 699–704.

Soil Survey Staff, USDA. (1960). *Soil Classification, 7th Approximation.* US Government printing office Washington, DC: 265 pages.

Thornes, J. B. (1973). Markov Chains and Slope Series. *Geographical Analysis,* **5,** 322–328.

Vauclin, M., Vieira, S. R., Vachaud, G., and Nielsen, D. R. (1983). The use of Co-Kriging with limiting field soil observations. *Soil Sci. Soc. Amer. J.,* **47,** 175–184.

Vieira, S. R., Nielsen, D. R., and Biggar, J. W. (1981). Spatial variability of field measured infiltration rate. *Soil Sci. Soc. Am. J.,* **45,** 1040–1048.

Vreeken, W. J. (1973). Soil variability in small loss watersheds: clay and organic matter content. *Catena,* **1,** 181–196.

Wagenet, R. J. (1984). *Measurement and Interpretation of Spatially Variable Leaching Processes,* pp. 209–236. Proc. Int. Soc. Soil Sci. and Soil Sci. Soc. Amer. workshop. Las Vegas, USA. PUDOC publishers, Wageningen, Netherlands.

Warrick, A. W., Myers, D. E., and Nielsen, D. R. (1986). Geostastistical methods applied to soil science. In Klute (ed.) *Methods of Soil Analysis Part 1*. 2nd edn. *Agronomy,* **9**, 53–82.

Webster, R., and Beckett, P. H. T. (1968). Quality and usefulness of soil maps. *Nature,* **219,** 680–682.

Webster, R., and Burgess, T. M. (1980). Optimal interpolation and isarithmic mapping of soil properties. III. Changing drift and universal Kriging. *J. Soil Sci.,* **31,** 505–524.

Wilding, L. P. (1984). *Spatial Variability: its Documentation Accommodation and Implecation to Soil Survey,* pp. 166–195. Proc. Int. Soc. Soil Sci. and Soil Sci. Soc. Amer. workshop. Las Vegas, USA. PUDOC publishers, Wageningen, Netherlands.

Young, M., and Evans, I. S. (1978). *Statistical Characterization of Altitude Matrices Report*. No. 5, Grant DA-ERO-591-73-G0040. Dept. of Geography, University of Durham, England: 26 pages.

Scales and Global Change
Edited by T. Rosswall, R. G. Woodmansee and P. G. Risser
© 1988 Scientific Committee on Problems of the Environment (SCOPE)
Published by John Wiley & Sons Ltd.

CHAPTER 10

Scale and the Measurement of Nitrogen-Gas Fluxes from Terrestrial Ecosystems

DAVID S. SCHIMEL,[1] STEPHEN SIMKINS,[2] THOMAS ROSSWALL,[3]
ARVIN R. MOSIER[4] and WILLIAM J. PARTON[5]

ABSTRACT

A key problem in the measurement of N gas fluxes from terrestrial ecosystems is their typically high degree of spatial and temporal heterogeneity. Gas fluxes vary at both fine and coarse scales of resolution. In this paper we review several studies of spatial variability in N gas fluxes. In the first example, fine-scale variation in N_2O production was examined using geostatistical techniques. A significant amount of the variation in N_2O production could be explained by spatial autocorrelation and by correlation with soil water. In the second example, seasonal and edaphic variations in N_2O production from the shortgrass steppe were measured. The results were incorporated into a simulation model which related N_2O production from nitrification and denitrification to soil moisture, temperature, and NH_4^+ and NO_3^- production. A simplified version of this model was developed to calculate annual N_2O production from a heterogeneous landscape. Spatial variations in NH_3^0 production from cattle urine were determined at the same site. NH_3^0 production is controlled by the joint spatial distributions of cattle urine deposition and of soil properties which control NH_3^0 evolution rate. The effect of N transport by cattle is to reduce mean NH_3^0 volatilization rates below maximal rates. A common technique to all of these studies was to relate hard-to-measure gas flux rates to more readily measured soil and landscape properties.

[1]Natural Resource Ecology Laboratory, Colorado State University, Fort Collins, Co 80523, USA
[2]Department of Plant and Soil Sciences, University of Massachussetts at Amherst, Stockbridge Hall, Amherst, MA 01003, USA
[3]Department of Water in Environment and Society, University of Linköping, S-581 83, Linköping, Sweden
[4]U.S. Department of Agriculture-Agricultural Research Service, Fort Collins, Co 80523, USA.
[5]Natural Resource Ecology Laboratory, Colorado State University, Fort Collins, Co 80523, USA

INTRODUCTION

Nitrogen gases play significant roles in atmospheric biogeochemistry at local and global scales (Crutzen, 1983). Nitrogen gases influence the climate as greenhouse gases, participate in the formation and destruction of O_3, influence atmospheric acidity, and are significant vectors for loss and gain of nitrogen from terrestrial ecosystems (Lacis *et al.*, 1981; Crutzen, 1983; Bolin *et al.*, 1983). Biogenic sources of N gases are currently significant, and may become more so with changes in climate and land use. Despite the importance of N gases in the atmosphere, fluxes in and out of terrestrial ecosystems are not well known. Data are particularly scanty for N species and from ecosystems whose emissions are not significant to N budgets. Such data may be important in defining global balances.

Although there are many problems with measurement of N gas fluxes, a key difficulty is the high degree of spatial and temporal variability characteristic of N gas fluxes. High variability is found at small scales, within experimental units and between sites which vary in vegetation, soil properties, or water balance. Folorunso and Rolston (1984) reported extreme small scale variability, with coefficients of variation for N_2O flux from 3×36 m plots of 161 to 508%. Mosier *et al.* (submitted) reported coarser-scale differences in N_2O flux from 80 to $160 \, g \, N \, ha^{-1} y^{-1}$ between a slope and an adjacent swale. Similar variations in NH_3^0 flux from the same pair of sites were reported Schimel *et al.* (1986). Accurate calculation of flux rates requires techniques for reducing random variation within experimental units, and knowledge of the factors that result in systematic variation between ecosystems. Considerable progress must be made in the measurement and modelling of fluxes in order to calculate fluxes over large areas and to predict how those will change with climate and land-use changes.

In this chapter we present three examples where gas flux rates were calculated for areas of considerable heterogeneity. The first example is of a study of within-site heterogeneity and demonstrates a class of techniques for minimizing experimental error in spatially variable data. The second example is of a study of systematic variation in gas flux rates within a landscape. The third study considers the impact of spatial variability mediated by transport processes. In these examples, gas flux rates, which are hard to measure, are related to other, more readily measured variables using statistical or modelling techniques. Our intent in this paper is not to present definitive numbers for specific ecosystems but, rather, to demonstrate that spatial heterogeneity is tractable and that its inclusion into experimental designs can improve both the accuracy of flux estimates and contribute to knowledge of rate controls.

GEOSTATISTICAL ANALYSIS OF WITHIN-PLOT VARIATION

Two approaches to the explanation of some of the small-scale variability associated with denitrification measurements are examined in this section.

First, the possibility is examined that a portion of the variability of denitrification rates may be spatially autocorrelated. Autocorrelation would permit improved estimation of the rates through the use of techniques from regionalized variable theory, such as kriging and cokriging (Journel and Huijbregts, 1978). Second, the dependence of denitrification rates in the field on soil water contents is examined. Estimates of denitrification rates over large areas would be greatly improved if a strong relationship were found with a variable such as soil moisture whose spatial distribution can be easily measured in the field, sensed remotely, or modelled more readily than denitrification.

The example used in this paper consists of denitrification rates measured in 31 groups of four soil cores taken at 2 m intervals along a 60 m transect at an agricultural field site in central Sweden. The cores in each group of four were taken at 8 cm intervals. Details of sampling and denitrification measurement are presented in Svensson *et al.* (1985). The measured denitrification rates varied widely about the mean rate (6.4 ng h^{-1}g^{-1} D.W. of soil) with a coefficient of variation of 146%.

A sample semivariogram (Webster, 1985) was calculated to obtain a quantitative expression of the relationship between the expected variance of two measurements and the distance separating the locations where the measurements were made (Figure 10.1). The lowest semivariance was found for measurements made at the closest spacing (2 m) between sampling points (Figure 10.1). The expected variance for two samples appeared to vary

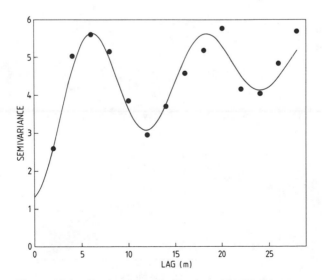

Figure 10.1   Semivariogram for logarithmically transformed denitrification rates. The theoretical model shown by the smooth curve was fit by nonlinear regression. The vertical axis is unitless

periodically with distance (Figure 10.1); measurements taken 12 and 24 m apart tended to be more similar to one another than measurements made at 6 and 18 m spacings. The smooth curve shown corresponds to the following function:

$$S = a_1\{1 - 0.5[1 + \cos(a_2 h)]\exp(-a_3 h)\} + a_4, \tag{1}$$

where S is semivariance, h is lag, and $a_1, \ldots, a_4$ are empirical constants determined by nonlinear regression.

Given a semivariogram, estimates for the variable at unsampled locations can be obtained through kriging (Journel and Huijbregts, 1978). Kriging was used to provide estimated rates at sampled locations for comparison to the measured rates. The measured rate at the location to be estimated was temporarily deleted from the data set. Then the estimated rate was obtained as a weighted average of the rates at the 14 locations nearest to the location where the rate was to be estimated, with weights based upon the semivariogram model. If the measured rate at the location where the rate is to be estimated is not deleted from the data set, then kriging will predict a rate equal to the measured rate (Journel and Huijbregts, 1978). The process of temporary deletion of a measurement at a location in order to obtain by kriging a meaningful prediction of the rate at that location is often referred to as jackknifing (Vieira *et al.,* 1983). The jackknifed rates of denitrification for each of the 31 groups of soil cores are shown in Figure 10.2 plotted against the rates measured at the same locations. The slope of the line of best fit (0.58)

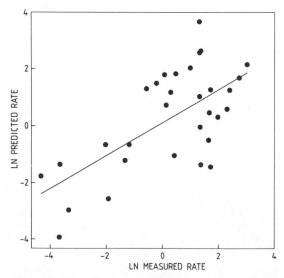

Figure 10.2 Denitrification rates predicted using kriging plotted against the actually measured rates

through the points in Figure 10.2 differs significantly from 0 at the $p < 0.01$ level. Kriging appears to provide meaningful estimates for the rates of denitrification along the transect, and the autocorrelation of denitrification rate suggested by Figure 10.1 is probably not illusory.

Although meaningful estimates of denitrification rates can be obtained with kriging, ideally one would wish the line of best fit (Figure 10.2) to have a slope of one with the points clustered tightly about it, which is not the case with the jackknifed estimates shown. Observed correlations suggested that cokriging using water content variables might be employed to improve the kriged estimates. Variables useful for cokriging must exhibit spatial autocorrelation before they can be used to improve the estimates for another variable with which they are correlated. Both NO_3^- concentration and CO_2-evolution rates had semivariograms so flat as to suggest very little spatial autocorrelation for these variables, Water fraction (g H_2O/g sample) did exhibit strong spatial autocorrelation. Using this model of the semivariogram (Figure 10.3), jack-knifed water fractions agreed with measured water fractions more closely than did jackknifed and measured denitrification rates. The slope (0.71) of the regression line of best fit for jackknifed vs. measured water fractions differed significantly from 0 at the $P < 0.0001$ level.

Cokriging is rarely worth the extra analytical effort unless the variable of interest has been undersampled in comparison with correlated variables

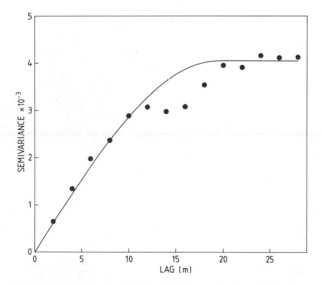

Figure 10.3 Semivariogram for water fraction. The spherical model shown by the smooth curve was fit by nonlinear regression after deletion of the points at lags of 14, 16, and 18 m, as explained in the text

(Journel and Huijbregts, 1978). Because soil water contents can be measured easily and inexpensively, we determined whether cokriging would outperform kriging in predicting denitrification rates when many more measurements of water fraction are available than of denitrification. If the use of soil water contents for cokriging substantially improves the quality of the estimates, then it may be possible to develop an optimal combination of water content and denitrification measurements that maximizes the accuracy of the predictions that may be obtained for a given investment of resources. Kriging and cokriging were used to generate predicted values for the sampling locations; cokriging outperformed kriging on these data sets with undersampled denitrification rates.

The cokriging procedure using jackknifing required temporary deletion of both the denitrification rate and the water fraction measured at the location where a denitrification rate was to be estimated. Thus, the water fraction actually measured at the location of interest was ignored in predicting the denitrification rate. This suggested the possibility of using the water fraction at the location where denitrification was to be estimated to obtain a prediction independent of the kriged estimate. Rates of denitrification are shown plotted against water fractions in Figure 10.4. Five of the locations along the transect were much drier than the others (Figure 10.4). At the 26 locations with water fractions greater than 0.2, the correlation between denitrification rate and water fraction appeared reasonably good. However, the data from the five

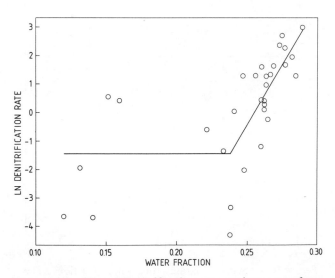

Figure 10.4   Plot of denitrification rate against water fraction. The bilinear model shown was fit by nonlinear regression

drier locations did not cluster about the same regression line with the rest of the data. The response of denitrification rate to increasing water fraction was modelled as shown by the two intersecting lines in Figure 10.4. The shape of the model shown in Figure 10.5 is in agreement with the results of Nömmik (1956), who found that significant denitrification did not occur until a certain moisture level was reached in the soil.

The cokriged and nonlinear regression estimates were combined using the following formula:

$$Z = (S_r Z_r + S_c Z_c)/S_r + S_c),$$ (2)

where Z is a predicted denitrification rate, S is the variance of a prediction, and subscripts r and c indicate regression and cokriging, respectively. This formula was derived from one in Granger and Newbold (1977) under the assumption that cokriging and the regression model give independent predictions. Cokriging provides an expected variance for each estimate, and this was used for S_c. The mean sum of squares left unexplained by the regression model was used for S_r. The predicted denitrification rates obtained from the combined estimator were the best of all the estimates produced during this study and left an unexplained sum of squares of only 41 out of 121. For comparison, the unexpected sum of squares left by cokriging and the regression model were 75 and 59, respectively. By using an optimal combination of the rates of denitrification predicted by cokriging and from a bilinear regression model of the effects of soil water contents on denitrification rates, it was possible to explain nearly two-thirds of the variation in the rates of denitrification along the transect.

Spatial variation in denitrification rates did not appear to be entirely random but appeared to contain a component that could be reduced by such techniques as kriging. In addition, much of the variability of denitrification was found to be associated with the variability of soil moisture, a quantity much more easily measured or remotely sensed than denitrification rates. The great disadvantage of both kriging and cokriging is that these techniques require at least one actual measurement of the variable to be estimated within the region for which the variable is to be estimated (Journel and Huijbregts, 1978). Although the data requirements of kriging and cokriging render these techniques unsuitable for use in estimating the rates of denitrification over truly large areas, such as continents, they are likely to prove ideally suited for estimating rates over areas of about 1 ha.

RATES AND PATHWAYS OF N_2O PRODUCTION

In the second example we consider a study of spatial and temporal variability in the rates and pathways of N_2O production. In this study, field and laboratory flux measurements were used to develop a simple model of N_2O

production from grassland soils (Mosier and Parton, 1985; Mosier *et al.,* submitted). The objective of this study was to develop a model that could be used to estimate fluxes over large areas, accounting for spatial and interannual variability, without requiring intensive data from each site modelled. The model partitions N_2O production into components resulting from nitrification and denitrification. Parameters allowing this partitioning were developed from experiments in which N_2O flux was measured with and without C_2H_2, and by relating N_2O flux to moisture and NO_3^- production.

Our current model is shown in Equation (3). Production of N_2O from nitrification and denitrification are represented separately as functions of temperature (M_t), soil water (M_d, M_n), NO_3, and NH_4. M_d and M_n are the effects of relative water content (actual available water/maximum available water) on denitrification and nitrification, respectively (Figure 10.5). S is an integrated parameter that allows the model to be fit for sites with varying textures and N availabilities.

$$\text{Denitrification} \qquad\qquad \text{Nitrification}$$
$$P_{N_2O} = (a \cdot NO_3 \cdot M_T \cdot M_d + b \cdot NH_4 \cdot M_t M_N + C) \cdot S, \qquad (3)$$

Parameters were estimated for two sites that represented much of the range of variability in soils and vegetation encountered in the shortgrass ecosystem (Schimel *et al.,* 1985).

The full model was not readily applicable to the target scale of study. A simplified version of the model was developed that did not include separate terms for NO_3^- and NH_4^+, because these variables require intensive time series data that are not usually available over large areas and are difficult to model with sufficient accuracy for use in the N_2O model. The simplified model uses an empirical multiplier related to the nitrogen availability of a given site and is otherwise the same. While the simplified model does not fit the data as well as the full model, it is much more suitable for coarse-scale applications.

The N_2O model was linked to soil temperature and water flow models driven by time series weather data from 1950 to 1984 to simulate interannual variability. Two sites were simulated, a midslope and a swale. Difference in mean plant production and soil texture were used to simulate the differences between the sites studied. Daily soil water and temperature were used as input into the simplified model to simulate interannual and intersite variations in N_2O production rates. Good agreement was obtained between observed and simulated results for those years when N_2O flux observations were made (Figure 10.6a, b).

Figure 10.5   The effect of soil NH_4^+ level on nitrification (a), the effect of relative water content on nitrification (M_n) and denitrification (M_d) (b), and the impact of soil temperature on nitrification and denitrification (c)

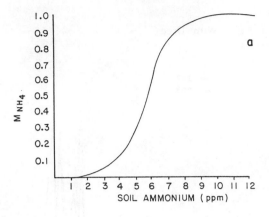

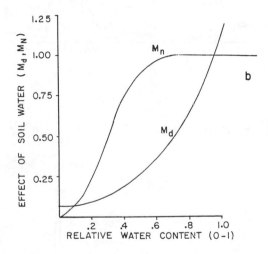

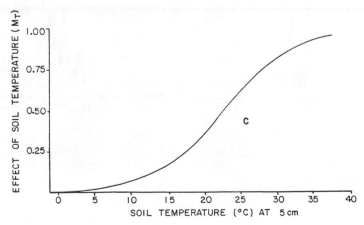

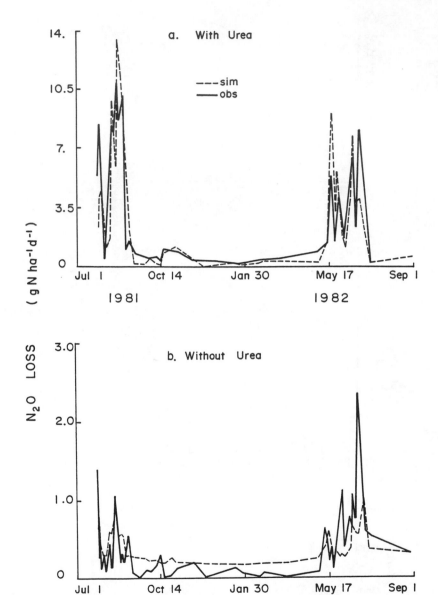

Figure 10.6   (a) Simulated and observed N_2O production for an unfertilized swale soil; (b) predicted and observed N_2O production for an unfertilized midslope soil

Table 10.1 Simulated annual N_2O loss $(g\,N\,ha^{-1}\,y^{-1})$ for two soil types with and without N fertilizer (from Mosier *et al.*, submitted)

Site	Mean loss $\pm$ SD	Highest annual loss	Lowest annual loss
Midslope	82 ± 7	100	73
Midslope + N	295 ± 112	614	85
Swale	161 ± 9	178	151
Swale + N	690 ± 290	1432	192

Results showed that mean simulated annual N_2O loss ranged from $82\,g\,N\,ha^{-1}\,y^{-1}$ for the midslope to $161\,g\,N\,ha^{-1}\,y^{-1}$ for the swale (Table 10.1). The variation between years is relatively small, with N_2O flux ranging from a minimum of $73\,g\,N\,ha^{-1}\,y^{-1}$ to a maximum of $100\,g\,N\,ha^{-1}\,y^{-1}$ for the midslope site (Table 10.1). As expected, N_2O production was highest during the wet years and lowest during dry years. Flux rates were highest in 1979, when the precipitation was 40 cm, and lowest in 1964, when precipitation was 10 cm. A reference point for the importance of N_2O flux rates to a site's N budget is the amount of N deposited in rainfall. Wetfall atmospheric N deposition has been observed at the shortgrass steppe site for the last six years by the National Atmospheric Deposition Program, with the total N deposited ranging from $2.0\,kg\,N\,ha^{-1}\,y^{-1}$ to $3.3\,kg\,N\,ha^{-1}\,y^{-1}$ with 60% of the N as NH_4^+. The simulated N_2O loss from the two sites ranged from 5.1 to 8.7% for the swale and from 2.4 to 4.8% of atmospheric-N inputs for the midslope. Measured losses of N from the system account for less than 20% of the annual inputs, and suggests that N may be accumulating in the system.

The fractions of N_2O produced by nitrification (N_F) were calculated with the full model for N-amended treatments at both swale and midslope sites. The average fraction of the total N_2O loss produced by nitrification was 85% for the swale site and the percentage ranged from 93% in a very wet year (59 cm annual rainfall) to 70% during a very dry year (10 to 11 cm annual rainfall). Higher nitrification N_2O losses during wet years result from the greater number of large rainfall events during wet years. This results in an increase in the length of time when soil water contents are at intermediate levels, while the number of days when the soil water content is near field capacity (conditions needed for denitrification) are only slightly increased. The increase in the time period with intermediate water content greatly enhances nitrification because nitrification rates stay high until the relative water content drops below 0.4 (see Figure 10.5b), while denitrification rates drop very rapidly as water content drops below field capacity (see Figure 10.5b). The higher fraction of N_2O produced by denitrification during the dry year is caused by a drop in the number of days with intermediate soil water content, rather than in the number of rainfall events, each of which is followed by more or less the same period of time at field capacity. Thus, N_2O production by denitrification is less temporally variable than is N_2O production by nitrification.

Differences in N_2O production seasonally and between sites were closely coupled to mineral-N dynamics. Soil NO_3^- concentration, nitrification, and mineralization rates all peaked in June, along with the N_2O production rate (Figure 10.6)(Schimel, 1982; Schimel *et al.*, 1985; Schimel and Parton, 1986). These data suggest that nitrification is the dominant vector for N_2O production in the shortgrass steppe, further suggesting that patterns of temperature and water availability have their effect on N_2O production by controlling N mineralization and nitrification. The formation of anaerobic microsites during wet periods is of secondary importance in the soils studied. While the difference in texture suggested that the difference in N_2O production between midslope and swale was due to porosity-related differences in water relations, the above argument suggests that differences in inorganic N turnover rate may be more significant. The swale has a significantly higher *in situ* N mineralization rate than did the midslope (55 vs 41 $kg\,N\,ha^{-1}\,y^{-1}$; Schimel *et al.*, 1985), which may contribute to the higher N_2O fluxes observed.

The sites chosen in this study were chosen to represent the range of variability in soil and vegetation properties found in the shortgrass steppe. Previous quantitative studies showed only two quantitatively separable major landscape units, "uplands and swales," based on either soil or vegetation analysis (Anderson, 1983; Yonker *et al.*, submitted). The processes leading to the strong differentiation between these landscape units are discussed in Schimel *et al.* (1985). Techniques such as those proposed in the first part of this paper, applied to each landscape unit, would further improve the precision of such an estimate.

AMMONIA PRODUCTION FROM CATTLE URINE

Production of NH_3^0 from wild and domestic animal urine is significant to the global NH_3^0 budget (Crutzen, 1983). In this example, we describe a study of NH_3^0 release from cattle urine that considered the correlated spatial distributions of cattle-urine deposition and soil properties that control NH_3^0 volatilization.

Flux of NH_3^0 from simulated urine patches on several soils (adjacent to those used in N_2O studies described above) was monitored by direct collection of NH_3^0 and by mass balance using ^{15}N (Schimel *et al.*, 1986). Urine was applied at several times of the year to characterize the effects of temperature and moisture on NH_3^0 flux. Total urine deposition was determined by monthly urine collection from catheterized, free-ranging cattle. Behavioural observations of uncatheterized free-ranging cattle were used to calculate spatial partitioning of urine, assuming that each urination represented an equal

Table 10.2 Spatial and seasonal variations in the deposition and subsequent volatilization of urine N (adapted from Schimel *et al.*, 1986)

Season	Urine deposition (g N/ha) Lowlands (swales)	Uplands	Loss (%) Lowlands	Uplands	Loss (g N/ha)
Growing (Apr–Oct)	2110	590	0	27	159
Dormant (Nov–Mar)	360	110	0	12	13
Total					172

portion of the total urine produced during a collection interval (Senft, 1983). These field studies were supplemented with extensive laboratory studies on moisture and water loss rate effects. Flux rates varied with soil type and time of year and the proportions of deposition lost as NH_3^0 shown in Table 10.2 were used in all subsequent calculations. NH_3^0 emission rates were low in swale soils, which had high rates of immobilization, low pHs and greater plant uptake. Emission rates were higher in upslope soils, which had the opposite properties. Interestingly, NH_3^0 emissions were high where N_2O emissions were low and *vice versa*.

Cattle deposition of urine was not uniform with respect to the soil properties identified as important in flux studies. Results from studies on cattle grazing and urine deposition behaviour, stratified by season and landscape position, showed deposition to occur disproportionately in areas of low potential loss (Table 10.2)(Senft, 1983; Stillwell, 1983). The product of the seasonally and spatially stratified deposition and volatilization data, assuming a pasture divided 70 : 30 between uplands and swales, yields an ecosystem-level estimate of loss of only $110 \, g \, N \, ha^{-1} \, y^{-1}$. Emission of NH_3^0 from pastures composed entirely of upland soils and at the same stocking rate would be $760 \, g \, N \, ha^{-1} \, y^{-1}$, a significantly higher value. A similar calculation for N_2O emission, using the rates presented above, results in an estimated ecosystem-level flux of $104 \, g \, N \, ha^{-1} \, y^{-1}$. N_2O emission from an upland pasture would be $80 \, g \, N \, ha^{-1} \, y^{-1}$. Any ecosystem or landscape in which N is transferred between areas of low and high emission potential will require careful study, particularly if either rates of transport or of emission are modified by anthropogenic activity.

CONCLUSIONS

Spatial heterogeneity is high for measurements of gas flux from terrestrial ecosystems. Both fine (as in the first example) and coarse grained (as in the last two examples) variability must be considered in computing estimates of gas fluxes from large areas. While no single class of techniques will resolve problems of measuring gas fluxes across all levels of scale, certain commonalities are evident in the several examples presented in this paper. First, identification of variables which are easy to measure or obtain from regional data sources and which are predictors of gas flux rates is critical. These predictor variables may be used in either statistical or simulation models to obtain spatially or temporally integrated estimates of gas flux rates. The variables chosen must be measurable in some way at the desired scale. In the study of N_2O flux modelling, variables which improved the fit of the model (NO_3^-, NH_4^+) were dropped from the model because they were difficult to measure or model at the chosen scale. Instead, an empirical variable, estimable from soil texture and organic matter content was substituted. Second, choice of appropriate predictor variables must be guided by thorough knowledge of the processes governing the emission rate. Without such mechanistic knowledge at a level of scale below the scale of integration, choice of predictor variables becomes empirical and extrapolation out of the original universe of study becomes problematic. Thus, successful attempts to estimate flux rates over large areas must be based on careful intensive studies to guide choice of predictor variables. Operating at multiple 'levels of resolution' becomes particularly important when controls over gas fluxes behave nonlinearly, as was the case for the effect of water on denitrification (example 1), and on the denitrification/nitrification ratio (example 2). While coarse-scale models need not, and usually must not, include detailed mechanism, oversimplification can lead to serious error when nonlinear functions are encountered.

The approaches presented here are not new but, rather, are rooted in systems analysis and are consistent with recent developments in hierarchy theory (O'Neill, 1988). Our purpose here is to give concrete examples of studies which take advantage of multiple scales of investigation and to illustrate by example techniques for extrapolating small-scale studies of gas flux to larger spatial domains.

REFERENCES

Anderson, M. (1983). *Soil and Vegetation Pattern on Shortgrass Catenas.* M.S. Thesis, Colorado State University, Fort Collins: 79 pages

Bolin, B., Crutzen, P. J., Vitousek, P. M., Woodmansee, R. G., Goldberg, E. D., and Cook, R. B. (1983). Interactions of biogeochemical cycles. In Bolin, B., and Cook, R. B. (Eds.) *The Major Biogeochemical Cycles and Their Interactions,* pp. 1–40. John Wiley & Sons, New York.

Crutzen, P. J. (1983). Atmospheric interactions—homogeneous gas reactions of C, N, and S containing compounds. In Bolin, B., and Cook, R. B. (Eds.) *The Major Biogeochemical Cycles and Their Interactions,* pp. 67–114. John Wiley & Sons, New York.

Folorunso, O. A., and Rolston, D. E. (1984). Spatial variability of field measured denitrification gas fluxes. *Soil Sci. Soc. Am. J., 48,* 1214–1219.

Granger, C. W. J., and Newbold, P. (1977). *Forecasting Economic Time Series.* Academic Press, Inc., New York.

Journel, A. G., and Huijbregts, C. J. (1978). *Mining Geostatistics.* Academic Press, Inc., New York.

Lacis, A., Hanson, G., Lee, P., Mitchell, T., and Lebedeff, S. (1981). Greenhouse effect of trace gases, 1970–1980. *Geophys. Res. Lett.,* **8,** 1035–1038.

Leffelaar, P. A. (1979). Simulation of partial anaerobiosis in a model soil in respect to denitrification. *Soil Sci.,* **128,** 110–120.

McGill, W. B., Hunt, H. W., Woodmansee, R. G., and Reuss, J. O. (1981). PHOENIX—a model of the dynamics of carbon and nitrogen in grassland soils. In Clark, F. E., and Rosswall, T. (Eds.) *Terrestrial Nitrogen Cycles: Processes, Ecosystem Strategies, and Management Impacts. Ecol. Bull. (Stockholm),* **33,** 49–115.

Mosier, A. R., and Parton, W. J. (1985). Denitrification in a shortgrass prairie: A modelling approach. In Caldwell, D. E., Brierly, J. A., and Brierly, C. L. (Eds.) *Planetary Ecology: Selected Papers From the Sixth International Symposium on Environmental Biogeochemistry,* pp. 441–452. Van Nostrand Reinhold, New York.

Mosier, A. R., Parton, W. J., and Schimel, D. S. (in press). Spatial and temporal variability in rates and pathways of nitrous oxide flux in a shortgrass steppe. *Biogeochemistry* (in press).

Nömmik, N. (1956). Investigations on denitrification in soil. *Acta Agric. Scand.,* **6,** 195–228

O'Neill, R. V. (1988). Hierarchy theory and global change. (Chapter 3, this volume).

Schimel, D. S. (1982). *The Effects of Hillslope Processes on Nutrient and Organic Matter Dynamics in a Shortgrass Steppe.* Ph.D. Dissertation, Colorado State Univ., Fort Collins, CO, USA.

Schimel, D. S., and Parton, W. J. (1986). Microclimatic controls of nitrogen mineralization and nitrification in a shortgrass steppe. *Plant and Soil,* **93,** 347–357.

Schimel, D., Stillwell, M. A., and Woodmansee, R. G. (1985). Biogeochemistry of C, N, and P in a soil catena of the shortgrass steppe. *Ecology,* **66**(1), 276–282.

Schimel, D. S., Parton, W. J., Adamsen, F. J., Woodmansee, R. G., Senft, R. L., and Stillwell, M. A. (1986). The role of cattle in the volatile loss of nitrogen from a shortgrass steppe. *Biogeochemistry,* **2,** 39–52.

Senft, R. L. (1983). *The Redistribution of Nitrogen by Cattle.* Ph.D. dissertation. Colorado State University, Fort Collins.

Stillwell, M. A. (1983). *The Effect of Bovine Urine on the Nitrogen Cycle of a Shortgrass Prairie.* Ph.D. dissertation. Colorado State University, Fort Collins.

Svensson, B. H., Klemedtsson, L., and Rosswall, T. (1985). Preliminary field denitrification studies of nitrate-fertilized and nitrogen-fixing crops. In Golberman, H. L. (Ed.) *Denitrification in the Nitrogen Cycle,* pp. 157–169. Plenum Press, London.

Vieira, S. R., Hatfield, J. L., Nielsen, D. R., and Biggar, J. W. (1983). Geostatistical theory and application to variability of some agronomic properties. *Hilgardia,* **51,** 1–75.

Webster, R. (1985). Quantitative spatial analysis of soil in the field. *Adv. Soil. Sci.,* **3,** 1–70.

Yonker, C. M., Schimel, D. S., Parnoussis, E., and Heil, R. D. (in press). Patterns of organic carbon accumulation in a semiarid shortgrass steppe. *Soil Sci. Soc. Am. J.* (in press).

Scales and Global Change
Edited by T. Rosswall, R. G. Woodmansee and P. G. Risser
© 1988 Scientific Committee on Problems of the Environment (SCOPE)
Published by John Wiley & Sons Ltd.

CHAPTER 11

Spatial and Temporal Scales in Groundwater Modelling

PAUL K. M. VAN DER HEIJDE

Holcomb Research Institute,
Butler University, Indianapolis,
Indiana 46208 USA

INTRODUCTION

The International Geosphere-Biosphere Programme (IGBP) of the International Council of Scientific Unions (ICSU) is aimed at 'describing and understanding the interactive physical, chemical, and biological processes that regulate the total Earth system, the unique environment for life, the changes that are occurring in this system and the manner in which they are influenced by human actions.' Each discipline involved has used its own spatial and temporal scales in describing the relevant processes quantitatively, thus generating its own specific resolution of the questions being asked. To succeed in obtaining an integrated approach to modelling the earth's resource systems, a consistent and coordinated use of scales is necessary across the various disciplines. This chapter describes the existing approaches to scale requirements in groundwater modelling and identifies further research to facilitate integration of groundwater systems into large-scale models of global processes.

Groundwater is a subsurface element of the hydrosphere, which is generally understood to encompass all the waters beneath, on, and above the earth's surface. Because groundwater systems occur in an irregular pattern all over the earth, they must be considered in any global analysis of the hydrosphere.

Many solar-powered processes occur in the hydrosphere, resulting in a continuous movement of water. This dynamic system is referred to as the hydrologic cycle. Its major elements are atmospheric water, surface water, water in the subsoil, groundwater, stream networks, lakes and ocean basins, and the water in the lithosphere (see Figure 11.1).

Movement of water occurs both within each element of the hydrologic cycle and as exchanges between the elements, and results in the dynamic character of

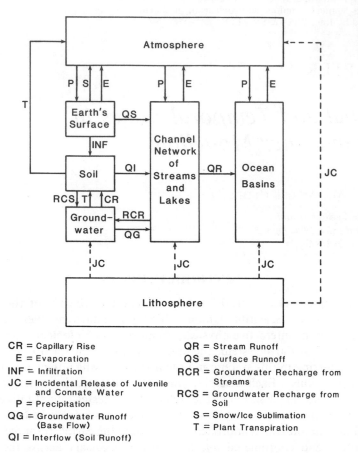

Figure 11.1   Elements of the hydrological cycle (after Domenico, 1972)

this relatively closed system. The exchange processes between the surface subsystem and the atmosphere include evaporation, precipitation (rainfall, fog, dew, hail, and snowfall), plant transpiration, and sublimation of snow. Infiltration, groundwater recharge from streams, and subsurface discharge into lakes and streams (both interflow and baseflow) are inter-element processes between surface and subsurface. Surface runoff forms the link between the stream network subsystem and the oceans (Domenico, 1972). Finally, juvenile, rejuvenated, and connate water is released from the lithosphere and becomes an active part of the hydrologic cycle. In addition, interactions take place between the subsurface hydrosphere and elements of the earth's biological environment.

Each of these physical and biological elements can constitute a system and might consist of various subsystems, each with its own spatial and temporal characteristics. In studying the interactions among such systems, as well as in aggregating smaller systems into larger ones, the problem of scale arises.

This chapter explores the variability of spatial and temporal scales in modelling the dynamics of groundwater systems. Special attention is given to scaling aspects of modelling the subsurface hydrosphere relevant to interactions with terrestrial and aquatic ecosystems.

Scales in groundwater hydrology can be discussed from two perspectives. First is the physical scale on which the hydrological processes take place. These processes provide the physical setting in which human interaction can be studied, as they occur in unintentional or managed alterations in the natural system. Another perspective is that of resource management, where socioeconomic and political conditions are paired with the hydrological and engineering aspects of a groundwater system. Modelling is an instrument to organize the data collection and to structure the analysis of the studied system. In groundwater hydrology, modelling has become the principle tool in the management of the resource. The extensive scientific foundation of groundwater modelling and the large database available indicate that groundwater systems might be modelled successfully in order to study their role in the earth system. Further research is required to integrate quantitative descriptions of groundwater systems into large-scale models of global processes, and to facilitate proper scaling of intersystem hydrologic and biologic processes.

To provide a basis for the discussion of scales in groundwater modelling, this chapter begins with a description of major aspects of groundwater systems, including definitions, relevant orders of magnitudes of processes, subsystems, and classifications.

CHARACTERISTICS OF GROUNDWATER SYSTEMS

A groundwater system is an aggregate of rock in which water enters and moves, and which is bounded by rock that does not allow any water movement, and by zones of interaction with the earth's surface and with surface water systems (Domenico, 1972). In such a system the water may transport solutes and biota; interactions of both water and dissolved constituents with the solid phase (rock) often occur.

Water in the earth occurs in spaces within solid matter. Various types of such spaces or pores exist, and include fractures. The interstices may vary in size from large limestone caverns to miniscule capillary openings in which water is held primarily by adhesive forces (Bear, 1979).

Subsurface water or underground water occurs in four zones. Directly beneath the land surface is the zone of aeration, or unsaturated zone, which contains both water and air. Its thickness may vary widely in time and space. In wetlands this zone may be absent, while in arid areas the thickness of this zone (sometimes called the vadose zone) can exceed 1000 m (Bouwer, 1978). That part of the unsaturated zone that supports plant growth is the root zone; it generally extends to a maximum depth of 2 m beneath the land surface (Heath, 1983). The soil zone is a major interaction area between the subsurface hydrosphere and the biospheric elements of terrestrial ecosystems.

The unsaturated zone is almost always underlain by rock layers that are fully saturated with water. This is the saturated zone, and the water in it is commonly referred to as groundwater. Groundwater forms an important fraction of global fresh water resources. It is estimated that 4000 km^3 of fresh water is stored up to a depth of 800 m under the land surface, comprising about 14% of the total amount of fresh water (L'vovich, 1979). After glaciers, which contain 85% of all fresh water, groundwater is the second largest global source of fresh water.

At the boundary zone between the unsaturated and saturated zone, the attraction forces between water and rocks are balanced against the pull of gravity. As a result, the smaller pores are water-saturated while the larger pores contain both water and air. This boundary between groundwater and vadose or soil water is known as the water table or the phreatic surface.

A special type of subsurface water occurs as a zone of permafrost. In addition to ice, this frozen zone may also contain liquid water at subzero temperatures. This free water is often highly saline. Permafrost is present in the northern regions of Eurasia, of North America, all of Antarctica, and the high mountain areas within orogenic belts (Klimentov, 1983). The thickness of the frozen zone may exceed 100 m.

A primary unit in groundwater systems is the aquifer, a lithologic unit or combination of units capable of transmitting relatively significant amounts of water (Figure 11.2). An aquifer may coincide with a geologic formation; it may cover a group of such formations, or it may be a part of a formation or a group of formations. It is often bounded by poorly transmitting layers called confining beds.

A water-table or phreatic aquifer is characterized by a freely moving upper boundary. The location of this water table follows the variations in hydraulic head. Although the hydraulic or piezometric head in a confined aquifer may fluctuate, its upper and lower physical boundaries are fixed.

The largest hydrogeologic unit is a groundwater basin. It is a system containing the entire network of flow paths taken by all the water recharging the basin (Freeze and Witherspoon, 1966). A groundwater basin consists of a single aquifer or several connected and interrelated aquifers. The water divide between two adjacent groundwater basins is not necessarily the same as that between the surface water drainage basins overlying them. Watersheds can lose part of their water to neighbouring watersheds through subsurface interbasin transfers. In a valley between mountain ranges, the drainage basin of the surface stream coincides closely with the groundwater basin. In

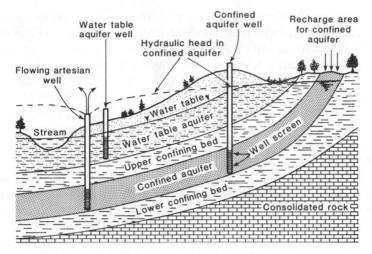

Figure 11.2 Groundwater system units (after Johnson, 1975)

limestone areas and large alluvial basins, the drainage and groundwater basins may have entirely different configurations.

As an illustration of the distribution of groundwater systems, areas with significant groundwater resources in the United States (areas with freshwater yields of more than 3.15 l/sec) are shown in Figure 11.3.

The largest groundwater systems in the United States are formed by major alluvial basins such as that in the High Plains groundwater region where the

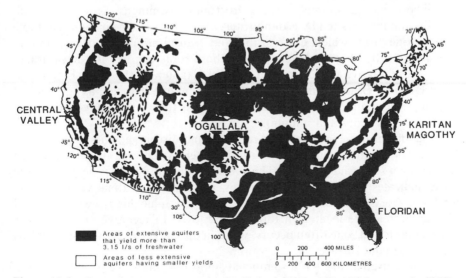

Figure 11.3 Groundwater resources in the United States (after Rickert *et al.*, 1979)

Ogallala aquifer extends over more than 20 000 km^2, and that in California's Central Valley. Other large regional aquifer systems exist in the Coastal Plain region, e.g. the Karitan and Magothy formations of unconsolidated sands and gravels underlying large regions in the south part of the northeastern and mid-Atlantic states, and the carbonate rock aquifer system formed by the Floridan formation in Florida and southeastern Georgia (Figure 11.3). Many smaller aquifer systems occur in these and other regions, and range from hundreds to a few thousand square kilometres. Numerous more localized aquifers of less than 100 km^2 are found in many parts of the country.

Water enters the groundwater system in recharge zones and leaves the system in discharge areas. In a humid climate, the major source of aquifer recharge is the infiltration of water and its subsequent percolation through the soil into the groundwater subsystem. This type of recharge occurs in all instream areas except along streams and their adjoining floodplains, which are discharge areas. In arid parts of the world, recharge is often restricted to mountain ranges, to alluvial fans bordering these mountain ranges, and along the channels of major streams underlain by thick and permeable alluvial deposits.

In addition to these natural recharge processes, artificial or man-made recharge can be significant. This type of recharge includes injection wells, induced infiltration from surface water bodies, and irrigation.

Outflows from groundwater systems are normally the result of a combination of inflows from various recharge sources. Groundwater loss appears as interflow to streams (rapid near-surface runoff); as groundwater discharge into streams (resulting in baseflow); as springs and small seeps in hillsides and valley bottoms, quicksands, geysers, frost mounds, pingos; as wetlands such as lakes and marshes fed by groundwater; as capillary rise near the water table into a zone from which evaporation and transpiration can occur; and as transpiration by phreatophytes (plants whose roots can live in the saturated zone or can survive fluctuations of the water table) (Toth, 1971; Freeze and Cherry, 1979). Other outflows are artificial or human-induced, as agricultural drainage (tile-drains, furrows, ditches) and wells for water supply or dewatering (e.g. excavations and mining).

A significant difference between recharge and discharge zones is their areal extent. The discharge areas of a groundwater system are in general much smaller than the recharge areas. The concentration of flowlines near discharge areas indicates that horizontal flow is more efficient than vertical flow. Recharge areas are commonly found in topographic highs, while natural discharge areas are located in topographic lows. In recharge areas a rather thick unsaturated zone often occurs, while in discharge areas the water table is close to the land surface.

The unsaturated zone has a significant smoothing influence on the temporal characteristics of the recharge of groundwater systems. Highly variable

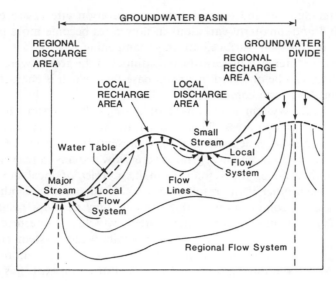

Figure 11.4 Schematic diagram of a regional groundwater flow system (after Toth, 1983)

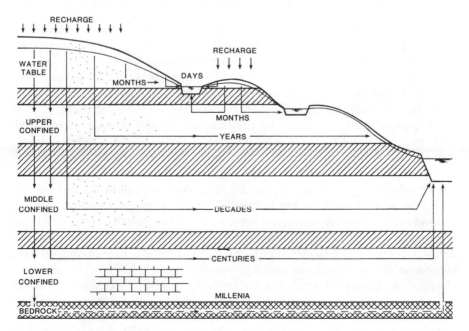

Figure 11.5 Schematic overview of groundwater residence times in large regional systems

(hourly) precipitation and diurnal evapotranspiration effects are dampened and seasonal and long-term variations in flow rates become more prominent further from the soil surface. In this dampening the higher frequency fluctuations are filtered, a proces which continues in the groundwater zone. Its ultimate effect can be observed in stream base flow which is characterized by seasonal and long-term components.

A groundwater system may consist of a single flow system between its recharge and discharge areas. This is generally the case when local relief is negligible and only a gentle regional slope is present. If the relief of the surface becomes more pronounced, local groundwater flow systems can develop. If the depth-to-width ratio of the system is small, a series of local flow systems adjacent to each other is the result. However, if the aquifer depth-to-width ratio increases, a combination of flow systems may develop, resulting in a hierarchically structured groundwater system with local, intermediate, and regional components (Figure 11.4). If the groundwater system consists of multiple aquifers, this hierarchical structure is even more evident (Figure 11.5). The notion that such a hierarchical structure exists has improved the effectiveness of modelling groundwater systems significantly (e.g. Freeze and Witherspoon, 1966).

MODELLING THE DYNAMICS OF GROUNDWATER

A groundwater system has two basic hydraulic functions: in storing water it acts as a reservoir, and in transmitting water from recharge to discharge areas it serves a conduit. A groundwater system can be considered as a reservoir that integrates various inputs (through mixing, among others) and dampens and delays the propagation of changes in inputs. The water movement is dictated by hydraulic gradients and geology-dependent hydraulic conductivity. In turn, these gradients are influenced by groundwater system boundary conditions such as those resulting from human-induced stresses on the system, climatic effects, and topography (land-surface and stream-related boundary conditions).

Groundwater systems are characterized by complex inflow-outflow-storage relations. System outflows are influenced by the origin and pathways of the groundwater. These relations are difficult to define directly from input and response data because of the dampening effect of storage on inflow, the lag or delay between the time water enters and exits the system, the variable rate and sometimes diffuse manner of recharge and discharge, and the heterogeneity of the geology. Therefore, mathematical models, based on a mechanistic description of the physical and chemical processes internal to the groundwater system, are widely used in groundwater hydrology.

The mathematical model for groundwater flow is derived by applying principles of mass conservation (resulting in the continuity equation) and conservation of momentum (resulting in the equation of motion). The generally applicable equation of motion in groundwater is Darcy's law, which originated in the mid-nineteenth century as an empirical relationship. Later, a mechanistic approach related this equation to the basic laws of fluid dynamics (Bear, 1972). Variability of parameters in space and time and uncertainty in data are often incorporated in these models.

The rate of groundwater movement can be expressed in terms of time required for groundwater to move from a recharge area to a discharge zone. This time ranges from a few days in zones adjacent to discharge areas in small local systems, to thousands of years for water that moves through deeper parts of the groundwater system (Figure 11.5). The large residence time in groundwater basins gives relatively slow chemical processes a chance to influence the composition of the water.

The groundwater transport of dissolved chemicals and biota such as bacteria and viruses is directly related to the flow of water in the subsurface. Many of the constituents occurring in groundwater can interact physically and chemically with solid phases such as clay particles, and with various dissolved chemicals. As a consequence, their displacement is both a function of mechanical transport processes such as advection and dispersion, and of physicochemical interactions such as adsorption–desorption, ion exchange, dissolution–precipitation, reduction–oxidation, complexation, and radioactive decay. Biotransformations taking place during the transport can alter the composition of the groundwater significantly.

In modelling the transport of dissolved chemicals, the principle of mass conservation is applied to each of the chemical constituents present. The resulting equations include physical and chemical interactions, as between the dissolved constituents and the solid subsurface matrix, and among the various solutes themselves. These equations might include the effects of biotic processes. To complete the mathematical formulation of a solute transport problem, equations are added describing groundwater flow and chemical interactions, as between the dissolved constituents and the solid subsurface matrix, and among the various solutes themselves. In some cases equations of state are added to describe the influence of temperature variations and the changing concentrations on the fluid flow through the effect of these variations on density and viscosity.

In a stochastic system the relationships governing the behaviour of the system are described in a probabilistic manner. If one element of a system is stochastic in nature, the system as a whole behaves stochastically, although the relationships themselves might be defined in a deterministic sense.

The solution of the equations describing a deterministic system is approached in three ways. If the solution is continuous in both time and space, it is called an analytical solution or model. For a solution that is discrete in either time or space, the term semianalytical model is used. A numerical model is discrete in both time and space and uses approximations for the derivatives in the governing equations. Spatial and temporal resolution in applying such models is a function of study objectives and availability of data.

No universal model can solve all kinds of groundwater problems; different types of models are appropriate for solving different types of problems. It is important to realize that comprehensiveness and complexity in a simulation do not necessarily equate with accuracy. An extensive discussion of the status of groundwater models is presented by Bachmat *et al.* (1985).

MULTIPLE SCALES IN MODELLING GROUNDWATER SYSTEMS

A wide range of both spatial and temporal scales is involved in the study of groundwater problems. Spatial scales range from less than a nanometre, for studying such phenomena as the interactions between water molecules and dissolved chemicals (Cusham, 1985), to hundreds of kilometres, for the assessment and management of regional groundwater systems (Toth, 1963). For temporal scales, two major categories can be distinguished: steady-state or average state, and a time-varying or transient state. Periodic fluctuations on a diurnal or seasonal scale are frequent in hydrogeology. Other processes display certain trends or occur rather randomly in nature (Table 11.1). Many processes exhibit a strong temporal effect immediately after their initiation but become stable after a while, moving to a steady-state. Other processes fluctuate on a scale that is often much smaller than necessary to include in the analysis of such systems. An averaging approach is then taken, resulting in steady-state analysis. The steady-state is also assumed when the analysis period is so short that temporal effects are not noticeable.

Dimensions in the time domain range from millenia in palaeohydrological simulations and risk-analysis for long-term isolation of radioactive waste through year-by-year, seasonal, monthly, weekly, daily, and hourly scales for field systems—to modelling of real-time systems on a basis of minutes and even seconds in certain laboratory experiments.

An important aspect of the scaling problem is related to the difference between the scale on which processes are mathematically described, and the subsequent aggregation into larger-scale formulations amenable to field analytical procedures. Small-scale descriptions are aggregated into large-scale models by applying averaging procedures. Such averaging applied to a statistical description of microscopic processes is commonly used to obtain continuous hydrodynamic field equations on the macroscopic scale (e.g. Bear, 1972, 1979). Although the resulting model requires less supporting field data than is required for a problem of the same physical extent, a certain amount of information regarding the real physical systems is lost. Also, in going to larger spatial and temporal scales, variations in system characteristics that could be ignored on the smaller scale may become important. Examples are the increasing importance of heterogeneities and anisotropy as related to the

Table 11.1 Summary of mechanisms tending to produce fluctuations in groundwater levels (after Freeze and Cherry, 1979)

	unconfined	confined	natural	man-made	short-lived	diurnal	seasonal	long-term	climate influence
Groundwater recharge (infiltration to water table)	×		×				×		×
Air entrapment during groundwater recharge	×		×		×				×
Evapotranspiration and phreatophytic consumption	×		×			×			×
Bank storage effects near streams	×		×				×		×
Tidal effects near oceans	×	×	×			×			
Atmospheric pressure effects	×	×	×			×			×
External loading of confined aquifers		×		×	×				
Earthquakes		×	×		×				
Groundwater pumpage	×	×		×				×	
Deep well injection		×		×				×	
Artificial recharge: leakage from ponds, lagoons, landfills	×			×				×	
Agricultural irrigation and drainage	×			×			×	×	×
Geotechnical drainage of open pit mines, slopes, tunnels, excavation sites	×			×				×	

geology of the system for larger spatial scales, and the effects of long-term recharge variations on the water balance of a system for long time periods. A major problem in this averaging process lies in evaluating the effects of assumptions made on the microscopic scale and the effects on the level of uncertainty in the modelling of a groundwater system. If such assumptions have to be incorporated in the macroscopic description, their formulation may be problematic. Another problem that may arise as a result of an averaging approach is that of defining the physical meaning of the resulting state variables and system parameters. Thusfar, no systematic evaluation of the consequences of this aggregation process in groundwater has been published, although an extensive database is available to carry out such a study.

To understand the effect of the various processes active in groundwater systems, each of these processes must be described on an appropriate scale.

The optimal scale for an individual process can be quite small. However, for the management of groundwater systems, such small scales are not required. The result is that the scales used in groundwater modelling range from microscopic to macroscopic (Bonnet, 1982; Table 11.2; Figure 11.6). On the microscopic scale, processes occur on atomic and molecular scales, such as the interactions between water molecules, organics, and clay surfaces and between organics and microbes, or processes on a particular scale involving large numbers of molecules. The granular scale represents the processes that occur on the scale of individual pore spaces and their adjacent grains. Significant for the formulation of most groundwater flow and transport equations is the scale of continuous equivalent media. The scale unit is defined as the Representative Equivalent Volume (REV) and contains many pores and grains (Bear, 1979). It is also the smallest macroscopic scale. Going to even larger scales, Bonnet (1982) defines the geologic or structural scale as a part of an aquifer containing sedimentological and elementary stratigraphical heterogeneities, and the hydrogeological or regional scale as an aquifer or groundwater basin. In groundwater modelling, the scales of interest for the study of interrelationships between the geohydrosphere and ecosystems range from site and local to

Table 11.2   Principal scales in hydraulics of subsurface porous media (after Bonnet, 1982)

Scale	Reference Volume	Order of Magnitude	Comments
atomic	atom	0.1 nm	too small for hydrogeology
molecular	molecule	1 nm	
particular	particle	μm	group of molecules which deform only by molecular diffusion
granular of capillary	pore space and adjacent grains	mm	
continuous equivalent medium	many pores and grains	m	representative equivalent volume (REV)
geologic or structural	part of aquifer	km	including sedimentological elementary stratigraphical heterogeneities
hydrogeological or regional	hydrogeologic system or aquifer	10's–100's km	groundwater basin

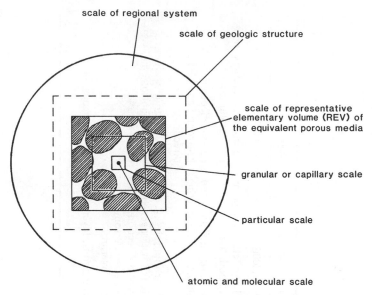

Figure 11.6   Scales pertinent to hydrology (Bonnet, 1982)

intermediate and regional (Table 11.3). Significant characteristics on each scale vary for different hydrogeological interests.

The processes of water movement and the transport of energy and dissolved constituents through porous media are well understood and are mathematically described on a macroscopic scale, using a representative equivalent volume (Bear, 1979). Such an approach can also be applied to flow and transport in fractured rock (Wang and Narasimhan, 1984). One way to make a system of fractures of varying size and orientation accessible to quantitative analysis is through the concept of equivalent porous media. Another approach often taken is based on applying stochastic principles to obtain representative parameters. Such approaches can be used to extend the results of small-scale studies to larger-scale problems.

Some of these approaches involve a direct relation between the spatial and temporal scales. This is especially the case in researching the physical principles underlying the responses of groundwater systems to varying stresses. In other cases the choice of temporal scales is related more to the type of problem under study and the materials of interest, such as the long-term storage of radioactive waste, where time scales up to thousands of years apply. Simulation of such large time periods may be limited by stability criteria inherent to the mathematical solution procedures included in the model.

The essential scaling problem is how to distinguish between the variables that can be considered as constants or as being uniform across discrete intervals of pertinent dimension (space, time), and the variables that cannot be

Table 11.3 Scales in groundwater modelling

	Site	Local	Intermediate	Regional
area	<100m	100–1000m	1000–10,000m	>10,000m
examples	tracer test, pumping test	point source, pollution, small well fields	small aquifers, large point-source pollution	basins large aquifers, non-point pollution
geology	homogeneous	single horizontal unit; some vertical layering	a few horizontal units and significant vertical layering	heterogeneous in both horizontal and vertical directions
flow	single aquifer or part of aquifers; homogeneous, possibly anisotropic	single aquifer or part of aquifers; homogeneous, possibly anisotropic	single or multiple aquifer(s); heterogeneous in horizontal direction, anisotropic, possible some vertical layering	multiple aquifers; heterogeneous, anisotropic, vertically layered
solute transport	homogeneous	homogeneous in horizontal direction, layered	heterogeneous, layered	heterogeneous, layered

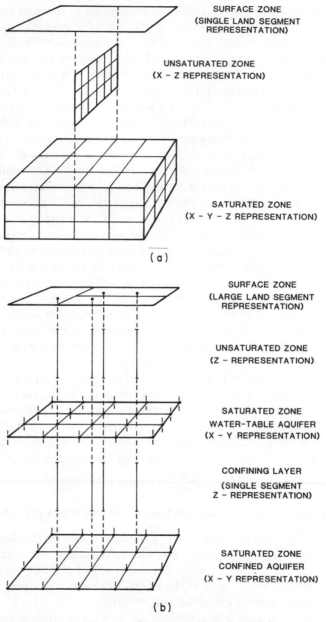

SURFACE ZONE
(SINGLE LAND SEGMENT
REPRESENTATION)

UNSATURATED ZONE
(X – Z REPRESENTATION)

SATURATED ZONE
(X – Y – Z REPRESENTATION)

(a)

SURFACE ZONE
(LARGE LAND SEGMENT
REPRESENTATION)

UNSATURATED ZONE
(Z – REPRESENTATION)

SATURATED ZONE
WATER–TABLE AQUIFER
(X – Y REPRESENTATION)

CONFINING LAYER
(SINGLE SEGMENT
Z – REPRESENTATION)

SATURATED ZONE
CONFINED AQUIFER
(X – Y REPRESENTATION)

(b)

Figure 11.7(a) Typical dimensionalities used to represent surface, unsaturated, and saturated zones in local-scale groundwater models (after Boutwell *et al.,* 1985). (b) Typical dimensionalities used to represent surface, unsaturated, and saturated zones in regional groundwater models

so considered (Beck, 1985). Problem decomposition in space or time is often applied to obtain optimal resolution in relation to computational efficiency. An example of such spatial and temporal decomposition is found in the modelling of infiltration into the soil and subsequent percolation toward the saturated zone. A distinction has been made between spatial discretization and connectiveness for local (Figure 11.7a) and for regional (Figure 11.7b) scales. Runoff from precipitation is split into a surface component (lumped horizontal segment) and infiltration (one- or two-dimensional, vertical). The infiltrated water percolates to the groundwater where a two-dimensional horizontal or three-dimensional model is used. For each of the submodels a different timestep is used, from hourly for the surface runoff and daily for the percolation, to weekly or monthly for the flow in the saturated zone.

In groundwater models, a significant distinction exists between local and regional discretization of the surface zone. This distinction reflects the difference in physiographic character between the subsurface and the surface, resulting in different approaches in aggregating small-scale phenomena into large-scale models. Also, limitations in presently existing data acquisition techniques influence the resolution used in modelling the surface zone.

Note the difference in treatment of the vertical components in groundwater models. In the regional models the flow in soils and between aquifers is mainly one-dimensional and vertical, to reduce the computational load. In the local model, second-order effects may be important enough to warrant the use of two-dimensional vertical simulation in the soil zone.

Another example can be found in simulating solute transport in fractured porous media where the movement of the solute in the fractures can be two orders of magnitude greater than in the porous matrix. A split-time approach increases the efficiency of the simulations (DeAngelis *et al.*, 1984).

With the increasing capacity and decreasing cost of computers, a trend prevails toward using smaller time scales for the same types of problems, resulting in higher temporal resolution.

SCALES RELEVANT TO GROUNDWATER MANAGEMENT SCALES

The scales used in groundwater modelling are determined by the characteristics of the groundwater system, by the availability of data, and by the nature of the system's management. These influences often include both natural and human-induced influences, such as the effects of climate, pumping, deep-well injection, and agricultural irrigation and drainage.

A discussion of spatial and temporal scales with respect to groundwater quality runs parallel with that of groundwater quantity. In general, human-induced influences affect local and intermediate scale, while large, regional-scale phenomena are of natural origin. Some human-induced quality changes are also on a regional scale, such as the amalgamated effect on water levels of groundwater withdrawal for irrigation; nonpoint pollution caused by use of fertilizers, herbicides, and pesticides in agriculture; acidification of groundwater as a result of acid precipitation; and the change in quality resulting from urbanization.

From a management point of view a system can be hierarchical, divided into administrative elements such as townships, counties, states, and river basins. If modelling is intended to provide optimal courses of action in the management of the water resources, an approach based on administrative elements can be successful. However, such as approach does not follow natural boundaries and elements.

In many management situations the selection of the scale of analysis is influenced by the restrictions in data availability. In part, such restrictions are imposed by the lack of techniques for obtaining higher-resolution data. For example, in groundwater modelling the recharge term of the subsurface water balance is directly related, through precipitation and evaporation, to atmospheric processes and conditions. These exchange variables between atmosphere and soil surface are obtained either through direct measurement (rainfall) or indirectly through calculation (evaporation), using various atmospheric variables. Such data were formerly available only for a limited number of sampling stations. Consequently, such stations are considered to represent a rather large area (Figure 11.7b), thus limiting severely the accuracy with which the recharge can be determined. In recent years remote sensing has developed as a promising means for obtaining areal values for a number of significant parameters (e.g. soil moisture and snow cover). As atmospheric scientists discuss simulations of global atmospheric exchanges based on a grid with cells circa 100 by 100 km (Dickinson, 1988), the possibility of using remote sensing to obtain important data for modelling groundwater systems and addressing problems of scale, becomes increasingly interesting.

The following discusses spatial and temporal scales from a phenomenological point of view.

Spatial scales

Water supply problems are generally related to availability of sufficient water to cover water needs, and to drawdown of groundwater levels and reduction of pressures and storage as a result of the exploitation of the resource. Industrial and municipal water supply requires wellfields with the wells relatively closely spaced (50–200 m) in order to obtain an efficient connection with the distribution network. Their area of influence can range from less than 1 km^2 to more than 10 km^2. Private, single-household wells have a small area of influence (often less than 100 m), but in some areas the combined drawdown of a large number of private wells can cause serious aquifer depletion. This is especially true with agricultural water use. Although individual irrigation wells may have a significant influence on a system (up to a few thousand metres), the total effect of large-scale irrigation from groundwater may have serious detrimental effects on the water levels and the storage in large aquifer systems, e.g. the long-term depletion of the Central Valley aquifers in California, and of the Ogallala aquifer in the High Plain region. This problem occurs most often in areas with low to moderate recharge from precipitation.

Groundwater pumpage aimed at lowering water levels may assume large-scale proportions, as with dewatering for mining operations. An example is the open-pit mining of lignite in the northern part of the Rhenish lignite mining district in West Germany, where drawdowns of more than 100 m occur to keep the pit dry. The influence of this dewatering is felt more than 20 km off site, and the affected area is still expanding as a result of continuing dewatering (Boehm, 1983).

Operation of groundwater systems and conjunctive management of coupled groundwater-surface water systems have their special scale requirements, ranging from the scale of a major watershed or river basin (for policy decisions) to that of sections of the aquifer or stretches of the river (for local planning and engineering purposes).

So-called human-induced point-source groundwater contamination, as from spills, leaching from landfills or lagoons, and underground tank failures, often occurs on a much more local scale (100–1000 m). However, if nothing is done about such groundwater deterioration, the affected area can become quite extensive (1000–10 000 m).

Some basins are affected by a large group of individual point sources such as septic tanks or landfills and dumps. The aggregated effect of these is comparable with that of nonpoint pollution. In such cases, individual mitigation has no effect and regulatory action for the entire basin is required.

Temporal scales

For water management purposes, temporal scales are important. Incidental local situations, such as construction site dewatering and chemical spills, have mainly short-term effects—weeks or months. Seasonal effects are related to agricultural uses and the use of aquifers as thermal energy sources or storage. Mid- to long-term scales (1–20 years) apply to many wellfield operations, dewatering of mining sites, and local pollution problems. Long-term effects (20–100 years) are of special interest in regional water resource development, hazardous waste displacement, and regional nonpoint pollution. Historical periods (100 years to millions of years) are of interest for palaeohydrogeological studies and for isolation of highly toxic, nondegradable chemicals and long-living radionuclides.

A typical example of temporal scales as applied in groundwater models is the study of the South Platte River in Colorado (Morel-Seytoux and Restrepo,

1985). This model currently simulates about a 100-mile stretch of river and hydraulically connected aquifer. The model is used for two types of analysis:

(1) daily operation of the conjunctive use river–aquifer system aimed at allocation of irrigation water according to availability and water rights

(2) evaluation of policies and legislation.

In the operational mode a daily simulation time-step is used for the surface water system and a weekly simulation is used for the groundwater away from the river. To account for the more rapid responses of the groundwater near the river, a correction is made to the results of the weekly simulations for the parts of the aquifer along the river. The scale for the use of the model in the development of policies and in evaluating new legislation, as in the formulation of new water-distribution rules, is much larger because long-term effects are of interest. In the South Platte River study, a weekly timestep is used for surface water, a monthly timestep for groundwater.

In planning remedial action, temporal scales are directly dependent on the extent of the polluted area, the geology, the important hydrological and biochemical processes, and the remedial action itself. For example, remedial actions designed for control of erosion and runoff, such as grading and surface water diversion, could require transient simulation with short timesteps for the rainfall and runoff processes that fluctuate rapidly (daily scale). In the saturated zone the flow is more regular and changes occur within a time frame of months or even years. Dynamically linked submodels, each with its own time scale, are then required for efficient simulation. To evaluate the threat of pollution on humans and the environment, or to analyse the effects of remedial action, simulation periods of 20–100 years are common. Much longer-term effects may have to be included, as in the case of long-living radionuclides and chemically inert toxic organics.

An example of temporal scale is radioactive waste storage in unsaturated systems, where effects must be evaluated for time periods up to ten thousand years (US EPA, 1982). Because of the strongly nonlinear character of the flow and transport equations for the unsaturated zone, the timesteps cannot be too large (Reisenauer *et al.*, 1982; Tripathi and Yeh, 1985). Tripathi and Yeh (1985) used variable time steps up to 20 000 years to simulate an unsaturated system for such an extended time.

GROUNDWATER–ECOSYSTEM RELATIONSHIPS

Ecology is concerned with the influence of external factors on organisms and with the way organisms modify their surroundings. The study of the interaction between water systems and ecological systems, or ecosystems, is important because water is a major environmental factor in most ecosystems.

For example, soil moisture influences the type of vegetation present; without soil moisture autotrophic plant covers could not exist (Budyko, 1977). Three areas of interaction between ecosystems and groundwater systems can be distinguished: terrestrial ecosystems in groundwater recharge areas, terrestrial ecosystems in groundwater discharge areas, and wetland aquatic ecosystems having complex interactions with groundwater.

Groundwater is influenced by the surface ecosystem through which it is recharged. For example, water consumption by plants can control the amount of water percolating into the saturated zone, and the chemical composition of groundwater can be affected significantly by transformation processes occurring in the litter and soil. A secondary interaction occurs as groundwater moves from its area of recharge to its point of discharge, often traversing distances greater than the extent of the ecosystem from which it originated. During its subsurface movement groundwater mixes with water originating from other recharge areas. Thus, ecosystems in groundwater discharge areas are indirectly influenced by remote ecosystems in recharge areas, in the same way as downstream ecosystems can be influenced by the headwaters of drainage systems (Loucks, 1983).

Numerous studies have shown groundwater to be a major component of stream and wetland ecosystems (Loucks, 1983). As groundwater discharges into streams and lakes, it tends to stabilize discharge and water levels, as well as water quality and biological productivity. Interactions between groundwater and terrestrial ecosystems also take place in various subsurface layers, mostly the unsaturated zone. Here, groundwater often functions as a source of transpiration water for plants, as a transporting agent of dissolved nutrients and other chemicals, and as the essential environment for various subsurface organisms (Loucks, 1983; Gerba, 1985).

Changes in natural conditions and human activities can significantly alter the physical and chemical characteristics of groundwater and ecosystem interactions. For example, the widespread problems of soil salinization and desertification are promoted by irrigation and other practices that alter the water balance.

Ecosytsems in recharge areas

Interactions between ecosystems and groundwater systems in recharge areas are rather one-sided. Plants take water from soil moisture fed by infiltrating precipitation, a process that shows wide daily and seasonal fluctuations. Water thus removed from the soil decreases the amount of water available for recharge of the groundwater system. The water requirements of the plants can be so great that little or no water is left for recharging the groundwater, and a seasonal depression in the water table results (Figure 11.8), especially in areas with a low precipitation–evaporation ratio. As groundwater in recharge zones

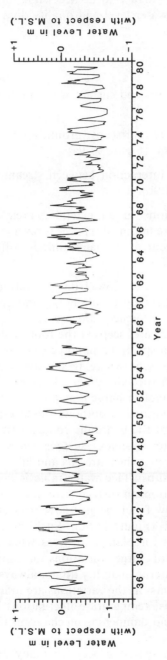

Figure 11.8 Seasonal fluctuations in groundwater levels; groundwater hydrograph of observation well 16-D, The Netherlands, for the period 1935–1980. Bottom of filter is at 14.15 m below surface; ground surface is 4.65 m above sea level

is generally too far below the surface to be accessible directly or indirectly through capillary rise, groundwater level and quality does not affect ecosystems in these zones.

Ecosystems in discharge areas

The contact between groundwater and surface water in discharge areas takes three forms:

(1) through free-flowing springs and seeps, groundwater collects in surface channels and feeds into a stream (Figure 11.9a)

(2) through direct subsurface interaction between stream, lake, or sea and groundwater system (Figure 11.9b)

(3) through complex interactions in an extended area such as wetlands (Figure 11.9c) and in lakes where in time discharge is reversed into recharge, or discharge and recharge can occur at the same time in different parts of the contact zone (Figure 11.9d).

As indicated, the several types of outflows from groundwater systems induce direct relationships between the type of discharge and specific ecosystems; often, however, other indirect influences are present. Such distinctive relationships occur between groundwater seeps at the foot of slopes and phreatophyte-dominated ecosystems; in springfed pond ecosystems and streams; and in estuarine systems and shelf seas where freshwater seeps create a local brackish environment hosting a specialized aquatic biota.

Periodic fluctuations in regional recharge may affect surface water systems in discharge areas through variations in outflow; this is especially important for stream base flow. Because of the hydraulic contact often present between surface water and groundwater systems, surface water levels can cause fluctuations in groundwater levels; when streams and lakes are not subject to large variations in levels, the subsurface water table is not subject to large fluctuations. In turn, as fluctuations in surface water levels affect groundwater levels in discharge areas, outflow from the groundwater system to the surface water system is also affected (Figure 11.9e). Such fluctuations result in variations in water availability for plants. In areas where such a regime is present, ecosystems tolerant of large variations in moisture may exist. However, such variations in water availability do not always coincide with the seasonal water needs of the plants, as the variations are influenced by climatic cycles (e.g. wet and dry periods, temperature fluctuations in recharge areas) and by the wave propagation and dampening mechanisms in the groundwater system.

Evapotranspiration is an important process in many discharge areas. A major component of evapotranspiration is plant transpiration during the day

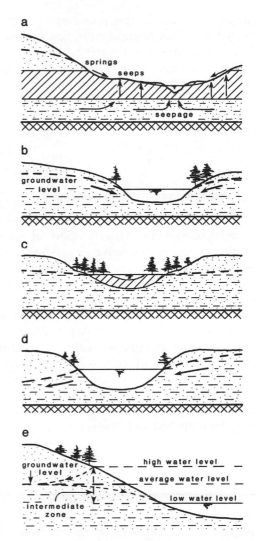

a - springs, seeps, seepage
b - effluent or gaining lake or
    stream
c - wetland
d - seepage with throughflow
e - reversing seepage from fluctu-
    ating surface water levels

Figure 11.9 Various types of surface water–groundwater contacts in discharge areas

with subsequent depletion of soil moisture and decline of groundwater levels resulting from capillary rise, as the water tables in these areas are close to the ground surface. During the night, recovery of groundwater levels from regional groundwater flow toward the discharge area, and increase of soil moisture content from capillary rise, can take place as the transpiration process is slowed. In some situations, when the water needs of the plants exceed the daily supply from groundwater and the discharge area extends away from the draining stream, a seasonal water table depression may result, (although it is often less pronounced than in recharge areas). In turn, plant growth may then be impeded by restricted water availability resulting from the extended decline of the water table.

Phreatophytes, such as cottonwood trees along southwestern mountain streams, root in or near the saturated zone; their roots are accommodated to a low-oxygen environment. Thus they are able to survive high water table levels that may submerge their roots temporarily. However, they also can survive long periods of low precipitation because they are not dependent on local infiltration (Miller, 1977).

Groundwater and wetland ecosystems

Probably the most complex relationships exist between groundwater and wetland ecosystems. The transport of water and nutrients in such systems results in quite distinctive ecosystem types and unique biota. Changed conditions in the groundwater recharge system often result in stresses on the receiving system, as when springs become seasonally or permanently dry, and when streams receive polluted groundwater discharge. Conversely, wetlands and lake systems can have a regulating function if they are in the path of groundwater flow (Figure 11.9d). The interactions in such systems work in two directions and at multiple scales: groundwater from upstream sources discharging into the surface water system affects aquatic ecosystem processes and productivity, while surface water infiltrating into the subsurface affects groundwater quality over large areas. These processes can be altered greatly by various natural or human-induced conditions, as in the case of wetlands that have been ditched and drained.

Soil salinization

A major environmental problem occurring at the interface of groundwater systems and the soil surface is soil salinization, which results from evaporation of mineralized capillary moisture from the upper horizons of the soil. Depending on the chemical composition of the soil moisture, saline soils of sodium chloride, sulphate, and nitrate occur in the United States, the USSR, Egypt, Iraq, Tunisia, Syria, Pakistan, India, and China, among others (Kovda, 1983). Saline soils, which are found in both recharge and discharge areas, generally result from natural or man-made changes in the hydrological regime. Changes in drainage patterns, such as those caused by irrigation under poor drainage and by construction of reservoirs and canals, may cause rising groundwater levels, so that water logging of previous relatively dry surface areas may occur. Another example of soil salinization is the aridization of dry land, sometimes in combination with the changing natural drainage patterns that may result from the relative lowering of sea levels (Kovda, 1983). Wherever the flow of mineralized water is more or less constant, as in discharge areas in lowlands near the sea, salinization continues over a long time period.

Inadequate drainage conditions from man-made recharge often cause the water table to rise from significant depth to within a metre of the soil surface. Minerals present in irrigation water and leachate formed by dissolving agricultural chemicals in the irrigated soil can aggravate the soil salinization problem. In Egypt, for example, almost all irrigated land is potentially affected by this process and at least half of it is now salt-affected (Kishk, 1986).

Quantification of exchanges through budgets

Water and chemicals are stored within an ecosystem and move from one ecosystem to another. The fluxes involved can be quantified by analysing transport paths and water, energy, or mass budgets (Miller, 1977; Loucks, 1983). Balancing water system outflows with inflows provides a tool for understanding the relative importance of various transport and exchange mechanisms, and also assures that all the important mechanisms are accounted for. By comparing the budgets for neighbouring ecosystems, inconsistencies in the budgets may surface. However, the components of such budgets cannot always be measured in the field, and the unmeasured component is taken as the residual of the budget calculation. Uncertainties inherent in field determination of certain budget components may reduce the accuracy of such an analysis.

Groundwater–ecosystem interaction: a matter of scale

Ecosystems vary widely in spatial scale; some are as large as groundwater systems, while others are much smaller and may even coincide with soil units. In the latter case, a single groundwater system may be a common element shared by many different ecosystems. Thus, ecosystem research across a range of scales must recognize the potential differences among the spatial and temporal scales of the ecosystem and the associated groundwater systems.

The prerequisite for quantifying relationships between groundwater systems and ecosystems is a quantitative understanding of each type of system. Conceptualization of the interactions between the systems can then proceed through interdisciplinary study and is a first step toward integral analysis of the combined system. In many instances the scales on which the systems are analysed are quite different (Figure 11.10); but resolving these differences is a crucial part of modelling the interactions between the surface and subsurface water, and the ecosystems influenced by the water system.

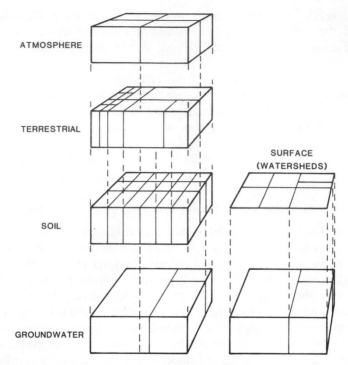

Figure 11.10   Scales and relative sizes of various hydrological
systems and ecosystems

DISCUSSION AND RESEARCH NEEDS

This paper has focused on the principles of groundwater modelling and
surveyed the problems of scale that are pertinent to such modelling. It has also
addressed the linkages between groundwater systems and aquatic and terrestrial ecosystems. These relationships often take the form of stresses put on one
system by the other, at various scales. To quantify the stresses and their effects
on the systems being considered, several scales for data collection and
modelling must be selected and defined. However, the scales on which
interfacing groundwater–ecosystem processes are usually modelled are not
identical. Hence, research on the available approaches to this problem is
needed, including averaging over larger elements, spatial or temporal decomposition, and selection of the smallest relevant scale as the basic element of
measurement and modelling. This latter approach requires extensive computer
resources, now increasingly available.

Many other questions implicit in the previous sections also need to be
answered so that physical and biological subsystems can be integrated into
large-scale models, with all under a single scale. Studies are necessary to

quantify interrelationships at various scales between hydrologic and biological subsystems. One element of such studies is improved definition of the most elementary units appropriate for work on the various subsystems. These basic units may be required as the foundation on which hierarchical modelling and integration can take place, leading to further definition of units at the scale required for regional modelling. In quantifying subsystem interactions, an illustrative study of the information loss and changing information requirements among groundwater systems aggregated at different scales from the basic units should be helpful. Such studies will be particularly important for modelling ecosystems influenced by groundwater discharge. Such research on aggregation could draw on the vast amount of existing regional groundwater data.

Another problem concerns selection of the appropriate scales for modelling studies. Here, the appropriate scales are determined by the objectives and by the level of resolution currently feasible for collecting and summarizing data on regional and global processes. In this context, the dynamic behaviour of the atmosphere, the hydrosphere, and the chemicals transported in air and water must be considered. The arbitrary scales associated with conventional modelling approaches in the various disciplines pose challenges for research on the integration between physical and biological subsystems.

Adapting the grid requirements from atmospheric modelling (on the scale of $10\,000$ km^2) into corresponding groundwater units may involve either parts of larger aquifers, or some number of smaller aquifers. In the latter case, representation at the prescribed scale may require calculation of equivalent variables describing storage throughput, transformations, and mixing in a large groundwater system. To unify optimally the modelling of the associated hydrological and biological systems, further research is necessary into the methodology of aggregation and decomposition of basic units. As part of such research, questions need to be asked concerning the relationships between spatial and temporal resolution and accuracy at the larger, aggregated scales. In this context, considerations as to the availability and quality of the databases will be significant. If we are to evaluate theoretical solutions to the indicated scaling problems of linked hydrologic and biologic systems, explicit criteria may be needed for selection of both small and large study basins and watersheds.

ACKNOWLEDGEMENTS

This paper has been prepared with support of the Holcomb Research Institute (HRI). The author acknowledges the resourceful discussions on large-scale ecological modelling with Orie L. Loucks and Richard A. Park of HRI and their constructive comments during the preparation of this paper.

REFERENCES

Bachmat, Y., Bredehoeft, J. D., Andrews, B., Holtz, D., and Sebastian, S. (1985). *Groundwater Management: The Use of Numerical Models.* 2nd Edition, edited by P. K. M. van der Heijde. Water Resources Monograph 5, American Geophysical Union, Washington, D.C.

Bear, J. (1979). *Hydraulics of Groundwater.* McGraw-Hill, New York.

Bear, J. (1972). *Dynamics of Fluids in Porous Media.* American Elsevier, New York.

Beck, M. B. (1985). *Water Quality Management: A Review of the Development and Application of Mathematical Models.* IIASA 11, Springer-Verlag, Berlin.

Boehm, B. (1983). *Conceptual Plan for Water Management Schemes in the Northern Part of the Rhenish Lignite Mining District*, pp. 571–579. Publication N142, Intern. Assoc. Hydrol. Sc.

Bonnet, M. (1982). *Methodologie de Modeles de Simuation en Hydrologie.* Document 34, Bureau de Recherches Geologique et Minieres, Orleans, France.

Boutwell, S. H., Brown, S. M., Roberts, B. R., and Atwood, D. F. (1985). *Modeling Remedial Actions at Uncontrolled Hazardous Waste Sites.* EPA/540/2–85/001, U.S. Environmental Protection Agency, Cincinnati, Ohio.

Bouwer, H. (1978). *Groundwater Hydrology.* McGraw-Hill, New York.

Budyko, M. I. (1977). *Global Ecology.* English translation 1980. Progress Publishers, Moscow.

Cusham, J. H. (1985). *The Need for Supercomputers in Clay–Water–Organic Mixtures.* Presented at DOE Seminar on Supercomputers in Hydrology, Water Resources Research Center, Purdue University, West Lafayette, Indiana, September 10–12.

DeAngelis, D. L., Yeh, G. T., and Huff, D. D. (1984). *An Integrated Compartment Model for Describing the Transport of Solute in a Fractured Porous Medium.* ORNL/TM-8983, Oak Ridge National Laboratory, Oak Ridge, Tennessee.

Dickinson, R. E. (1988). *Atmospheric systems and global change.* (Chapter 5, this volume.)

Domenico, P. (1972). *Concepts and Models in Groundwater Hydrology.* McGraw-Hill, New York.

Freeze, R. A., and Cherry, J. A. (1979). *Groundwater.* Prentice-Hall, Englewood Cliffs, New Jersey.

Freeze, R. A., and Witherspoon, P. A. (1966). Theoretical analysis of regional groundwater flow: 1. Analytical and numerical solutions to the mathematical model. *Water Resources Research*, **2**, 641–656.

Gerba, C. P. (1985). Microbial Contamination of the Subsurface. In Ward, C. H., McCarty, P. L., and Giger, W. (Eds.) Ground Water Quality. John Wiley and Sons, New York.

Heath, R. C. (1983). *Basic Ground-Water Hydrology.* Water Supply Paper 2220, U.S. Geological Survey, Reston, Virginia.

Johnson Division. (1975). *Ground Water and Wells.* Johnson Division, UOP Inc., Saint Paul, Minnesota.

Kishk, M. A. (1986). Land degradation in the Nile valley. *Ambio*, **15** (4), 226–230.

Klimentov, P. P. (1983). *General Hydrogeology.* English Translation, MIR Publishers, Moscow, USSR.

Kovda, V. A. (1983). Loss of productive land due to salinization. *Ambio*, **12**(2), 91–93.

Loucks, O. L. (1983). *Wetland Characteristics—Their Land–Water Interactions.*

Proceedings 2nd. SCOPE Workshop on Wetland Management, Songkla, Thailand, April 1983 (in press).

L'vovich, M. I. (1979). *World Water Resources and Their Future*. English Translation, American Geophys. Union, Washington, D.C.

Miller, D. H. (1977). *Water at the Surface of the Earth*. Academic Press, New York.

Morel-Seytoux, H. J., and Restrepo, J. I. (1985). *SAMSON Computer System*. HYDROWAR Program, Colorado State University, Fort Collins, Colorado.

NRC, (1981). *Coal Mining and Ground-Water Resources in the United States*. National Academy Press, Washington, D.C.

Reisenauer, A. E., Key, K. T., Narasimhan, T. N., and Nelson, R. W. (1982). *TRUST: A Computer Program for Variable Saturated Flow in Multidimensional, Deformable Media*. NUREG/CR-2360, PNL-3975, Battelle Pacific NW Lab, Richland, Washington.

Rickert, D. A., Ulman, W. J., and Hampton, E. R. (1979). *Synthetic Fuels Development, Earth-Sciences Considerations*. U.S. Geological Survey, Reston, Virginia.

Toth, J. (1963). A Theoretical Analysis of Groundwater Flow in Small Drainage Basins. *J. Geophys. Res.*, **68**, 4795–4811.

Toth, J. (1971). Groundwater Discharge: a Common Generator of Diverse Geologic and Morphologic Phenomena. *Int. Assoc. Hydrol. Sci. Bull.*, **6**, 7–24.

Tripathi, S., and Yeh, G. T. (1985). *HYDROGEOCHEM on the Hypercube Computer*. Presented at DOE Seminar on Supercomputers in Hydrology, Water Resources Research Center, Purdue University, West Lafayette, Indiana, September 10–12, 1985.

U.S. Environmental Protection Agency. (1982). *Environmental Standards for the Management and Disposal of Spent Nuclear Fuel, High-level and Transuranic Radioactive Wastes* (40 *CFR Part* 191). Federal Register (47 FR 58196), Washington, DC.

Wang, J. S. Y., and Narasimhan, T. N. (1984). *Hydrologic Mechanisms Governing Fluid Flow in Partially Saturated, Porous Tuff at Yucca Mountain*. LBL-18473, Lawrence Berkeley Lab., Berkeley, California.

Scales and Global Change
Edited by T. Rosswall, R. G. Woodmansee and P. G. Risser
© 1988 Scientific Committee on Problems of the Environment (SCOPE)
Published by John Wiley & Sons Ltd.

CHAPTER 12

Variability of the Fluvial System in Space and Time

STANLEY A. SCHUMM

Department of Earth Resources,
Colorado State University,
Fort Collins, Colorado 80523, *USA*

ABSTRACT

The fluvial system is a process–response system that includes the morphologic component of channels, floodplains, hillslopes, and the cascading component of water and sediment. The system changes progressively through geologic time, as a result of normal erosional and depositional processes, and it responds to changes of climate, baselevel, and tectonics. Therefore, there can be considerable variability of fluvial system morphology and dynamics through time. In addition, there is great variability in space or location as a result of different geologic, climatic, and relief conditions.

The prediction and postdiction of fluvial system behaviour is greatly complicated by this variability, and there are seven potential problems that must be considered in any attempt to extrapolate empirical relations in space and time.

INTRODUCTION

The term fluvial is from the Latin word *fluvius,* a river. When carried to its broadest interpretation a fluvial system not only involves stream channels but also entire drainage networks. The size of fluvial systems ranges from that of the vast Mississippi, Missouri, and Ohio river system to small badland watersheds of a few square metres. The time periods that are of interest to the student of the fluvial system can range from a few minutes of present-day activity, to channel changes of the past century, to the geologic time periods required for the development of the billion-year-old gold-bearing palaeo-channels of the Witwatersrand conglomerate and the even older and spectacular channels and drainage networks on Mars. Therefore, the range of temporal and spatial dimensions of the fluvial system is very large (Carey, 1962; Fisher, 1969).

In order to simplify discussion of the complex assemblage of landforms that comprise a fluvial system, it can be divided into three zones (Figure 12.1A). Zone 1 is the drainage basin, watershed, or sediment-source area. This is the area from which water and sediment are derived. It is primarily a zone of sediment production, although sediment storage does occur there in important ways. Zone 2 is the transfer zone, where, for a stable channel, input of sediment can equal output. Zone 3 is the sediment sink or area of deposition (delta, alluvial fan). These three subdivisions of the fluvial system may appear artificial because obviously sediments are stored, eroded, and transported in all the zones; nevertheless, within each zone one process is dominant.

Each zone, as defined above, is an open system. Each has its own set of morphological attributes, which can be related to water discharge and sediment movement. For example, the divides, slopes, floodplains, and channels of Zone 1 form a morphological system. The components of which are related statistically one to the other (e.g. valley-side slopes and stream gradients are directly related to the abundance and spacing of channels). In addition, the energy and materials flow form another system, a cascading system. Components of the morphological system (channel width, depth, drainage density) can be related statistically to the cascading system (water and sediment movement, shear forces, etc.) to produce a fluvial process-response system.

The fluvial system can be considered at different scales and in greater or lesser detail depending upon the objective of the observer. For example, a large segment, the dendritic drainage pattern is a component of obvious interest to the geologist and geomorphologist (Figure 12.1A). At a finer scale is the river reach of Figure 12.1B, which is of interest to those who are concerned with what the channel pattern reveals about river history and behaviour and to engineers who are charged with maintaining navigation and preventing channel erosion. A single meander can be the dominant feature of interest (Figure 12.1C), which is studied by geomorphologists and hydraulic engineers for information that it provides on flow hydraulics, sediment transport, and rate of bend shift. Within the channel itself is a sand bar (Figure 12.1C), the composition of which is of concern to the sedimentologist, as are the bed forms (ripples and dunes) on the surface of the bar (Figure 12.1D) and the details of their sedimentary structure (Figure 12.1E). These, of course, are composed of the individual grains of sediment (Figure 12.1F) which can provide information on sediment sources, sediment loads, and the feasibility of mining the sediment for construction purposes.

As the above demonstrates, a variety of components of the fluvial system can be investigated at many scales, but no component can be totally isolated

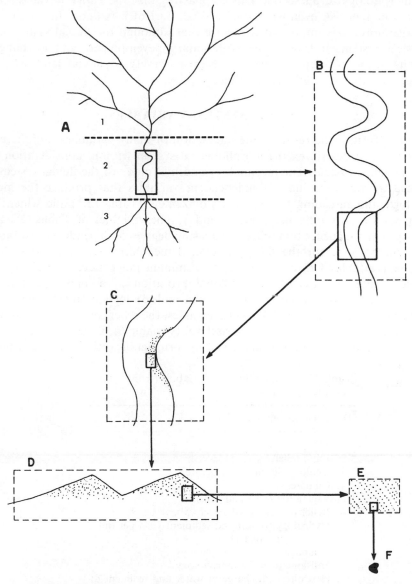

Figure 12.1 Idealized sketch showing the components of a fluvial system. See text for discussion

because there is an interaction of hydrology, hydraulics, geology, and geomorphology at all scales. This emphasizes that the entire fluvial system cannot be ignored, even when only a small part of it is under investigation. Furthermore, it is important to realize that although the fluvial system is a physical system, it follows an evolutionary development, and it changes through time. Therefore, there are a great variety of fluvial landforms in space, and they change through time.

FLUVIAL SYSTEM VARIABLES

Any landform is the result of the interaction of many variables, and there is little value in describing the morphological system without consideration of the cascading system. The morphology and hydrology of the fluvial system is related to the controlling of independent variables that produce the morphologic and cascading characteristics of Zone 1 (Table 12.1) and which, in turn, significantly influence Zones 2 and 3. The variables of Table 12.1 are arranged in a sequence that reflects increasing degrees of dependence insofar as this can be done for the fluvial system. Time, initial relief, geology, and climate (variables 1 through 4) are the dominant independent variables that influence the progress of the erosional evolution of a landscape and its hydrology. Vegetation type and density (variable 5) depend on lithology (soil) and climate (variables 3 and 4). As time passes, the relief, or the volume of the drainage system remaining above baselevel (variable 6), is determined by the factors above it in the table, and relief in turn significantly influences runoff

Table 12.1 Fluvial System Variables (from Schumm and Lichty, 1965)

Drainage system variables

---

 1. Time
 2. Intial relief
 3. Geology (lithology, structure)
 4. Climate
 5. Vegetation (type and density)
 6. Relief or volume of system above baselevel
 7. Hydrology (runoff and sediment yield per unit area within Zone 1)
 8. Drainage network morphology
 9. Hillslope network morphology
10. Hydrology (discharge of water and sediment to Zones 2 and 3)
11. Channel and valley morphology and sediment characteristics (Zone 2)
12. Depositional system morphology and sediment charactieristics (Zone 3)

---

and sediment yield per unit area within the drainage basin (variable 7). Runoff acting on the soil and geologic materials produces a characteristic drainage-network morphology (variable 8, drainage density, channel shape, gradient, and pattern) and hillslope morphology (variable 9, slope angle, length, and profile form). These morphological variables in turn strongly influence the cascading system, the volumes of runoff and sediment that are eventually discharged from Zone 1 (variable 10). It is the volume and type of sediment and discharge volume and flow character that largely determines channel morphology and the nature of fluvial deposits that form in Zones 2 and 3 (variables 11 and 12).

In Table 12.1 only upstream controls are listed, but the fluvial system can also be significantly influenced by variations in the downstream baselevel (sea level and lake level change, dam construction). Lowering of baselevel will rejuvenate the drainage system. The effect on Zones 1 and 2 will be significant, with feedback to Zone 3 of greatly increased sediment production and a change of sediment characteristics (Figure 12.1A). For further discussion of these variables and their influence on the drainage system see Schumm (1977) or Richards (1982).

The variability of a fluvial system under the influence of only three controls; stage of development, relief, and climate (variables 1, 2, 4) is summarized in Figure 12.2. Only two examples are shown, and geologic conditions are the same for each example. Example 1 is either a young, high-relief, or dry-climate drainage basin. Example 2 is either an old, low-relief or humid-climate drainage basin. The variables time, relief, and climate act in the same general way on the landscape.

For the youthful, high-relief, dry and sparsely-vegetated drainage basin the drainage density (D, ratio of total channel length to drainage area) will be high, and both hillslope inclination and stream gradient (S) will be steep. A fully-developed drainage network will produce high discharge per unit area (Q), high peak discharge (Qp), relatively low base flow (Qb), and sediment loads and yield (Qs) will be high. The fine-textured (high drainage density) drainage network will permit the rapid movement of water and sediment from Zone 1 to Zone 2. In Zone 2 the high sediment load (sand and gravel), the high bed load, and the highly variable (flashy) nature of the discharge will produce a bed-load channel of steep gradient, large width to depth ratio, low sinuosity (P, ratio of channel length to valley length) and a braided pattern. Channel shifting and change will be common. Downstream (Zone 3) the large quantity of coarse sediment may form an alluvial fan or fan delta. Deposition will be rapid, and the sedimentary deposit will contain many discontinuities and numerous sand bodies.

At the other extreme is Example 2, an old, low-relief, humid, well-vegetated drainage basin that has a low drainage density (D), gentle slopes, and low discharge per unit area (Q). A high percentage of the precipitation infiltrates

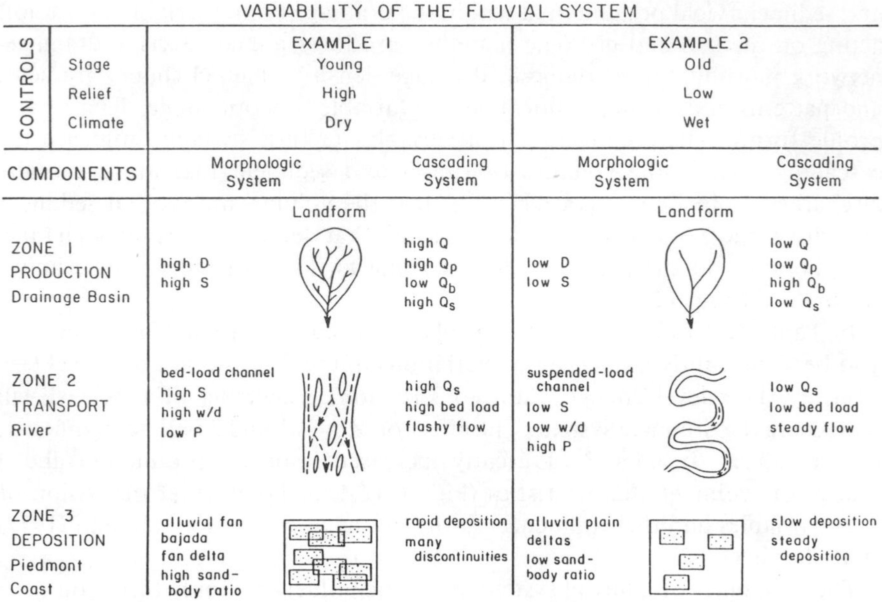

VARIABILITY OF THE FLUVIAL SYSTEM

CONTROLS		EXAMPLE 1		EXAMPLE 2	
Stage		Young		Old	
Relief		High		Low	
Climate		Dry		Wet	
COMPONENTS		Morphologic System	Cascading System	Morphologic System	Cascading System
		Landform		Landform	
ZONE 1 PRODUCTION Drainage Basin		high D, high S	high Q, high Q_p, low Q_b, high Q_s	low D, low S	low Q, low Q_p, high Q_b, low Q_s
ZONE 2 TRANSPORT River		bed-load channel, high S, high w/d, low P	high Q_s, high bed load, flashy flow	suspended-load channel, low S, low w/d, high P	low Q_s, low bed load, steady flow
ZONE 3 DEPOSITION Piedmont Coast		alluvial fan, bajada, fan delta, high sand-body ratio	rapid deposition, many discontinuities	alluvial plain, deltas, low sand-body ratio	slow deposition, steady deposition

Figure 12.2   Two examples of very different fluvial systems, showing the variability of the morphologic and cascading components of the examples in the three geomorphic zones (Figure. 1): D, drainage density; S, gradient; w/d, width–depth ratio; P, sinuosity; Q, water discharge per unit area; Qb, base flow; Qp, peak discharge; Qs, sediment load

or is lost to evapotranspiration. Peak discharge (Qp) will be relatively low, and groundwater will be abundant, leading to high base flow (Qb). Sediment loads and sediment yields will be low. This produces a suspended-load channel in Zone 2, which transports relatively fine sediments (fine sands, silt, and clay) at a low slope in a channel with high sinuosity and a low ratio of width to depth. Discharge will be relatively steady, although during major precipitation events large floods will move through the valley. The fine sediment and the steady nature of the flow will cause slower rates of deposition in Zone 3, and there will be a few sand bodies and an alluvial plain or delta will form.

A change of climate can transform Example 1 to Example 2 or vice versa or to some intermediate stage. The character and volume of the sediments delivered to Zones 2 and 3 will also change, and significant channel adjustments will result (Schumm, 1977). Without tectonic interruptions through geological time, the erosional evolution of a landscape should result in a transition from drainage basins and channels like Example 1 to those more similar to Example 2 (Figure 12.2). As the relief of the drainage basin is reduced during the erosional evolution, drainage density will decrease, slopes will decline, and the amount and grain size of the sediment will decrease. The result will be a transition from a braided to a meandering channel in Zone 2 and to finer grained, more uniform deposits in Zone 3.

The relationships displayed in Figure 2 are straightforward and well known. They demonstrate clearly that, because of the number of variables acting, the fluvial system will have a complex history as it adjusts to climatic changes and human influences through time. In addition, at any one time the range of geology, relief, and climate will guarantee that a great range of river characteristics exist. The following discussion will, for simplicity, be limited to river channels. In order to consider fluvial-system variability, first the great variety of existing river channels will be described (variation in space), and then it will be possible to consider how they change with time.

VARIABILITY IN SPACE

Rivers usually increase in size in a downstream direction, as a larger channel is required to convey the increasing discharge. In general, channel width and meander dimensions increase as about the 0.5 power of discharge, and channel width increases as about the 0.4 power of discharge. However, when river discharge decreases in a downstream direction, channel dimensions also decrease. For example, the Finke River in central Australia flows from its source in the McDonald Range as a wide sandy channel, becoming smaller as the river flows into the Simpson Desert where it eventually disappears.

Rivers exhibit an astonishing diversity that can be related to the variations of water discharge and sediment load, as well as to the presence of bedrock outcrops, man's activities, and tectonic influences. Pattern is a simple means of classifying alluvial channels. Five basic channel patterns exist (Figure 12.3): straight channels with either migrating sand waves (pattern 1) or migrating alternate bars with sinuous thalweg (pattern 2), two types of channels, a highly sinuous channel of equal width (pattern 3a) and channels that are wider at bends than in crossings (pattern 3b), the meandering-braided transition (pattern 4) and a typical braided-stream (pattern 5). The relative stability of these channels and their shape and gradient as related to relative sediment size, load, velocity of flow, and stream power, are also indicated on Figure 12.3. It has been possible to develop these patterns experimentally by varying the gradient, sediment load, stream power, and the type of sediment load transported by the channel (Schumm and Khan, 1972).

Alluvial channels also have been classified according to the type of sediment load moved through the channels as suspended-load, mixed-load, and bed-load channels. Water discharge determines the dimensions of the channel (width, depth, meander dimensions), but the relative proportions of bed load (sand and gravel), and suspended load (silts and clays) determine not only the shape of the channel but width–depth ratio and channel pattern. A suspended load channel has been defined as one that transports less than 3% bed load

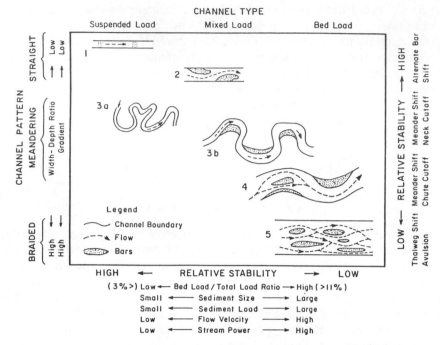

Figure 12.3   Channel classification based on pattern and type of sediment load with associated variables and relative stability indicated (from Schumm, 1981)

(sand and gravel) and a bed-load channel as one transporting more than 11% bed load. The mixed-load channel lies between these two (Figure 12.3).

Figure 12.3 suggests that the range of channels from straight through braided forms a continuum, but experimental work and field studies have indicated that the pattern changes between braided, meandering, and straight occur at river-pattern thresholds (Figure 12.4). The changes of pattern take place at critical values of stream power, gradient and sediment load (Schumm and Khan, 1972).

As anyone knows who has looked out of the window of an aircraft or who has studied aerial photographs or maps of rivers, one river can reveal a great variability in its pattern in a downstream direction. For example, meander cutoffs may convert a meandering reach of channel to a relatively straight reach that with time will regain its original sinuosity. However, if the type of sediment load transported by a stream is modified by the contibutions of a tributary, the channel morphology can change completely. For example, a sinuous suspended-load channel can braid, or a braided channel can meander, when there is a sufficiently great change in the type of sediment load transported through the channel.

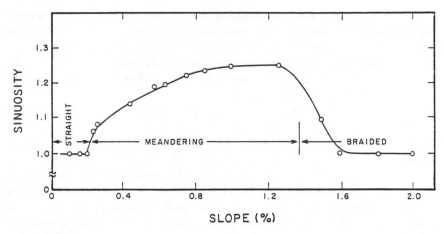

Figure 12.4 Relation between flume slope and sinuosity during experiments at constant water discharge. Sediment load, stream power, velocity increase with flume slope and a similar relation can be developed with these variables (from Schumm and Khan, 1972)

Variations of sinuosity are also influenced by the slope of the valley floor, as the river adjusts its pattern to maintain a constant gradient over the changing valley-floor slope (Adams, 1980; Burnett and Schumm, 1983). For example, as valley floor slope increases, sinuosity increases in order to maintain a constant channel gradient (Figure 12.4). However, if the change is too great, the large increase of slope (stream power) may cause a change from meandering to braided or channel incision.

Although the above discussion concentrates on channel patterns, the classification of Figure 12.3 indicates that channel dimensions, hydraulic characteristics, shape (width–depth ratio) and gradient all change as pattern changes. Therefore, as sediment loads or valley slopes change there can be significant variations of channel morphology. The degree of change depends on channel sensitivity, and it is important to realize that channels that are near a pattern threshold may change their characteristics dramatically with only a slight change in the controlling variable. For example, some rivers that are meandering and that are near pattern thresholds become braided with only a small addition of bed load, whereas other less sensitive channels will not be affected significantly (Figure 12.4).

VARIABILITY THROUGH TIME

Geomorphologists and engineers often have different concepts of the change in fluvial systems, depending upon the time span under consideration. On the one hand, historical studies involving long periods show the fluvial system evolving progressively with major changes occurring as a result of climatic change or tectonism. On the other hand studies of channel process are made during short time spans when the landform may not change significantly. Because of this disparate approach, the historically oriented geologist and geomorphologist frequently find communication difficult with the process-oriented geomorphologist and engineer.

The passage of time or the evolutionary stage of landform development is important to the geologist, but it need not be to engineers. This leads to very different perspectives. For example, is stream gradient (variable 11, Table 12.1) an independent or dependent variable? During a long time span, it is clearly dependent upon the discharge of water and sediment that passes through a channel (variable 10, Table 12.1). It is equally clear, however, when studying sediment transport in river channels and flumes, that the steeper the channel or flume the higher is sediment transport. For example, a meander cutoff will steepen the gradient of the affected reach of channel and increase sediment transport. Obviously gradient can be both an independent and a dependent variable, depending upon the objective and the time span considered during an investigation.

An investigator needs to consider how to view a fluvial system for his particular needs; either as a purely physical system during a short span of time, or as a historical system that changes with time, because the different viewpoints strongly influence conceptions of the landscape and the conclusions relating to cause and effect. For example, Schumm and Lichty (1965) suggest that landform evolution can be considered during three time spans of different duration: cyclic, graded, and steady (Figure 12.5). Cyclic-time span encompasses a major period of geologic time, perhaps that involving an erosion cycle. Over the long span of cyclic time a removal of material (potential energy) occurs, and the characteristics of the system progressively change (Figure 12.5a). When viewed from this perspective a fluvial system is undergoing continual change (dynamic equilibrium, Figure 12.6c).

The graded-time span refers to a short period of cyclic time (decades or centuries). When the system is viewed from this perspective, there is a continuous adjustment among its components. There may be slight progressive change of the landforms, but this is masked by the fluctuations about the average values (Figure 12.5b). In other words, the progressive change during cyclic time is seen during a shorter span of time as a series of fluctuations about or approaches to an equilibrium (steady-state equilibrium, Figure 12.6b). Seasonal and other short-term fluctuations mask any slow and progressive change.

During the brief steady-time span of a few hours, or days, a static equilibrium (Figure 12.6a) may exist, in contrast to the steady-state equilibrium of graded time. The landforms during this time span are truly time-independent, because they do not change.

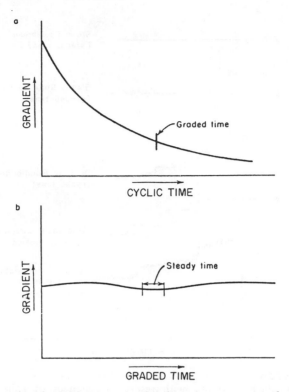

Figure 12.5 Diagram showing the concepts of cyclic, graded, and steady time as reflected in changes of stream gradient through time. (a) Progressive reduction of channel gradient during cyclic time. During graded time (b), a small fraction of cyclic time, the gradient remains relatively constant, but there are fluctuations of gradient about the mean. Gradient is constant during the brief span of steady time (from Schumm and Lichty, 1965)

Dynamic, steady-state, and static equilibrium define the types of landform behaviour assumed for cyclic, graded, and steady time spans, but metastable and dynamic metastable equilibria should also be included because they reflect the influence of thresholds that can cause abrupt episodes of system adjustment. As an example, the slow progressive reduction of sediment size and quantity delivered from the eroding Zone 1, during cyclic time, is expected to produce a progressive decrease of stream gradient in Zone 2. However, the channels need not respond to slight changes of sediment load, but, in fact, there can be a lag until a threshold is exceeded. In this case, the channel responds abruptly to the slow cumulative effects.

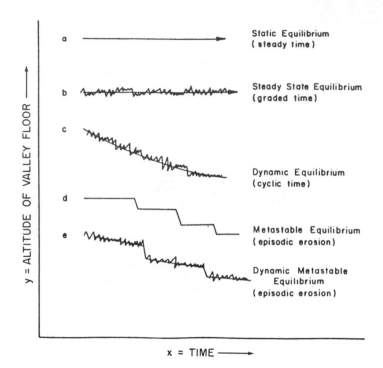

Figure 12.6 Types of equilibria based on Chorely and Kennedy (1971). Each is shown with respect to changes of valley-floor elevation with time. (a) Static eqilibrium—no change with time (steady time). (b) Steady-state equilibrium—variations about a constant average condition (graded time). (c) Dynamic equilibrium—variations about a changing average condition (cyclic time) (d) Metastable equilibrium—static equilibrium separated by episodes of change as thresholds are exceeded. (e) Dynamic metastable equilibrium—dynamic equilibrium with episodic change as thresholds are exceeded (from Schumm, 1977)

In summary, a river can be considered unchanging or static during a steady time span, it can undergo natural and expected variations during a graded-time span, and it can be viewed as undergoing progressive or episodic change during a cyclic time span. According to the scheme of Figure 12.6, these channels are open systems in static, steady state, or dynamic equilibrium, respectively. Of course, a channel can undergo alterations of gradient, shape, patterns, and dimensions during all three time spans if it is adjusting to major changes of climate, baselevel, or tectonics.

It is during steady time that an apparent reversal of cause and effect may occur. This is best demonstrated by comparing the conflicting conclusions that could result from studying fluvial processes in the hydraulic laboratory and in a natural stream. The quantity of sediment transported through a flume is dependent on the velocity and depth of the flowing water, which for given discharge depends on flume shape and slope. An increase in sediment transport will result from an increase in the slope of the flume or an increase in discharge. In a natural stream, however, over periods of time (graded time) it is apparent that mean water and sediment discharge, and sediment size are independent variables that determine the slope of the stream and, therefore, the hydraulic characteristics at a cross section. Furthermore, over a long period of geologic (cyclic) time the independent variables of geology, relief, and climate determine the discharge of water and sediment, with all other morphologic and hydraulic variables being dependent. Confusion can only be avoided if discussion is restricted to a consideration of a single time span. This

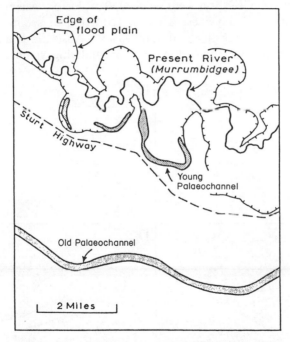

Figure 12.7 Diagram made from an aerial photograph of a portion of the Riverine Plain near Darlington Point, New South Wales Australia. The sinuous Murrumbidgee River, which is about 200 ft wide, flows to the left across the top of the figure (upper arrow). It is confined to an irregular floodplain on which a large oxbow lake (youngest palaeochannel, middle arrow) is preserved. The oldest palaeochannel (lower arrow) crosses the lower part of the figure (from Schumm, 1977)

will prevent unprofitable arguments between engineer and earth scientist, and
between historical and process-oriented geomorphologists. This is, of course,
the reason that the relations of Figures 12.5 and 12.6 are stressed here.

During graded and steady time, channel morphology reflects the influence of
a complex series of independent variables, but the discharge of water and
sediment integrates most of these, and it is the quantity of sediment and water
moving through the channel that largely determines the morphology of the
alluvial channels. As these variables change through time, a result of long-term
erosional reduction of a sediment source area (Zone 1) or climatic change or
land-use change, the river channel will respond.

The world wide climate changes of the Pleistocene had profound effects on
rivers. The Mississippi River provides an excellent example of the change of a
braided river to a meandering channel as sediment loads delivered from the
retreating Pleistocene ice sheet were reduced. This history was repeated in the
Ohio River valley and anywhere a major reduction of sediment load took
place. The Murrumbidgee River of southeastern Australia is an example of
river metamorphosis from braided to a large meandering channel and finally to

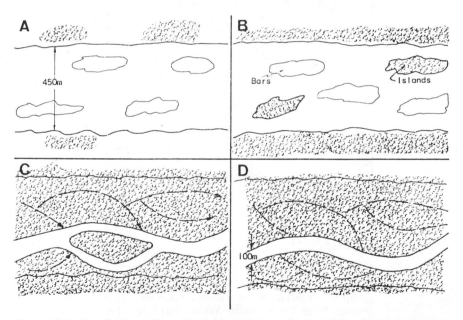

Figure 12.8   South Platte River metamorphosis. A. Early 1800s, discharge is intermittent, bars are transient. B. Late 1800s, discharge is perennial, vegatation is thicker
on flood-plain and islands are forming. C. Early 1900s, droughts allow vegetation to
become established below mean annual high water level, bars become islands, and a
single channel forms. D. Modern channel, islands are attached to the floodplain.
Braided patterns on floodplain are vestiges of historic channels (from Nadler and
Schumm, 1981)

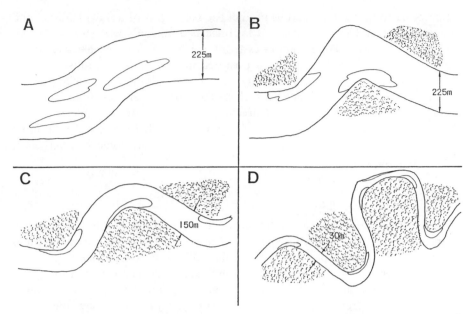

Figure 12.9   Arkansas River metamorphosis at Bent's Old Fort. A. Pre-1900 channel. B. 1926 channel. C. Channel between 1926 and 1953. D. Modern channel (from Nadler and Schumm, 1981)

a small meandering channel as climate varied from relatively dry to wet to semiarid (Figure 12.7). Similar changes of channel pattern have been documented for the northern Polish Plain and the Gulf Coast of the United States (Starkel, 1983; Baker and Penteado-Orellana, 1977).

Some of the most recent and drastic changes of river channels have taken place during the past 100 years in the semiarid western USA. These result from climatic fluctuations or from reduction of peak discharge by water storage behind dams. For example, the South Platte River in northwestern Colorado changed from a braided river 450 m wide in 1930 to a single channel in 1970, as a result of reduced peak discharge and increased base flow as a result of increased irrigation (Figure 12.8). Similar changes have occurred along the North Platte and Arkansas Rivers (Nadler and Schumm, 1981). However, the Arkansas River, as a result of greatly increased suspended-sediment load from a rejuvenated tributary and a reduction of peak discharge, changed from braided to sinuous (Figure 12.9).

SEVEN PROBLEMS

Although this discussion should concentrate on problems of scale (size and time) there are other important problems that must be considered when attempting to understand and to predict the behaviour of a fluvial system and its components. The earth scientist is concerned not only with prediction but postdiction as well. He extrapolates from the present to the past and to the future. However, any prediction or postdiction (extrapolation) must be made with caution because there are at least seven problems associated with extrapolation in the earth sciences. Two have been mentioned earlier, scale (space and time) and sensitivity (thresholds), but in order to understand the hazards of prediction, especially at a regional or global scale, five additional problems should be considered as follows: location, convergence, divergence, singularity, and complexity.

Scale

This review has already dealt with scale problems involving both time and size. In general, the shorter the time span, the smaller the space, and the more rapid the process the more specific can be explanation and extrapolation.

Time can be viewed as an index of the rate of energy expenditure, work done, or change of entropy. These variables cannot be measured through geologic time; therefore, time is used as a surrogate just as drainage basin area can be used as a surrogate for water discharge from ungauged watersheds. Time, of course, has always been of great concern to the earth scientist (Thornes and Brunsden, 1977; Cullingford *et al.*, 1980).

The basic problem with time is that the human individual does not live long enough to appreciate or understand the present sufficiently in order to apply it to the past and to the future. Many geomorphic processes operate too slowly to provide an adequate record for prediction. Another way of appreciating this problem is to consider a human individual with different life spans. A human with a life span of one day would likely conclude that the earth's surface was static; if it were 100 years, the conclusion would be that geomorphic processes are modifying the surface. A human who lived for 10,000 years would note climatic and tectonic instability and the evolutionary development of landforms. (For a botanical example see Gleason, 1926.)

Explanations or hypotheses may change as the size and complexity of the feature increases or decreases (Arnett, 1979). For example, Penning-Roswell and Townshend (1978) show that, although local variations of stream gradient can be explained by variations in size of bed-material, the gradient of a long segment of a river is better explained by water discharge. Morgan (1973) shows the relative importance of climate on drainage density in Malaysia. At a small scale (second-order drainage basin) both climate and lithology determine drainage density, but lithology dominates at meso-scale (Klang River basin) and climate at macro-scale (west Malaysia).

When dealing with larger and older landforms, less of their properties can be explained by present conditions and more must be inferred from the past.

Thus microfeatures and events such as sediment movement and bed forms in a river are understandable in the light of recent experience without historical information. However, channel morphology may reflect a sizeable historical component of explanation. For example, rivers flowing on valley or alluvial plain surfaces, which were formed under Pleistocene conditions, are significantly influenced by these conditions (Schumm, 1977, 1979). Large features such as drainage networks that are stucturally controlled (trellis, rectangular) and mountain ranges will obviously be explained predominantly by historical information.

In summary, scale is important for extrapolation. The longer the time span and the larger the area the less accurate will be predictions and postdictions that are based upon present conditions. Therefore, extrapolation of modern records is hazardous, and geomorphic predictions for periods in excess of perhaps 1000 years should be based upon worst-case conditions of climate change, tectonic activity and baselevel change. This, of course, involves an understanding of the past.

Location

Even the smallest components of a landscape such as a first-order stream may have a considerable range of potential energy from mouth to drainage divide. The morphology varies and the materials and energy flow varies from place to place (Graf, 1982). For example, sediment delivery ratios (ratio of sediment delivery to sediment production in a basin) are usually below 1.0 indicating that there is sediment in storage. In other words some portions of a geomorphic system are eroding whereas others are aggrading.

Not all components of a high energy landscape are functioning in phase. Some tributaries may be stable while others may be aggrading or degrading, depending on local circumstances or the rate at which they respond to changes in the main channel (Schumm and Hadley, 1957; Schumm, 1977). Therefore, conclusions about a river system may depend on the part of the system studied. For example, rejuvenation of a drainage basin by baselevel change will cause a wave of accelerated erosion to advance headward through the basin. The lag time for features near the drainage divide may be very long for large basins, and events in one part of the basin may be very different from those occurring elsewhere for hundreds or even thousands of years.

In summary, predictions based upon data from one location may not be useful elsewhere. Extrapolation in space is as hazardous as extrapolation in time.

Convergence (equifinality)

Convergence refers to the production of similar effects from different processes and causes. For example, braided streams result from aggradation, but they also can be 'stable' with the braided morphology being the effect of high bedload transport on steep gradients. Similarly flashy discharges may maintain a braided pattern in one channel, while similar but more uniform discharge in another channel will form meanders. Therefore, it is sometimes difficult to infer processes from form (Pitty, 1982, p. 44; Chorley and Kennedy, 1971, p. 294) and attempts to do so have been termed the *genetic fallacy* (Harvey, 1969, p. 80, p. 409).

In addition to similar forms, similar effects may also be produced by very different causes. For example, incision of a stream may be due to baselevel lowering, tectonic uplift or climate change. Therefore, a fragmentary record from the geologic past or limited observations at present may be an inadequate base upon which to postdict or predict. Obviously under these conditions extrapolation must rest upon a careful study of the system of interest.

Divergence

Divergence is the opposite of convergence and refers to similar processes and causes producing different effects. This depends on the nature of the landscape, climate, and geology. For example, a climate change may trigger massive landslides in one area, gullying in another, and a limited response elsewhere.

Hydrologic and geomorphic studies show that both sediment yield and drainage density are a maximum in semi-arid areas (Langbein and Schumm, 1958; Gregory and Gardiner, 1975). Therefore, a similar change in climate will have very different effects in arid, semi-arid and humid regions (Figure 12.10A). For example, an increase of precipitation in arid regions will significantly increase drainage density and the export of sediment from that area. A similar change of precipitation in semi-arid regions, due to the increased vegetative cover, will produce less sediment, and drainage density will decrease as vegetation obliterates low-order channels.

Another example is provided by the patterns of river channels. An increase in energy due to increased discharge or perhaps to active tectonic steepening of the valley floor may cause a straight stream to begin to develop a sinuous pattern (Figure 12.10B, a to c), a mildly meandering stream to become more sinuous (b to c), a highly sinuous stream to become braided (c to d), or have no effect on a braided stream (d to e).

Similar changes of the inputs of energy or matter to a system may produce very different effects; therefore, existing conditions must be thoroughly understood before extrapolations are made.

Singularity

Just as all people are the same but each has singular characteristics, landforms (river, hillslopes), when examined in detail, have sufficient differences so that

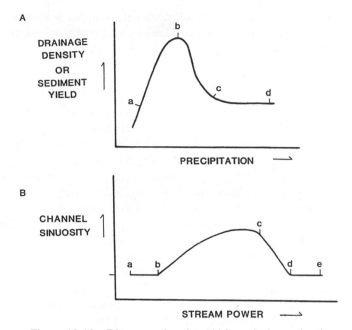

Figure 12.10   Diagrams showing (A) how drainage density (total length of stream channels divided by drainage area) or sediment yield varies with mean annual precipitation, and (B) how sinuosity (channel length divided by valley length) varies with stream power (tractive force times velocity of flow). With an increase of precipitation drainage density and sediment yield increase to maximum (a to b) in semiarid regions, decrease (b to c) and then remain relatively constant (c to d) in humid regions, with other variables remaining the same. With an increase of stream power or velocity, sinuosity remains constant at low values (a to b), increases with meandering (b to c), decreases through a transition from meandering to braided (c to d) and then remains braided (d to e) (From Schumm, 1985)

they can be considered to be singular. Hence, each singular landform will respond to a change in either slightly or significantly different ways and at different rates. This really is the key to the difficulty of short-term prediction. General relationships (laws) will be of only partial assistance in specific cases. For example, when a geomorphic variable is plotted against a controlling variable, the data usually scatter over half a log cycle or even over a full log cycle. This is a poor basis from which to predict individual response, and landform response to change may appear to be random.

Uncertainty of predictions pertains to all sciences, but accurate prediction in physics and chemistry are based upon large 'clean' samples, whereas samples of the earth scientist are very small, and each sample may be considered singular if not unique (Nairn, 1965). There is, therefore, singularity of form and process in location and time.

Sensitivity

One aspect of singularity noted earlier but which must be treated separately is the sensitivity of landscape components (Brunsden and Thornes, 1979; Wolman and Gerson, 1978). Sensitivity is the susceptibility of landforms to change. For example, mass movement, gully formation, and changes of river pattern may be triggered by relatively minor changes of a controlling variable if the geomorphic conditions have developed to conditions of incipient instability (Schumm, 1977). Therefore, a minor input may cause a major change at one location, but elsewhere it may have little or no effect (Begin and Schumm, 1984). The major change, when viewed by the geomorphologist, could be interpreted to be the result of a climatic or baselevel change, which would be incorrect.

The reason for such variable response is the existence of threshold conditions, which when exceeded produce a large change (Schumm, 1979). Thus, within a landscape composed of singular landforms there will be sensitive and less sensitive landforms. The sensitive landforms will respond significantly to perhaps even a minor change or even to a large hydrologic event (Begin and Schumm, 1984). For example, sediment may accumulate in a valley or channel until it is incised and removed during an apparently normal hydrologic event. The accumulation of sediment had reached a threshold of gradient instability. In a series of meanders one may increase in amplitude to a condition where cut off is inevitable. That meander was sensitive. The others were not, although all were singular.

At locations b, c, and d on Figure 12.10B there are river pattern or sinuosity thresholds; obviously a channel that plots on this diagram near b, c, or d is sensitive. Those plotting near a or e are insensitive to pattern changes.

Complexity

The final problem is the complexity of the response of a geomorphic system. The complex system itself when interfered with or modified is unable to adjust in a progressive and systematic fashion (Schumm, 1977). The movement of a wave of rejuvenation into a drainage system, for example, will affect downstream reaches long before the upstream reaches are affected. This will create a situation in which it is very difficult for a given reach of a channel or a given tributary to adjust progressively. In fact, there will be hunting for a new condition of stability (Figure 12.11A), which is referred to as complex response (Schumm, 1977). In high energy systems the behaviour can even be

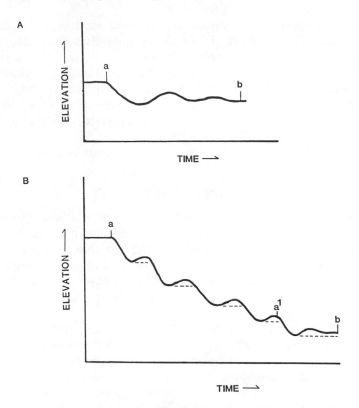

Figure 12.11 Diagrams illustrating (A) complex response and (B) episodic erosion. In each case a stream is affected at time a by a climate, baselevel or land-use change that induces degradation. When the impact is relatively small, (A), the stream degrades, aggrades, and degrades until a new condition of relative stability is achieved at time b. When the change is large (B), and large quantities of sediment are moved, the major degradation is episodic being interrupted by periods of aggradation until relative stability is achieved at time b. Area above dashed lines indicate extent of aggradation. The complex response of (A) occupies the space between a¹ and b on the diagram of episodic erosion (B) (from Schumm, 1985)

episodic with periods of aggradation interrupting degradation (Figure 12.11B) until a new condition of stability has been achieved (Schumm, 1977). This produces a very complex geomorphic and stratigraphic record, the details of which cannot be attributed to external influences but rather to the adjustment of the system itself.

For example, the incision of streams crossing the Canterbury Plain in New Zealand into glacial outwash, and the incision of Sierra Nevada streams into hydraulic mining debris has produced 'degradational terraces'. These unpaired terraces reflect pauses in down-cutting as the channels became clogged with sediment. For example, some tributaries of the Bear and Yuba rivers in California have as many as 12 unpaired terraces, all of which formed since 1880 (Wildman, 1981). These are only landscape details, but careful evaluation of a responding stream is required because aggradation may follow the expected degradation, as a natural result of increased sediment movement (Womack and Schumm, 1977). Episodic behaviour (Bergstrom and Schumm, 1981) may only be characteristic of high-energy fluvial systems.

DISCUSSION

The purpose of science is to accumulate knowledge with the ultimate goal being prediction. A historical science has the goal of both prediction and postdiction because both the future, present and past must be understood. Just as it is necessary to understand history in a social and political sense so it is necessary to understand landform change and response to change during the geologic past in order to understand existing landforms and to predict future change.

The understanding of river behaviour tells us that with an increase of discharge, the channel width, depth, and meander dimensions will increase and gradient will decrease. If bedload (sand and gravel) increases channel width and meander dimensions will increase but sinuosity will decrease as gradient and width-depth ratio increase. However, the seven problems demonstrate why prediction in geomorphology is of low resolution. This discussion of seven problems is not an attempt to discourage prediction. Rather, it can be construed as encouragement for it. If problems are recognized then they can be solved (Schumm, 1985). The assumption is that the uncertainty can largely be removed by careful study of landforms.

General principles and concepts developed from field and experimental studies should be brought to bear on every situation, but quantitative relations must be developed for each location, and usually they cannot be extrapolated from the source of data used to develop the relations. Therefore, this requires additional data collection, field work, surveying, sampling, and analyses. Consideration of the problems may, therefore, be time-consuming, difficult, and expensive, but never as expensive as failure. The development of the required understanding of landform morphology and behaviour will be intellectually rewarding, and it will be cost effective.

There is no obviously preferred scale that can be used for the study of the fluvial system. However, Klemes (1983) states that 'It is natural that we have the best grasp of things which are within the "human-scale", i.e., accessible to us directly through our unaided senses: roughly from 1/10 of a millimeter to a few kilometers in space and from 1/10 of a second to a few decades in time.

Table 12.2 Variables affecting geomorphic hazards and the site risk associated with each hazard. An X indicates a hazard of concern and the sites that may be at risk for a change of variables and through the passage of time (from Schumm *et al.*, 1982)

Geomorphic hazards	Time	Discharge +	–	Sediment load +	–	Base-level up	down	1000 years surface	1000 years sub-grade	10,000 years surface	10,000 years sub-grade	100,000 years sub-grade	deep burial
1 Drainage networks													
(a) Erosion													
(i) rejuvenation		X			X		X	X	X	X	X	X	X
(ii) extension		X			X		X	X	X	X	X	X	
(b) Deposition													
valley filling				X		X		X		X			
(c) pattern change	X												
capture					X	X	X	X	X	X	X	X	X
2 Slopes													
(a) Erosion													
(i) denudation-retreat	X	X			X	X	X	X		X		X	
(ii) dissection		X			X	X	X	X	X	X		X	
(iii) mass failure	X	X			X	X	X	X		X		X	
3 Channels													
(a) Erosion													
(i) degradation (incision)		X			X		X	X	X	X	X	X	X
(ii) nickpoint formation and migration	X	X			X		X	X	X	X	X	X	
(iii) bank erosion	X	X			X	X	X	X		X			
(b) Deposition													
(i) aggradation				X	X	X		X		X			
(ii) back and downfilling				X	X	X		X		X			
(iii) berming				X				X		X			
(c) Pattern change													
(i) meander growth and shift	X							X		X			
(ii) island and bar formation and shift	X			X				X		X			
(iii) cutoffs	X	X		X		X	X	X		X			
(iv) avulsion		X		X			X	X	X	X		X	X
(d) Metamorphosis													
(i) straight to meandering		X			X		X	X		X			
(ii) straight to braided				X	X	X	X	X		X			
(iii) braided to meandering		X			X	X	X	X		X			
(iv) braided to staight				X	X	X	X	X		X			
(v) meandering to straight		X			X	X	X	X		X			
(vi) meandering to braided				X	X		X	X		X			

Table 12.2 (*continued*)

	Variables							Risk					
Geomorphic hazards	Time	Discharge +	−	Sediment load +	−	Base-level up	down	1000 years sub-surface	1000 years sub-grade	10,000 years sub-surface	10,000 years sub-grade	100,000 years sub-grade	100,000 years deep burial
4 Piedmont and coastal plains													
(a) Erosion													
dissection		X		X	X	X		X	X	X	X	X	X
(b) Deposition													
(i) aggradation				X	X	X			X	X			
(ii) progradation	X					X			X	X			
(c) Pattern change													
(i) development of pattern		X				X		X	X	X	X	X	
(ii) avulsion	X	X		X		X			X	X			

For the processes at this scale we have an intuitive feel ...'. In fact, the scale used should depend upon the problem under consideration, which will determine the component of the fluvial system or the size of the unit studied. Therefore, many scales of size and time can be used depending upon what aspect of the fluvial system is being considered. Table 12.2 provides an example. It is an attempt to indicate landform change as time passes, as discharge increases or decreases, as sediment load increases or decreases, or as baselevel rises or falls. The purpose of the table is to suggest how important a variety of geomorphic hazards is for radioactive waste disposal during three time periods for three modes of disposal, on the surface, subsurface, and by deep burial. The hazards grouped under drainage networks and slopes are in Zone 1, channel hazards are Zone 2, and piedmont and coastal plain hazards are Zone 3. The table is presented here to provide final emphasis on the complexity and dynamics of the fluvial system in both space and time.

REFERENCES

Adams, J. (1980). Active tilting of the United States midcontinent: Geodetic and geomorphic evidence: *Geology,* **8,** 442–446.

Arnett, R. R. (1979). The use of differing scales to identify factors controlling denudation rates. In Pitty, A. F. (Ed.) *Geographical Approaches to Fluvial Processes*, pp. 127–147. Geobooks, Norwich UK: 300 pages

Baker, V. R., and Penteado-Orellana, M. M. (1977). Adjustment to Quaternary climatic change by the Colorado River in central Texas. *Jour. of Geol.,* **85,** 395–422.

Begin, Z. B., and Schumm, S. A. (1984). Gradational thresholds and landform singularity: Significance for Quaternary Studies. *Quaternary Research,* **31,** 267–274.

Bergstrom, F. W., and Schumm, S.A. (1981). Episodic behaviour in badlands. *Internal. Assoc. Hydrological Sci. Pub.,* **132,** 478–489.

Brunsden, D., and Thornes, J. B. (1979). Landscape sensitivity and change. *Inst. British Geog., Trans.,* **4,** 463–484.

Burnett, A. W., and Schumm, S. A. (1983). Neotectonics and alluvial river response. *Science,* **222**, 49–50.

Carey, S. W. (1962). Scale of geotectonic phenomena. *Geol. Soc. India Journal,* **3**, 97–105.

Chorley, R. J., and Kennedy, B. A. (1971). *Physical geography: A systems approach.* Prentice-Hall, NY: 370 pages.

Cullingford, R. A., Davidson, D. A., and Lewin, J. (1980). *Timescales in geomorphology.* John Wiley and Sons, New York: 360 pages.

Fisher, A. G., (1969). Geological time–distance rates: The Bubnoff unit. *Geol. soc. America, bull.,* **80**, 549–552.

Gleason, H. A. (1926). The individualistic concept of the plant association. *American Midland Naturalist,* **21**, 92–100.

Graf, W. L. (1982). Spatial variation of fluvial processes in semi-arid lands. In Thorn, C. E. (Ed.) *Space and Time in Geomorphology,* pp. 193–217. George Allen & Unwin, London.

Gregory, K. J., and Gardiner, V. (1975). Drainage density and climate. *Zeit. Geomorph.,* **19**, 287–298.

Harvey, D. (1969). *Explanation in geography.* Edward Arnold, London: 521 pages

Klemes, V. (1983). Conceptualization and scale in hydrology. *Jour. Hydrology,* **65**, 1–23.

Langbein, W. B., and Schumm, S. A. (1958). Yield of sediment in relation to mean annual precipitation. *American Geophys. Union Trans.,* **39**, 1076–1084.

Morgan, R. P. C. (1973). The influence of scale in climatic geomorphology. A case study of drainage density in West Malaysia. *Geogr. Annaler,* **55A**, 107–115.

Nadler, C. T., and Schumm, S. A. (1981). Metamorphosis of South Platte and Arkansas Rivers, eastern Colorado. Physical Geography, **2**, 95–115.

Nairn, A. E. M. (1965). Uniformitarianism and environment. *Palaeogeography, Palaeoclimatology, Palaeoecology,* **1**, 5–11.

Penning-Roswell, E. C., and Townshend, J. R. G. (1978). The influence of scale on the factors affecting stream channel slope. *Inst. British Geogr. Trans.,* **3**, 395–415.

Pitty, A. F. (1982). *The Nature of Geomorphology.* Methuen, London, 161 p.

Richards, K. (1982). *Rivers, Form and Process in Alluvial Channels.* Methuen, London: 358 pages.

Schumm, S. A. (1977). *The Fluvial System.* John Wiley and Sons, New York: 338 pages

Schumm, S. A. (1979). Geomorphic thresholds: the concept and its applications. *Inst. British Geogr. Trans.,* **4**, 485–515.

Schumm, S. A. (1981). Evolution and response of the fluvial system: Sedimentologic implications. *Soc.-Econ. Paleon. Mineral. Spec. Pub.,* **31**, 19–29, see p. 26.

Schumm, S. A. (1985). Explanation and extrapolation in geomorphology. Seven reasons for geologic uncertainty. *Japanese Geomorph. Union. Trans.,* **6**, 1–18.

Schumm, S. A., and Hadley, R. J. (1957). Arroyos and the semi-arid cycle of erosion. *Amer. Jour. Science,* **225**, 161–174.

Schumm, S. A., and Lichty, R. W. (1965). Time, space and causality in geomorphology. *American Jour. Science,* **263**, 110–119.

Schumm, S. A., and Khan, H. R. (1972). Experimental study of channel patterns. *Geol. Soc. America Bull.,* **83**, 1755–1770, see p. 1767.

Schumm, S. A., Costa, J. E., Toy, T. J., Knox, J. C., and Warner, R. F. (1982). *Geomorphic Hazards and Uranium Tailings Disposal in Management of Wastes from Uranium Mining and Millings.* International Atomic Energy Agency (Vienna) 1AEA-SM-262, p. 111–124.

Starkel, L. (1983). Climate change and fluvial response. In Gardner, R. and Scoging, H. (Eds.) *Mega-geomorphology*, pp. 195–211. Oxford Univ. Press, Oxford.

Thornes, J. B., and Brunsden, D. (1977). *Geomorphology and time*. Halstead Press, Wiley, NY: 208 pages.

Wildman, N. A. (1981). *Episodic Removal of Hydraulic-Mining Debris, Yuba and Bear River Basins, California*. Unpublished M.S. thesis, Colorado State Univ., 107 pages.

Wolman, M. G., and Gerson, R. (1978). Relative scales of time and effectiveness of climate in watershed geomorphology. *Earth Surface Processes,* **3**, 189–208.

Womack, W. R., and Schumm, S. A. (1977). Terraces of Douglas Creek, northwestern Colorado: An example of episodic erosion. *Geology,* **5**, 72–76.

CHAPTER 13

Estuaries—Their Natural and Anthropogenic Changes

STEPHAN KEMPE

SCOPE/UNEP International Carbon Unit,
Geological–Palaeontological Institute University of Hamburg,
Bundesstr. 55, D-2000 Hamburg 13,
Fed. Rep. of Germany

ABSTRACT

Estuaries are defined as embayments where freshwater and seawater mix. During mixing, chemical reactions may or may not occur, i.e. certain components mix conservatively from fresh to salt water, while others undergo complicated reactions, exchanges, sedimentation, or are consumed or liberated by biological activity. Estuaries are under constant change: this follows from tidal, diurnal, seasonal, or glacial–interglacial cycles. However, anthropogenic change is of more importance today. Not only is the chemistry, suspended matter load, and water discharge of the input to the estuary modified, but dredging and damming has changed many estuaries completely since the beginning of industrialization. Estuaries are extremely diverse phenomena, defying generalization. Thus one example was chosen to illustrate anthropogenic impact on an estuary: The Elbe River and the inner German Bight. Here the behaviour of nutrients, organic substances, and CO_2 pressure is discussed in more detail for the inner mixing zone and the long-term effects on processes and sediments in the adjacent North Sea are highlighted.

INTRODUCTION

Estuaries are of prime importance for mankind. Since the development of sea going ships in the second millenium B.C. they provided linkage of national and international commerce. In turn, the settlements near such harbours developed to become centres of power and wealth, such as Rome, Venice, Lisboa, Amsterdam, London, New York. Later, industrialization called for places of easy access where large amounts of goods could be stored, processed, and distributed. Thus many of the industrial and population centres of the world are found at estuaries today and economic forces have shaped almost any estuary.

For the the natural scientist estuaries are gigantic mixing vessels for waters of various biological, thermal, hydrochemical, and suspended matter characteristics. Nowhere else do waters of principally different origins mix as thoroughly and rapidly as in estuaries. The number of estuaries fringing the continents is large, every river outlet to the sea qualifies. Cameron and Pritchard (1963) define an estuary as a 'semi-enclosed coastal body of water which has a free connection with the open sea and within which seawater is measurably diluted with fresh water derived from land drainage'.

One can classify estuaries according to their geological history, their morphology, their hydrophysical mixing pattern, their climatic situation, their biological activity, their tidal response, their degree of pollution, or to other criteria depending on one's interest.

GENERAL CONSIDERATIONS

The shape of present day estuaries is essentially determined by the late Glacial and Holocene sea level history and by the amount of sediment the river or rivers feeding the respective estuaries carry (Figure 13.1). Rivers with relatively small sediment yields and high tidal energy will tend to form open channels which allow seawater to ingress far inland. Rivers with high sediment yields or which are exposed to low tides tend to form deltas. In deltas the shoreline is built outwards by newly deposited sediment much faster than coastal currents or waves can erode it. Also local subsidence of crust plays a role in determining the kind of estuary a river forms: rising areas will sustain delta growth, subsiding areas may cause ingression of seawater even to a point to drown former deltas.

The ratio of estuary volume to the headwater discharge determines the residence time of freshwater in the estuary. Residence times range from a few hours (large river deltas) to years (fjords with small tributary areas) and may vary during the seasons by an order of magnitude or more.

The physical mixing of seawater with fresh water in the estuary may produce various forms of density gradients (Figure 13.2). Depending on the height of the tidal wave, the depth of the channel and the river discharge, the estuary may either be vertically mixed, more or less stratified, or of a salt wedge character.

Mixing of dissolved or particulate components is called conservative, if their concentration changes in direct proportion to the increasing salinity between their fresh water and seawater concentrations (Figure 13.3, top). Deviation from this conservative mixing line indicates either subtraction from or

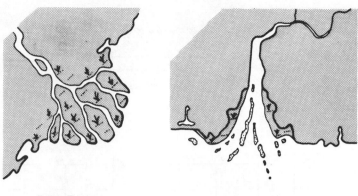

DELTA

- large sediment yield

- stable or rising coast

- sea level declining

- low tidal wave

ESTUARY

- low sediment yield

- subsiding coast

- sea level rising

- high tidal wave

- original channel formed

by glaciers or formerly

larger rivers

Figure 13.1   In geomorphology distributary river mouths are termed deltas. Those consisting only of a single channel are termed estuaries. Geochemically any enclosed reach in which river and seawater mix is an estuarine environment, thus deltas consist of many individual estuaries

addition to the phase under consideration. These additions or subtractions may be caused by thermodynamically governed mineral equilibria, by adsorption or desorption processes, by sedimentation, erosion or by biological activity. The same compound may behave differently in different estuaries. It may even alter its behaviour throughout the seasons in the same estuary. Such a compound is, for example, dissolved SiO_2 which in estuaries of large residence times and reasonable clear water may be depleted in summer by diatoms or silicoflagellates while it shows a truly conservative behaviour during the cold season or in estuaries which are too turbid to sustain a significant phytoplankton activity (Boyle *et al.*, 1974).

Concentration of suspended matter (SM) generally does not behave conservatively. Seawater carries much lower SM-concentrations than most river waters, thus its concentration should be diluted seawards (Figure 13.3, bottom). As the estuary widens up towards the open sea, one may expect a

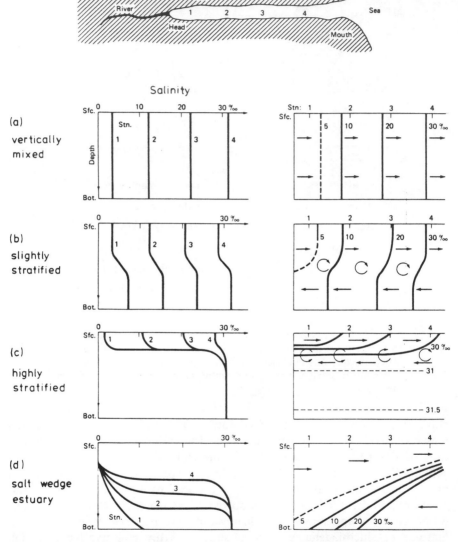

Figure 13.2   Salinity/depth diagram (left) and longitudinal salinity sections (right) for various hydrodynamic types of estuaries, illustrated for four sampling stations (after Pickard, 1975)

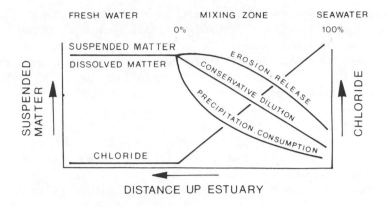

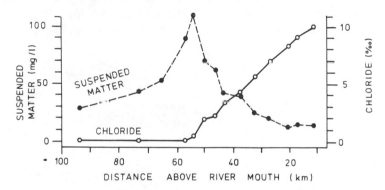

Figure 13.3 Top: General scheme of conservative and non-conservative mixing of a component carried by freshwater in a higher concentration than in seawater (the alternative being also possible). Bottom: The actually measured distribution of suspended matter (surface of York River, Virginia, 24 March 1960, modified after Nelson, 1960, quoted from Meade, 1972) does not follow any of the mixing lines, indicating that its concentration distribution is governed by hydrodynamical processes

decrease in flow velocity and thus a settling of particles. Furthermore, the increase of salinity, as has been shown by many laboratory experiments, should cause the flocculation of clay minerals very early in the mixing process (Krone, 1962; Whitehouse *et al.*, 1960). However, measurements very often show a turbidity maximum stradelling the point up to where a salinity increase can be measured. As Meade (1972) points out, salinity flocculation, even though often quoted as being the cause of the turbidity maximum, cannot explain this observed SM-concentration pattern. Also, rivers heavily polluted by waste salts should not show any turbidity maximum at all. The Weser in

Germany is such a river which carries a salinity of 4‰ but nevertheless displays a distinct turbidity maximum (Wellershaus, 1982). There is a very good reason why salt flocculation is not observed in the real estuary. This is due to the ubiquity of organic coatings. Rivers do not carry pure clay mineral suspensions, but rather the suspended material is lumped together in flocs. The glue of these flocs may be organic molecules such as mucopolysaccharides. Floc sizes can measure up to 1 mm and may be limited by the size of turbulence (Eisma, 1988). Flocs can be made visible by in-situ photography or by careful in-situ filtration avoiding shock of samples (Eisma *et al.*, 1983). Laboratory grainsize analysis therefore never gives useful information about size distribution of SM under natural conditions.

The causes of the turbidity maximum seem to be hydrodynamic in nature. Details of the hydrodynamics of the estuary with regards to SM-concentrations are given, for example, by Officer (1976, 1981). Because the fresh water barotropically flows out on top of the baroclinically intruding salt water, material which settles from the fresh water finds itself going inland again

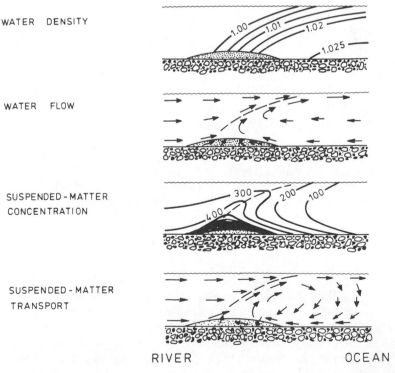

Figure 13.4   Scheme of hyrodynamic conditions in an estuary showing density, water flow, suspended matter concentration, and suspended matter transport (after NEDECO, 1965)

(Postma, 1967). Thus a sort of horizontal eddy is formed trapping both fresh water and seawater suspensions (Figure 13.4). Numerical modelling has indeed shown that this process can cause an enrichment of suspended matter at the tip of the salt wedge (Festa and Hansen, 1978; Officer and Nichols, 1980). At the turbidity maximum sedimentation of fines is to be expected. However, things seem to be even more complicated in tidally ifluenced estuaries. Observations show (Eisma *et al.*, 1983; 1985) that marine suspended matter moves upstream far beyond the tip of the salt wedge into the fresh water tidal reach. One reason for this transport is the asymmetry in flow velocity of the tides and the effect of neap and spring tide cycles (Postma, 1961; Officer, 1981; Allen *et al.*, 1980; Castaing and Allen, 1981). In Figure 13.5 the prolongation of the ebb period is shown as one proceeds upstream. Consequently the flood period becomes shorter but its flow velocity increases. Thus bottom sediments become more effectively eroded during flood than during ebb phases causing a successive deplacement of sediment upstream.

Even though suspension concentration behaves non conservatively throughout the estuarine mixing process, composition of SM may behave conservatively. The SM composition with regard to its mineral matter is mostly

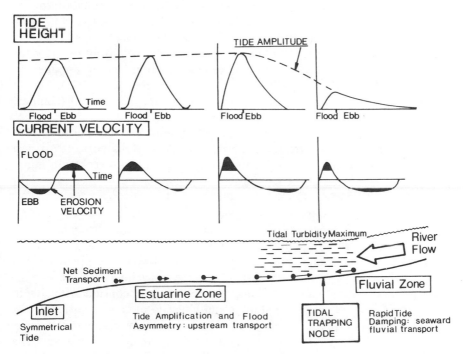

Figure 13.5   Scheme of asymmetry effect of tides in an estuary with high tidal wave on upstream transport of suspended matter and its accumulation in a turbidity maximum (from Allen *et al.,* 1980)

explained by a mixture of a fresh water and a marine source (Eisma, 1988). Clay minerals pass the estuary with hardly any changes. Also composition of coatings with regard to heavy metals and their speciation does not undergo marked changes apart from the mixing of two different sources (Schoer and Eggersgluess, 1982). These conclusions are substantiated by a survey of the sediments of the Delaware estuary (Bopp and Biggs, 1981). Factor analysis of 18 parameters measured on 119 samples (size fraction below 63 μm) indicated three sources of sediments: Factor I (50.9% of total variance) composed of Fe, Mg, K, Al, and Li signals the presence of Holocene marsh deposits. Factor II (24.1%) contains Sr, Mg, Ca, and Na and represents the marine fine material increasingly dominating seawards. Finally, Factor III groups Cu, Cr, Hg, Pb, and organic matter linked to river input into the bay. Total concentration of heavy metals (natural and anthropogenic) is depending on the grain size distribution (Schoer *et al.*, 1982). Most heavy metals are transported bound to clay surfaces or associated with grain coatings and are concentrated almost quantitatively in the size fraction below 6.3 μm. Thus, comparison of total concentrations is meaningful only if restricted to certain grain size fractions or normalized to certain background elements (Al, Sc).

Flocculation of dissolved iron has been observed both in the laboratory as well as by field studies in many estuaries causing non-conservative concentration curves indicating second order kinetics (see Fox and Wofsy, 1983, for further quotations). However, amounts thus removed from solution are small compared to the amount of iron present as grain coatings and do not alter the composition of the SM measurable (Schoer and Eggersgluess, 1982). Similar findings are reported for other elements such as Mn, Cu, Zn, Al (compare Sholkovitz, 1976).

Laboratory flocculation of organic matter is reported to amount to 3–11% for Scottish rivers which are rich in humic colloids (Sholkovitz, 1976). For most rivers flocculation of dissolved organic material seems to play a minor role only, i.e. its effect may be so small that it is masked by the accuracy of the dissolved organic carbon (DOC) determination. The mixing curve for DOC thus appears to be conservative (compare Figure 13.14) (Eisma *et al.*, 1982; Ittekkot *et al.*, 1982). However, the role of organic matter flocculation for removal of dissolved inorganic trace constituents may not be neglected.

Once the estuarine water, now of a brackish, intermediate salinity, has left the morphological confinements of the estuary it mixes into the waters of the adjacent sea. Depending on the volume of the river under consideration, depth and width of the shelf, wind conditions and intensity of coastal currents, and tides the plume of the river may vary considerably in size and structure. Ship-based studies can define such a plume only very roughly because it is a highly dynamic object, but satellite and aircraft images can give real time information of high resolution (Szekielda, 1982; Szekielda *et al.*, 1983a). It appears that mixing occurs by interfingering of water masses, formation of patches, and upwelling. The patches seem to show a continuous size distribution from entities smaller than 100 m to more than 50 km across with observed life times of up to several days (Szekielda and McGinnis, 1985a). Long life times of these plume structures enable phytoplankton to thrive. Thus one can discern in many plumes two zones, an inner one where sediment load from the river plays the more important role for the reflectivity of the water and an outer one where the chlorophyll of living plankton causes turbidity. Mathematical methods are now available to extract these optical properties from the reflectivity data of aircraft and satellite measurements and translate them into more common units like mg SM/l (Fischer, 1983). Such a double zonation has been shown to exist in the plume of the Amazon (Szekielda *et al.*, 1983a; Szekielda, 1985) and the Rio de la Plata (Szekielda *et al.*, 1983b). Satellite investigations can also yield estimates of overall seasonal variation of plumes and their relative productivity (Szekielda and McGinnis, 1985a, b).

SYNOPSIS OF CHANGES

Any density and velocity field in an estuary is but of very transient nature. Tidal forces induce a 12 hour cycle, seasons modify the mixing process with relation to discharge, temperature and chemical composition of inflow, and kind and magnitude of biological activity. Further, interannual and long-term processes occur such as erosion and deposition, sea level changes, natural changes in the hydrological and climatic settings, and anthropogenic changes with regard to dredging, construction of dams and harbours, and pollution of various sorts. If we could gather a data record of all possible measurable parameters in the estuarine system spaced at short term intervals for the life time of that estuary (i.e. in cases for large tropical rivers up to several million years) and then run a frequency analysis, we would obtain a diagram similar to Figure 13.6. This diagram illustrates that processes of very short-term frequencies and processes of long-term frequencies shape the estuaries. Most estuaries are not older than the Holocene sea level rise, but some of the large tropical rivers occupy the same outlet over consecutive glacial cycles. During Glacial times, they cut deep canyons which are filled up during Interglacial times (Amazon, Parana, Orinoco, Zaire, Indus, etc.). Thus more material reaches the shelf slope during the Glacials than during the Interglacials per unit of time. In the Figure the Holocene sea level rise is given with a dashed curve because it is for most estuaries a sort of mono event. This is especially true for those estuaries which formed in areas formerly covered by ice and where drainage patterns were established anew by the glaciation. Also, the impact of human activity which since about 150 years is felt to a larger extent in estuaries is not a cyclic event and thus dashed in the graph. True cyclic events are seasons, neap-spring tides, day-night cycles, tides, and velocity changes caused by tides. Storms also play an important role if one thinks about the

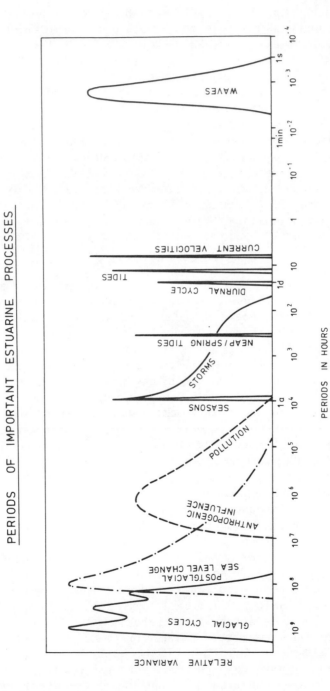

Figure 13.6 Synopsis of periods of short-term and long-term estuarine processes

catastrophes which struck the North Sea coast in former centuries drowning islands and cutting large bays within a few days. The same is true for tropical storms which devastate low-lands with a certain seasonal regularity (Bangladesh, 1985, for example). Tsunamis are another event which shape estuaries. They, however, do not occur at any preset periodicity, but strike irregularly. At the lower end of our scale we find waves which oscillate at the order of seconds. They are important for beach transport and for the gas exchange (Broecker *et al.*, 1978; Broecker *et al.*, 1980).

ANTHROPOGENIC CHANGES OF ESTUARIES

Changes of sea level

During the past century the sea level rose by about 12 cm. The long-term Holocene sea level trend amounts to 2 cm, which suggests that modern sea level rise is significantly different from natural background (Gornitz *et al.*, 1982) (Figure 13.7). This fact causes serious concern as one suspects that the rise is associated with the slight global warming observed for the same period. If model calculations prove to be correct, then the anthropogenic input of CO_2 and other trace gases to the atmosphere may cause a rise in temperature of about 2 °C until the middle of the next century (Grassl *et al.*, 1984). Gornitz *et*

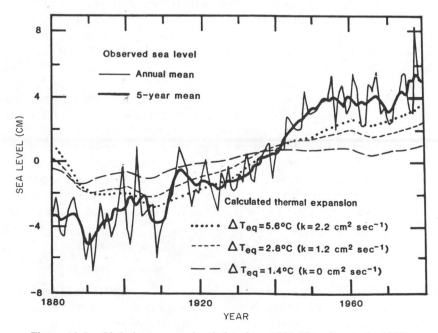

Figure 13.7   Global mean sea level rise since 1880 (Gornitz *et al.,* 1982)

al. (1982) suggest that the sea level may rise by about 40 to 60 cm until then. Half of this increase may be attributed to the thermal expansion of the ocean warm surface layer, half of it may be due to the melting of glaciers.

A rise of half a metre or more would cause serious problems in many low-land estuaries. The first counter measure to such an increase would be raising dykes. Industrialized countries, which already have substantial dykes along their shore lines, even defending agricultural areas below sea level successfully, will probably be able to cope with such a challenge. Along the North Sea coast of the Netherlands and Germany dykes have been raised to 6–8 m above mean sea level. Danger does probably not arise form an increase in sea level alone as only severe storm surges can break these battlements. Furthermore, changes in tide levels have been noticed along the German coast in a magnitude much larger than the global sea level changes. Since the last century the mean low water is falling (Siefert, 1982) but the amplitude of the tide and the frequency of storm surges is increasing in the estuaries (Plate, 1983). For the station St. Pauli, Hamburg, these two trends amount to roughly 80 cm and 140 cm since 1900, respectively. These trends are, in part, man-made. The deepening of the estuary for shipping, the dyking of marshes and other measures have contributed to these trends. However, climatic and stochastic causes must be assumed in addition because the trends are the same for all German rivers regardless of their development (Siefert, 1982). Numerical modelling shows that storm surges depend on direction and force of wind much stronger than on morphometric factors (Sündermann and Zielke, 1983).

Sea level changes will cause much higher damage in developing nations. Millions of people live in delta regions practically at sea level (Nile delta, Indus delta, Bangladesh, Thailand, Mekong delta, Changjiang delta). Building effective dykes will, in many cases, be impossible due to lack of capital. However, even if the capital could be found, it might not be economically feasible to build the dykes at all. The advance of brackish groundwater inland may cause salinization of the ground, thus preventing effective agricultural use of the saved land. Coastal settlements will also suffer, especially in those areas, where no tides occur and people are not used to defending themselves against storm surges and inundations and where half a metre of sea level difference would cause the loss of protective beaches, mangroves, or lagoon barriers. However, if the increase occurs slowly enough, settlements can be rearranged within the normal life time of such structures.

Changes in the riverine input to the estuary

There is hardly any river whose water shed is not severely affected by man's activity. Activities altering the sediment load of a river also alter the natural state of its estuary. Man has either increased the load of sediment or decreased it. Deforestation and the advance of agriculture are usually connected with large increases in the erosion rate: in the Black Sea sedimentation rate increased significantly starting 1500 years ago, when more and more of the Russian plains became farmlands (Degens *et al.*, 1980). Meade (1982) estimates, that the erosion in the east coast area of the United States increased by a factor of 10 at the arrival of European farmers. These increases lead to the advanced siltation of channels and to the accelerated growth of deltas.

The opposite effect is caused by building dams on major streams. The diversion of water for irrigation and industrial purposes not only decreases the discharge (the Columbia, for example, lost 20% of its runoff since 1874; Kempe, 1982a), but the reservoirs behind the dams also serve as sediment traps. The most spectacular example is the Assuan High Dam on the Nile (Figure 13.8). Since its closure in 1965 the ecology, hydrochemistry and SM loads of the Nile have dramatically altered (Kempe, 1983). Before the dam closure more than 50×10^6 t/yr of SM passed the Delta Barrage, while today only 4×10^6 t/yr are monitored (Schamp, 1983). Consequences for the Nile delta are grave: fishery declined, coastal erosion set in, inundations threatened, and fields were ruined by salts and changing drainage.

In most rivers, the hydrochemistry is similarly out of balance. Rivers which are chemically unaffected by civilization are rare today as can be seen from the three volumes of documentation on river chemistry produced by the

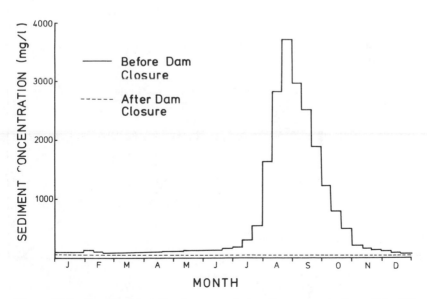

Figure 13.8 Seasonal variation in suspended matter concentration of the Nile at Gaafra (35 km below Assuan) before and after the closure of the Assuan High Dam in 1965 (Schamp, 1983)

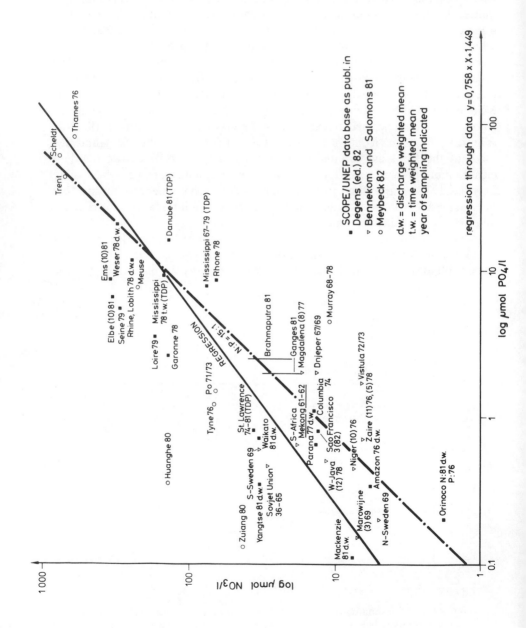

The plot axes and labels (rotated):

log μmol NO₃/l (vertical axis: 1000, 100, 10)

log μmol PO₄/l (horizontal axis: 0.1, 1, 10, 100)

Data points labelled: Thames 76, Scheldt, Trent, Ems (10) 81, Weser 78 d.w., Elbe (10) 81, Seine 79, Rhine, Lobith 78 d.w., Meuse, Danube 81 (TDP), Mississippi 78 t.w. (TDP), Loire 79, Garonne 78, Mississippi 67–79 (TDP), Rhone 78, Tyne 76, Po 71/73, REGRESSION, N:P = 15:1, Brahmaputra 81, Ganges 81, Magdalena (8) 77, Murray 68–78, Dnjeper 67/69, St. Lawrence 74–81 (TDP), Waikato 81 d.w., S-Africa, Mekong 61–62, Columbia 74, Vistula 72/73, Huanghe 80, Parana 77 d.w., Sao Francisco 3 (82), Zaire (11) 76, (5) 78, Zuiang 80, S-Sweden 69, W-Java (12) 78, Niger (10) 76, Amazon 76 d.w., Yangtse 81 d.w., Sovjet Union 36–65, Marowijne (3) 69, N-Sweden 69, Mackenzie 81 d.w., Orinoco N:81 d.w. P:76

■ SCOPE/UNEP data base as publ. in
   Degens (ed.) 82
▽ Bennekom and Salomons 81
○ Meybeck 82

d.w. = discharge weighted mean
t.w. = time weighted mean
year of sampling indicated

regression through data y = 0,758 × X + 1,449

SCOPE/UNEP Project 'Transport of Carbon and Minerals in Major World Rivers' (Degens, 1982; Degens *et al.*, 1983, 1985; Kempe, 1984). Spectacular is the increase in nutrient loads with the advance of intensive agricultural techniques in recent years. Figure 13.9 relates average phosphate and nitrate concentrations of major world rivers. Rivers from industrial regions display concentrations two orders of magnitudes higher than rivers from pristine tropical and arctic areas. It is interesting to note that, on average, pollution of nitrate and phosphate occurs at a mole ratio close to 15 : 1. This is the rate with which marine plankton uses these two elements. Thus rivers can fertilize the estuaries and coastal seas. The Mississippi has become the largest single source of nutrients to the ocean and the Rhine carries already half the nutrient load of the Amazon with only 1% of its discharge (Meybeck, 1982; Kempe, 1982a, 1984). The increase in nutrient load is by no means under control. Figure 13.10 shows the steep increase in the nitrate concentration of the Mississippi from 1963 to 1979. Similar steep increases are recorded for the Rhine where values of 20 $mgNO_3/l$ and 1 $mgPO_4/l$ are common.

Similar to nutrients, the load of organic matter has increased in rivers. Forced erosion delivers particulate humic matter into the rivers and industrial, agricultural, and municipal waste inputs add labile dissolved and particulate organic matter. Parameters such as BOD (biological oxygen demand), COD (chemical oxygen demand), TOC (total organic carbon) and others increased. Added load of organics fuels excess respiration in rivers, pushing their internal CO_2 pressure into ranges ten or twenty times that of the atmospheric pressure. Figure 13.11 displays the long-term longitudinal profile of the PCO_2 of the Rhine. One can note several features on the graph: The upstream source of the turbulent Rhine, Lake Constance, is of low PCO_2 because here phytoplankton consumes nutrients and hence CO_2 in the limnic environment. In the river itself, respiration takes over as phytoplankton cannot grow in the turbid and turbulent Rhine water. Consecutive input of organic matter fuels the respiration further and CO_2 is given off to the atmosphere. In fact, respiration runs at such a high rate, that all oxygen should be consumed were it not for the high nitrate concentration of the Rhine which serves as a second source for oxygen by means of the denitrification reaction (Kempe, 1982a). Thus estuaries of polluted rivers receive waters high in organic substance, high in PCO_2, high in nutrients, and low in oxygen not to talk about heavy metals, chlorinated hydrocarbons, radioactive substances, heat, or other wastes of human productivity.

Figure 13.9 Plot of mean nitrate versus mean phosphate concentrations for various world rivers. Exaplanation of signs: 76, average of 1976; (10)81, single sample October 1981; t.w., time weighted mean; d.w., discharge weighted mean; TDP, total dissolved phosphate; TDN, total dissolved nitrogen (inorganic); (Kempe, 1984)

266

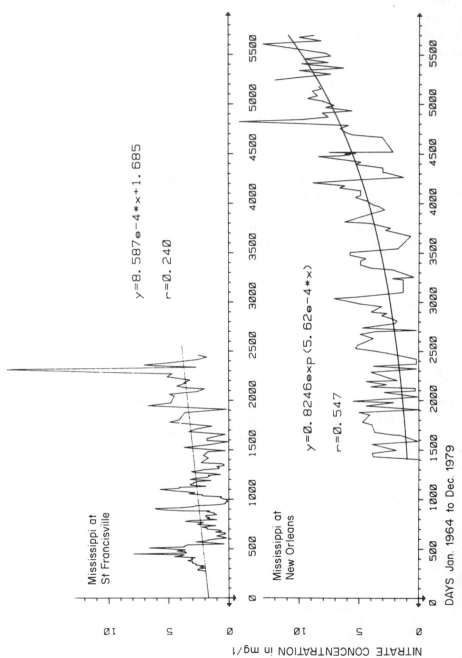

Figure 13.10 Detailed record of the nitrate concentration increase of the Mississippi of two stations shortly above its estuary (1963–1979) (Kempe, 1982a)

267

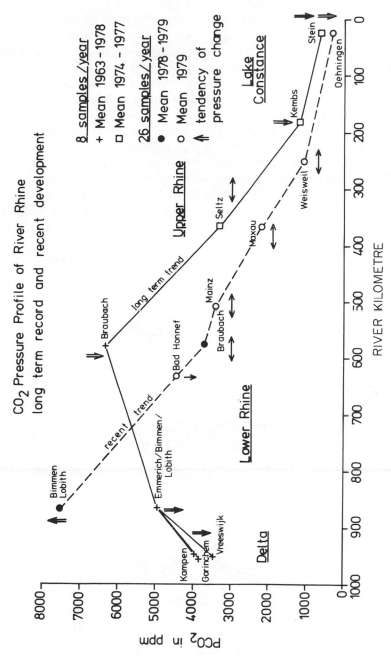

Figure 13.11 Longitudinal PCO₂ profile of the Rhine as calculated from long-term hydrochemical monitoring at 13 stations. Times for which means apply are given. River flow is from right (outflow of Lake Constance) to left (Rhine delta in the Netherlands)(Kempe, 1982a)

Direct alteration of the physiography of estuaries

Not only upstream alterations change the physiography of an estuary, but most estuaries are directly affected by civil engineering. Harbours necessitate the dredging of the main channel down to a depth which is not in equilibrium with sedimentation/erosion. Adjacent marshes are dyked to obtain land for cattle raising, agriculture, or industry and the river looses its storm flood plain. Consequently, tides travel further upstream and intensify. More dykes need to be built and more dredging is necessary to keep moving sediment at bay. Figure 13.12 displays the development of the Seine Estuary within the last 150 years from an open tidal flat estuary to a dyked channel (Avoine *et al.*, 1981). In the mouth of the Rhine annually 10.5×10^6 t of material are dredged out (Eisma, 1988) which is a far larger amount than the total load of the Rhine across the German/Dutch border (3.4×10^6 t SM/yr, average 1966–1973; Kempe *et al.*, 1981). In effect, the dredging battles mainly the net inflow of SM from the North Sea estimated to amount to 13×10^6 t/yr (Eisma, 1988). Such

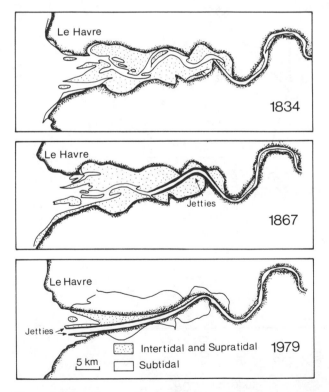

Figure 13.12   Change in the physiography of the Seine Estuary between 1834 and 1979 (from Avoine *et al.*, 1981)

figures illustrate the great magnitude of man's efforts to effectively manage the Rhine mouth, a statement which certainly applies to many estuaries all over the world.

The Elbe River and the German Bight, a case study

The Elbe River, which drains highly industrialized regions in Central Europe, is one of the most polluted rivers on Earth (compare Figure 13.9, data since 1978 available from Arbeitsgemeinschaft für die Reinhaltung der Elbe.) Thus it can serve as an extreme case of anthropogenic change. Thames, Rhine, Weser, and Elbe are the four most important riverine inputs to the North Sea. With additional inputs by dumping of waste and dredge material, oil spills, and input from air pollution the North Sea is probably the most polluted coastal sea (Weichart, 1973). Due to its shallowness (50–100 m) pollutants are moved about for a long period because almost no new sediment is formed to trap these substances. Transport of material is along the German and Danish coast to the north and suspended matter is finally deposited in the Skagerrak where water depths exceed 300 m (Jansen *et al.*, 1979).

Pollution of the Elbe with organic substances and nutrients seems to date back more than 30 years. Data of the Hamburgian Water Works, dating back to the early 1950s, already show high nutrient concentrations. Also the PCO_2, which can be calculated from standard hydrochemical records to give a measure of organic matter respiration if direct measurements of organics are missing, indicates quite high levels of organics throughout the past thirty years (Kempe, 1982a). This behaviour is unlike the Rhine, which shows significant trends of concentrations with time (increases for nutrients throughout the period, increase for organic matter until 1972 and decrease since then).

Figure 13.13 gives a simplified view of the transformations some of the pollutants suffer while passing through the estuarine system. Phosphate, nitrate, and dissolved organic matter seem to pass conservatively through the turbidity maximum zone. Figure 13.14 plots their concentrations with decreasing salinity for measurements made in October 1981. During other times of the year DOC, however, can behave non-conservatively (Seifert, 1985). Also, particulate organic carbon (POC in % of total suspended matter not as absolute content per litre) shows a conservative behaviour (Ittekkot *et al.*, 1982). This is even true for the particulate carbohydrates, which seem to vary conservatively between the fresh water and the seawater suspended matter composition (Lohse and Michaelis, 1983). Particulate carbohydrates suggest that terrestrial plant matter is only a small part of the organic matter present. Rather, living and dead plankton of riverine or marine provenience and bacterial cell walls seem to determine carbohydrate spectra. Seifert (1985) could show that the POC and particulate carbohydrate composition is influenced by seasons (primary productivity) and headwater volume of the

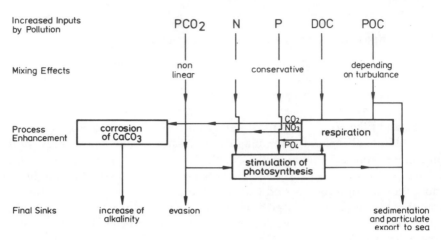

Figure 13.13   Scheme of fluxes and geobiochemical processes for the elements C, N, P in polluted estuaries (Kempe, 1982b)

river and may also—especially in spring—deviate from conservative mixing. Figure 13.15 (distribution of DOC in the Elbe–Weser Plume, Seifert, 1985) shows that the influence of the Elbe River reaches far into the inner German Bight and is directed towards the North (which is also known from satellite pictures). Distribution of salinity is very similar to this DOC pattern. This suggests, that most of the DOC is conservatively mixed into the North Sea water which one can see intruding from the west in the form of water masses low in DOC. How much of the DOC in the plume is derived from nutrient induced in-situ phytoplankton blooms still needs to be shown.

More complicated is the fate of the high PCO_2 of the Elbe water. In a numeric thermodynamic mixing experiment (using the WATMIX model of Wigley and Plummer, 1976) one can calculate how the PCO_2, the pH, and the saturation index of calcite should change if the mixing would be mass-conservative between the river water and the North Sea end members. Figure 13.16 gives the result: due to the high PCO_2 of the Elbe water the river is undersaturated with respect to calcite (the same can be calculated to be true for aragonite and dolomite) but the North Sea water is—as any seawater

Figure 13.14   Plot of concentrations of various parameters versus salinity for the Elbe Estuary. Data of October 1981, R. V. VALDIVIA Cruise (Ittekkot *et al.*, 1982; Kronberg *et al.*, 1982). Note conservative mixing for DOC, POC, phosphate and nitrate and non-conservative mixing for silica

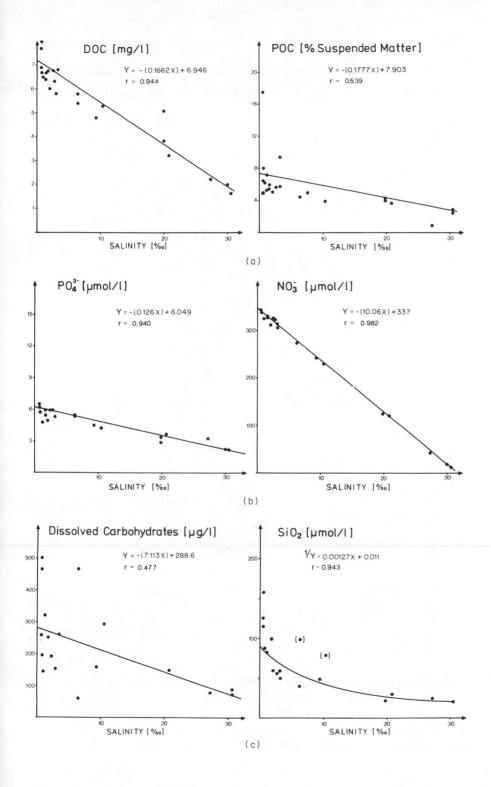

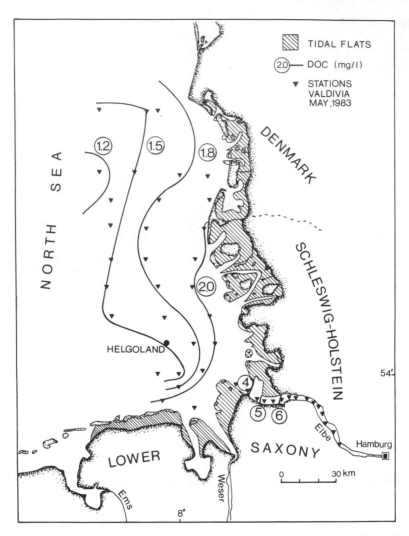

Figure 13.15   Distribution of dissolved organic carbon at the surface of
the eastern half of the German Bight in May 1983 (VALDIVIA Cruise 12).
Note the northward transport of the plume of the Elbe/Weser estuaries
along the coast and bulges of North Sea water lower in DOC intruding into
the plume from the west (Seifert, 1985)

—highly supersaturated. Undersaturation should persist up to a point where
more than 60% seawater has been admixed to the fresh water. This long
persistence of an undersaturation derives from non-linear redistributions
among species of the carbonate system (Wigley and Plummer, 1976) and leads
even to an increase in undersaturation in the innermost estuary, at least in the

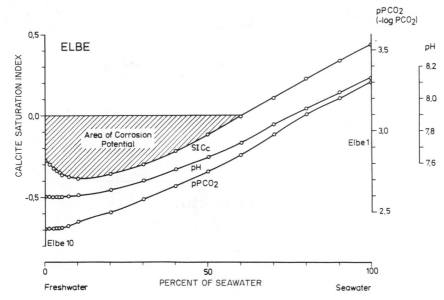

Figure 13.16 Thermodynamic model of various mixtures of Elbe fresh water (left) and of North Sea water (right, 30.5‰ salinity). End-member solutions as of composition as found in October 1981 R. V. VALDIVIA stations Elbe 1 and Elbe 10 (Kempe, 1982b). SI Cc = saturation index of calcite; $pPCO_2$ = log CO_2 pressure, calculated by programme WATMIX (Wigley and Plummer, 1976)

numeric calculation. In effect, additional CO_2 is liberated during the mixing process formerly locked up in ionic form. Nature, however, does not necessarily follow mass-conservation during the mixing process: Figure 13.17 gives the plots of the carbonate alkalinity and pH versus the salinity for October 1981 in the Elbe estuary. One can see that both parameters (which are independent measurements) display steep jumps very early in the mixing process. The increase in pH signals the loss of free CO_2 from the water, and the increase in alkalinity indicates dissolution of carbonates. What happens is that marine carbonates become dissolved and saturate the high undersaturation of the Elbe with respect to calcite very early in the mixing (Kempe, 1982b, 1984). Thus the high CO_2 pressure of the Elbe is not degassed to the atmosphere nor is it mixed mass conservatively into the North Sea water, rather it is titrated by marine carbonates to form excess alkalinity. In fact, the maximum of the undersaturation (or corrosion) potential in Figure 13.16 can be calculated to be equivalent to about 100 μmol $CaCO_3$/l which would result in an increase in the alkalinity of about 200 μmol/l which is the height of the observed jump. Measurements during other seasons show that this alkalinity jump occurs at times of low alkalinity in the head waters and is a rather normal feature in the hydrochemistry of the Elbe estuary. It must be concluded that

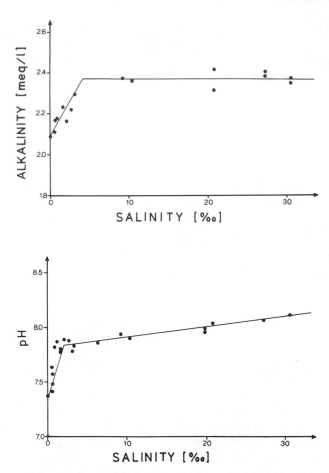

Figure 13.17   Actual alteration of alkalinity (top) and pH (bottom) in the Elbe Estuary with increasing salinity as measured in October 1981 R. V. VALDIVIA Cruise (Kempe, 1982b)

Figure 13.18   Long-term record of PO_4-P (top) and NO_3-N (bottom) concentrations of the inner German Bight at Helgoland, 1962–1984. Solid line = linear regression; small broken lines = 98% confidence interval; large broken line = overall average; solid line represents low-pass filter, periods greater than 2 years are shown. Equations of linear regressions: $Y = 0.548 + 0.00005*X$ (for PO_4-P) and $Y = 5.541 + 0.001299*X$ (for NO_3-N). Data by Harms and Mangelsdorf, statistics by Berg and Radach (1985)

275

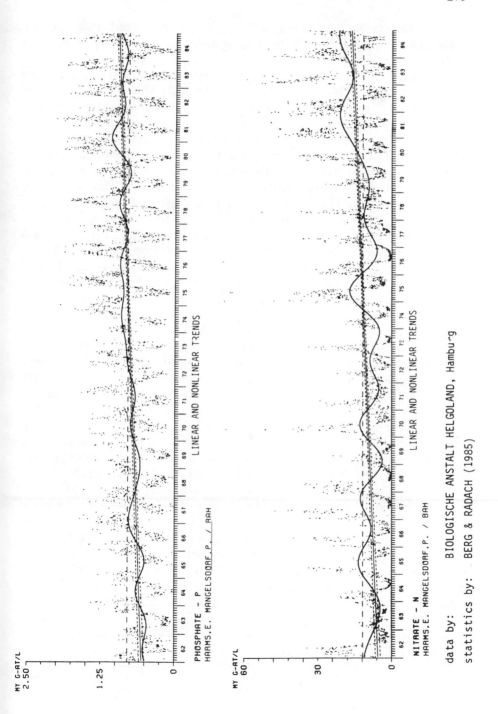

MY G-AT/L
2.50

1.25

0

PHOSPHATE - P
HARMS.E. MANGELSDORF.P. / BAH

LINEAR AND NONLINEAR TRENDS

MY G-AT/L
60

30

0

NITRATE - N
HARMS.E. MANGELSDORF.P. / BAH

LINEAR AND NONLINEAR TRENDS

data by: BIOLOGISCHE ANSTALT HELGOLAND, Hamburg

statistics by: BERG & RADACH (1985)

such active carbonate corrosion and therefore a buffering of the anthropogenically increased PCO_2 may occur in other rivers highly polluted with labile organics as well.

Respiration consumes oxygen from the water. However, due to the advance of waste treatment this seems not to be the main threat for the oxygen concentration of the Elbe estuary. Rather, release of ammonia from treatment plants in East and West Germany is important. This ammonia triggers bacterial nitrification reactions if, in early summer, temperatures rise above about $16\,°C$. Thus downstream of the sewage plant outlet of Hamburg (said to be the fourth largest tributary of the Elbe) oxygen is depleted regularly and anoxia can occur (Arbeitsgemeinschaft für die Reinhaltung der Elbe, 1978–1984). Fish are killed and drift to shore. These paroxysms of a heavily polluted system attract the interest of the news media much more than the overall pollution of the river. Cure for this illness is in sight as Hamburg will soon employ a nitrification unit at its sewage plant. Similar anoxic stretches are reported from other estuaries as well. The Scheldt is an especially well documented example (Wollast, 1983). Cynically, though, one may ask, what does it help to save the life of fish, if these fish are polluted by heavy metals to such an extent that professional fishery is prohibited since 1984 in the Elbe estuary and no fish from the Elbe may be sold?

How, if at all, eutrophication by rivers is affecting the coastal seas is a long standing debate. This debate mostly suffers from the fact that no long-term data are available for coastal seas. Ship measurements are infrequent and tend to cluster in times of fine, or at least quite fair, weather conditions. Winter and storm data are missing. In situ techniques have been developed to provide better data. However, these sets of data are not numerous and they are being utilized only now in a few places. For example, for the first time such a long-term record is available only now for discussion: it comprises data collected since 1962 by the Helgoland Biological Institute (Gillbricht, Treutner, Harms, Mangelsdorf, published in Biologische Anstalt Helgoland, 1962–1984) at their station on the Island of Helgoland in the Inner German Bight outside the Weser and Elbe estuary (compare Figure 13.15). At the station up to 50 parameters are measured daily during work days throughout the year. The data have been digitized, screened for outliers and tested for statistical significant trends by Berg and Radach (1985). Salinity at the station is on average around 32‰ but changes seasonally with discharge of the rivers by as much as 4‰. Figures 13.18 and 13.19 show four of the most interesting

---

Figure 13.19   Long-term record of diatom carbon and flagellate carbon of the inner German Bight at Station Helgoland, 1962–1984. For explanation of regressions see Figure 13.18. Equations of linear regressions: $\log(Y) = 0.786 + 0.000012*X$ (for diatom-C) and $\log(Y) = 0.393 + 0.00012*X$ (for flagellate-C). Data by Gillbricht, statistics by Berg and Radach (1985)

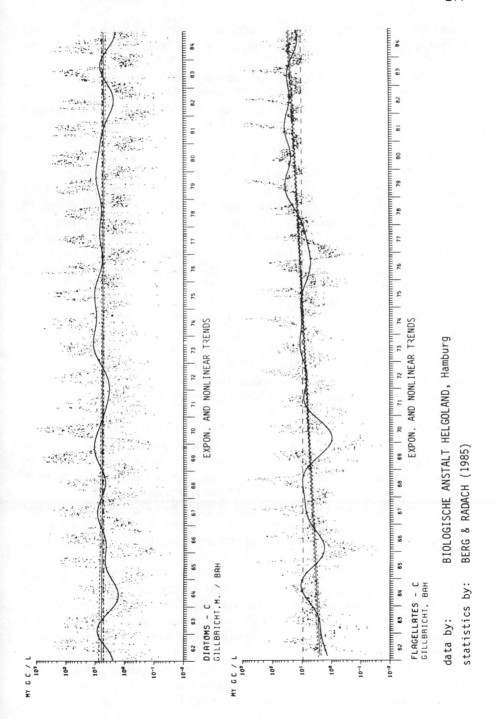

MY G C / L

DIATOMS - C
GILLBRICHT.M. / BAH

EXPON. AND NONLINEAR TRENDS

MY G C / L

FLAGELLATES - C
GILLBRICHT, BAH

EXPON. AND NONLINEAR TRENDS

data by: BIOLOGISCHE ANSTALT HELGOLAND, Hamburg

statistics by: BERG & RADACH (1985)

long-term curves: PO_4-P, NO_3-N, diatoms-C, and flagellates-C. For both PO_4-P and NO_3-N positive and highly significant linear trends were found. PO_4-P increased from 0.55 to 0.97 μg-at/l and NO_3-N from 5.56 to 16.45 μg at/l in 23 years. Apparently also the seasonal amplitude increased but final tests still have to be run. Even though more and more nutrients seem to become available, phytoplankton in summer still brings concentrations down to values near total depletion. As a consequence one must expect that standing phytoplankton-C must have increased. This is in fact what is found, phytoplankton-C increased exponentially by a factor of four (from 8.9 to 36.6 μgC/l). Most of this increase is due to flagellates (Figure 13.19, bottom). Diatoms, however, did not increase at all (Figure 13.19, top). Again, as one would expect because one of their nutrients, silica is not influenced by pollution and has remained relatively unchanged throughout the period. One must realize that there is no accumulation of nutrients taking place at the station as the residence time of both water and recent sediments in the North Sea is low. Thus the station probably monitors annual pollution events, i.e. those nutrients provided by rivers to the coastal current within the last few weeks, months, or during the last season if one assumes a small sediment pool as an interim buffer. This inference is substantiated, for example, for the nitrate curve, the long-term frequency curve of which has a high correlation with the discharge record of the Elbe (Berg and Radach, 1985). One conclusion of hope can be drawn from this: cutting the nutrient input by rivers to the North Sea may, within a few years, curb eutrophication. Speculation exists as to the final resting place of the exported nutrients: possibly the young sediments in the Skagerrak bury most of them.

Another part of the pollutants seems to be mixed into the sediments of tidal flats along the Dutch, German, and Danish coasts. Here lateral migration of tidal channels and heavy waves during storm tides move large quantities of sediments around and thereby mix younger substances with older sediments of longer residence times. Schwedhelm and Irion (1985) have recently shown that heavy metal and phosphorus (Figure 13.20) enrichment is quite evident throughout the German Wadden Sea. Concentrations are especially high in the immediate vicinity of the Elbe and Weser Estuaries. The heavy metal measurements show that in places these pollutants have been mixed down up to several metres. In this upper mixing zone no general concentration gradient can be defined, suggesting recent reworking of the column. The case for phosphorus is somewhat different. Its concentration profile does not show a break at the same erosion basis, rather the concentrations decline quickly within the first few centimetres, suggesting that most of the phosphate is remobilized quickly. In fact, most of the phosphorus is bound in easily reducible Fe and Al phosphates (Figure 13.20, inorg. fraction), thus allowing easy remobilization to pore waters. The pore fluids are released during low tides or when the sediment is reworked by benthic fauna or resuspended during

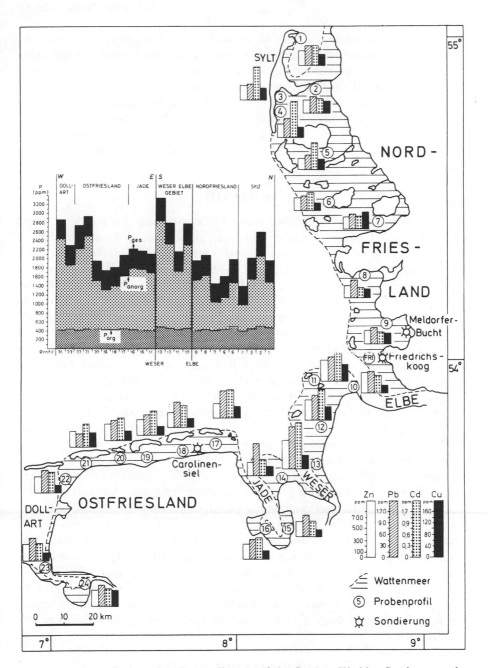

Figure 13.20 Pollution of surface sediments of the German Wadden Sea between the Dollart (south-west) and Sylt (North) for the sediment size fraction < 2 μm. Map shows distribution of Zn, Pb, Cd, and Cu and inset graphs contents of total phosphorus, inorganic phosphorus, and organic phosphorus (Schwedhelm and Irion, 1985)

storms. Thus phosphorus is kept in the biological cycle for several turnovers before finally enclosed in sediments or exported to the Atlantic Ocean. Also PCBs show large concentration increases in the North Sea and in the Wadden Sea (Boon *et al.*, 1985). Main input seems to be the River Rhine.

So far no geochemical model has been coupled to a circulation model of the North Sea (e.g. Backhaus, 1985) and calculation of fluxes, inputs, and outputs to the system is only at its beginning (e.g. Weichart, 1973; Postma, 1981; Eisma, 1981; Wollast 1983; Eisma and Kalf, 1987; Schwedhelm and Irion, 1985).

OUTLOOK

Even without the anthropogenic interference estuaries are places of complicated hydrodynamic, chemico-physical, geological, and biological processes. The state of the system changes with weather, tides, seasons, sea level, glacials, and tectonic movements over a wide range of frequencies. We are far from understanding all of these interdependencies. The example of the Elbe and the inner German Bight teaches that many conflicting uses alter the estuarine system often in an unwanted and unexpected way. Scientists only now begin to understand what the dredging of channels, the dyking of marshes and the digging of harbour basins means for the tidal behaviour of an estuary and its vulnerability to storm surges. Pollution is another unwanted effect of industrial and intensified agricultural activity. Quietly heavy metals and nutrients have spread from the river mouths into the coastal sea, its water, sediments, and biota. Adverse effects are now evident: the fauna and flora have changed and—in the case of the Elbe—it is forbidden to sell fish. Pollutants are measurable throughout the North Sea and it may be only a question of time until its fish cannot be sold any more.

Certainly, further research is needed to describe the system more completely and to derive models capable of predicting future changes. Data acquisition is poor, even for relatively well investigated estuaries like the Elbe. In-situ monitoring should be pushed especially to obtain data for storm surges, winter conditions, and extreme tides. Numerical hydrodynamic models need to be coupled to geochemical and ecological models. However, these activities should not obscure the fact that only stringent pollution control will effectively decrease current input to estuaries and that a full 'clean-up' may never be reached. Developing nations have the opportunity to 'learn' from the mistakes of industrialized regions and to organize their industrialization in a less harmful way.

ACKNOWLEDGEMENTS

The author is supported by a grant of the German Minister for Research and Technology (BMFT).

REFERENCES

Allen, G. P., Salomon, J. C., Bassoullet, P., Du Penhoat, Y., and De Grandpr, C. (1980). Effects of tides on mixing and suspended sediment transport in macrotidal estuaries. *Sediment. Geol.*, **26**, 69–70.

Arbeitsgemeinschaft für die Reinhaltung der Elbe (1978–1984). *Water quality data of the Elbe from Schnackenburg down to the North Sea.* Annual Rep., Hamburg (in German).

Avoine, J., Allen, G. P., Nichols, M., Salomon, J. C., and Larsonneur, C. (1981). Suspended-sediment transport in the Seine Estuary, France: effects of man-made modifications on estuary-shelf sedimentology. *Mar. Geol.*, **40**, 119–137.

Backhaus, J. O. (1985). *Estimates of the Variability of Low Frequency Currents and Flushing-Times of the North Sea.* Intern. Council Expl. Sea, C:24, Sess. Hydrog. Comm. : 23 pages.

Berg, J., and Radach, G. (1985). *Trends in Nutrient and Phytoplankton Concentrations at Helgoland Reede (German Bight) Since 1962.* Intern. Council Expl. Sea, L:2, Sess. R. Biol. Oceanogr. Com.: 16 pages.

Biologische Anstalt Helgoland (1962–1984). Annual Reports, Hamburg (in German).

Boon, J. P., van Zantvoort, M. B., Govaert, M. J. M. A., and Duinker, J. C. (1985). Organochlorines in benthic polychaetes (*Nephtys* spp.) and sediments from the southern North Sea. Identification of individual PCB components. *Netherl. J. Sea Res.*, **19**, 93–109.

Bopp, F., and Biggs, R. B. (1981). Metals in estuarine sediments: factor analysis and its environmental significance. *Science*, **214**, 441–443.

Doyle, E., Collier, R., Dengler, A. T., Edmond, J. M., Ng, A. C., and Stallard, R. (1974). On the chemical mass-balance in estuaries. *Geochim. Cosmochim. Acta*, **38**, 1719–1728.

Broecker, H. C., Petermann, J., and Siems, W. (1978). The influence of wind on CO_2-exchange in a wind-wave tunnel, including the effects of monolayers. *Journ. Mar. Res.*, **36**, 595–610.

Broecker, W. S., Peng, T. H., Mathieu, G., Hesslein, R., and Torgérsen, T. (1980). *Gas Exchange Rate Measurements in Natural Systems.* Symp. on Capillary Waves and Gas Exchange, Ber. Sonderforschungsbereich 94 'Meeresforschung', Univ. Hamburg, 17, 71–86.

Cameron, W. M., and Pritchard, D. W. (1963). Estuaries. In Hill, M. N. (Ed.) *The Sea. Vol.* 2, pp. 306–324. Wiley, New York.

Castaing, P., and Allen, G. P. (1981). Mechanisms controlling seaward escape of suspended sediment from the Gironde: A macrotidal estuary in France. *Mar. geol.*, **40**, 101–118.

Degens, E. T., Michaelis, W., Garrasi, C., Mopper, K., Kempe, S., and Ittekkot, V. (1980). Varve-chronology and early diagenetic alteration of organic substances of Holocene sediments of the Black Sea. *Neues Jahrb. Geol. Paläontol. Monatsh.*, **2**, 65–86 (in German).

Degens, E. T. (Ed.) (1982). *Transport of Carbon and Minerals in Major World Rivers, Vol. 1.* SCOPE/UNEP Sonderbd., Mitt. Geol. Paläont. Inst. Univ. Hamburg, **52**: 764 pages.

Degens, E. T., Kempe, S., and Soliman, H. (Eds.) (1983). *Transport of Carbon and Minerals in Major World Rivers, Vol. 2.* SCOPE/UNEP Sonderbd., Mitt. Geol. Paläont. Inst. Univ. Hamburg, **55**: 525 pages.

Degens, E. T., Kempe, S., and Herrera, R. (Eds.) (1985). Transport of Carbon and Minerals in Major World Rivers, Vol. 3. SCOPE/UNEP Sonderbd., Mitt. Geol. Paläont. Inst. Univ. Hamburg, **58**: 645 pages.

Eisma, D. *(1981). Supply and deposition of suspended matter in the North Sea. Spec. Publs. int. Ass. Sediment.*, **5**, 415–428.

Eisma, D. (1988). Transport and deposition of suspended matter in estuaries and the nearshore sea. In Degens, E. T., Kempe, S., and Gan W. B. (Eds.) *Transport of Carbon and Minerals in Major World Rivers, Vol. 4.* SCOPE/UNEP Sonderbd., Mitt. Geol. Paläont. Inst. Univ. Hamburg, **65** (in press).

Eisma, D., and Kalf, J. (1987). Dispersal, concentration and deposition of suspended matter in the North Sea. *J. Geol. Soc. London,* **144**, 161–178.

Eisma, D., Cadee, G. C., and Laane, R. (1982). Supply of suspended matter and particulate and dissolved organic carbon from the Rhine to the coastal North Sea. In Degens, E. T. (Ed.) *Transport of Carbon and Minerals in Major World Rivers, Vol.* 1, pp. 483–505. SCOPE/UNEP Sonderbd., Mitt. Geol. Paläont. Inst. Univ. Hamburg, **52**.

Eisma, D., Boon, J., Groenewegen, R., Ittekkot, V., Kalf, J., and Mook, W. G. (1983). Observation on macro-aggregates, particle size and organic composition of suspended matter in the Ems Estuary. In Degens, E. T., Kempe, S., and Soliman, H. (Eds.) *Transport of Carbon and Minerals in Major World Rivers, Vol.* 2, pp. 295–314. SCOPE/UNEP Sonderbd., Mitt. Geol. Paläont. Inst. Univ. Hamburg, **55**.

Eisma, D., Bernard, P., Boon, J. J., Van Grieken, R., Kalf, J., and Mook, W. G. (1985). Loss of particulate organic matter in estuaries as exemplified by the Ems and Gironde Estuaries. In Degens, E. T., Kempe, S., and Herrera, R. (Eds.) *Transport of Carbon and Minerals in Major World Rivers, Vol.* 3. pp. 397–412. SCOPE/UNEP Sonderbd., Mitt. Geol. Paläont. Inst. Univ. Hamburg, **58**.

Festa, J. F., and Hansen, D. V. (1978). Turbidity maxima in partially mixed estuaries: a two-dimensional numerical model. *Estuar. Coastal Mar. Sci.,* **7**, 347–359.

Fischer, J. (1983). Remote sensing of suspended matter, phytoplankton and yellow substances over coastal waters. Part I: Aircraft measurements. In Degens, E. T., Kempe, S., and Soliman, H. (Eds.) *Transport of Carbon and Minerals in Major World Rivers, Vol.* 2, pp. 85–95. SCOPE/UNEP Sonderbd., Mitt. Geol. Paläont. Inst. Univ. Hamburg, **55**.

Fox, L. E., and Wofsy, S. C. (1983). Kinetics of removal of iron colloids from estuaries. *Geochim. Cosmochim. Acta,* **47**, 211–216.

Gornitz, V., Lebedeff, S., and Hansen, J. (1982). Global sea level trend in the past century. *Science,* **215**, 1611–1614.

Grassl, H., Maier-Reimer, E., Degens, E. T., Kempe, S., and Spitzy, A. (1984). CO_2, carbon cycle and climate, I: Global carbon balance; II: Radiation balance and water budget. *Naturwissenschaften,* **71**, 129–136, 234–238 (in German).

Ittekkot, V., Spitzy, A., and Lammerz, U. (1982). VALDIVIA Cruise October 1981: dissolved organic matter in the Elbe, Weser and Ems Rivers and the German Bight. In Degens, E. T. (Ed.) *Transport of Carbon and Minerals in Major World Rivers, Vol.* 1, pp. 749–756. SCOPE/UNEP Sonderbd., Mitt. Geol. Paläont. Inst. Univ. Hamburg, **52**.

Jansen, J. H. F., Van Weering, T. C. E., and Eisma, D. (1979). Late Quaternary sedimentation in the North Sea. *Acta Univ. Upsala, Symp. Univ. Ups. Ann. Quing Cel.,* **2**, 175–187.

Kempe, S. (1982a). Long-term records of the CO_2 pressure fluctuations in fresh water. In Degens, E. T. (Ed.) *Transport of Carbon and Minerals in Major World Rivers, Vol.* 1, pp. 91–331. SCOPE/UNEP Sonderbd., Mitt. Geol. Paläont. Inst. Univ. Hamburg, **52**.

Kempe, S. (1982b). Valdivia Cruise, October 1981: carbonate equilibria in the estuaries of the Elbe, Weser, Ems, and in the Southern German Bight. In Degens, E. T. (Ed.)

Transport of Carbon and Minerals in Major World Rivers, Vol. 1, pp. 719–742. SCOPE/UNEP Sonderbd., Mitt. Geol. Paläont. Inst. Univ. Hamburg, **52**.

Kempe, S. (1983). Impact of the Assuan High Dam on water chemistry of the Nile. In Degens, E. T., Kempe, S., and Soliman, H. (Eds.) *Transport of Carbon and Minerals in Major World Rivers, Vol.* 2, pp. 401–423. SCOPE/UNEP Sonderbd., Mitt. Geol. Paläont. Inst. Univ. Hamburg, **55**.

Kempe, S. (1984). Sinks of the anthropogenically enhanced carbon cycle in surface fresh waters. *J. Geophys. Res.*, **89**, 4657–4676.

Kempe, S., Mycke, B., and Seeger, M. (1981). River loads and erosion rates in Central Europe 1966–1973. *Wasser und Boden*, **1981**, 126–131 (in German).

Kronberg, B. I., Lammerz, U., Schütt, M., and De Jonge, V. N. (1982). Valdivia Cruise October 1981: mineral nutrients in the Elbe, Weser, and Ems Rivers and the German Bight. In Degens, E. T. (Ed.) *Transport of Carbon and Minerals in Major World Rivers, Vol.* 1, pp. 655–666. SCOPE/UNEP Sonderbd., Mitt. Geol. Paläont. Inst. Univ. Hamburg, **52**.

Kronc, R. B., (1962). *Flume Studies of the Transport of Sediment in Estuarial Shoaling Processes—Final Report.* Univ. Calif. Hydr. Eng. Lab. and Sanitary Eng. Res. Lab., Berkeley: 110 pages.

Lohse, J., and Michaelis, W. (1983). Carbohydrates in particulate matter of the Elbe Estuary. In Degens, E. T., Kempe, S., and Soliman, H. (1983). *Transport of Carbon and Minerals in Major World Rivers, Vol.* 2, pp. 371–384. SCOPE/UNEP Sonderbd., Mitt. Geol. Paläont. Inst. Univ. Hamburg, **55**.

Meade, R. H. (1972). Transport and deposition of sediments in estuaries. *Geol. Soc. Amer. Mem.*, **133**, 91–120.

Meade, R. H. (1982). Sources, sinks, and storage of river sediments in the Atlantic drainage of the United States. *J. Geol.*, **90**, 235–252.

Meybeck, M. (1982). Carbon, nitrogen, and phosphorus transport by world rivers. *Am. J. Sci.*, **282**, 401–450.

NEDECO, (1965). *A Study of the Siltation of the Bangkok Port Channel, Vol.* 2, *the Field Investigations.* Netherlands Engineering Consultants, The Hague: 474 pages.

Officer, C. B. (1976). *Physical Oceanography of Estuaries.* Wiley, New York: 465 pages.

Officer, C. B. (1981). Physical dynamics of estuarine suspended sediments. *Mar. Geol.*, **40**, 1–14.

Officer, C. B., and Nichols, M. M. (1980). Box-model application to a study of suspended-sediment distributions and fluxes in partially mixed estuaries. In Kennedy, V. (Ed.) *Estuarine Perspectives*, pp. 329–340. Academic Press, New York.

Pickard, G. L. (1975). *Descriptive Physical Oceanography.* Pergamon Press, Oxford: 214 pages.

Plate, E. J. (1983). Zeitreihenuntersuchungen der Sturmfluten im Mündungsbereich der Elbe. *Die Küste*, **28**, 201–219 (in German).

Postma, H. (1961). Transport and accumulation of suspended matter in the Dutch Wadden Sea. *Neth. J. Sea Res.*, **1**, 148–190.

Postma, H. (1967). Sediment transport and sedimentation in the estuarine environment. In Lauff, G. H. (Ed.) *Estuaries*, pp. 158–179. Am. Assoc. Adv. Sci. Publ., **83**.

Postma, H. (1981). Exchange of materials between the North Sea and the Wadden Sea. *Mar. Geol.*, **40**, 199–213.

Schamp, H. (1983). Sadd el Ali, the High Dam of Assuan, I, II. *Geowiss. in unserer Zeit*, **1**, 51–59 and 73–85 (in German).

Schoer, J., and Eggersgluess, D. (1982). Chemical forms of heavy metals in sediments and suspended matter of Weser, Elbe, and Ems Rivers. In Degens, E. T. (Ed.)

Transport of Carbon and Minerals in Major World Rivers, Vol. 1, pp. 667–685. SCOPE/UNEP Sonderbd., Mitt. Geol. Paläont. Inst. Univ. Hamburg, **52**.

Schoer, J., Nagel, U., Eggersgluess, D., and Förstner, U. (1982). Metal contents in sediments from the Elbe, Weser, and Ems Estuaries and from the German Bight (Southeastern North Sea): grain size effects. In Degens, E. T. (Ed.) *Transport of Carbon and Minerals in Major World Rivers, Vol.* 1, pp. 687–702. SCOPE/UNEP Sonderbd., Mitt. Geol. Paläont. Inst. Univ. Hamburg, **52**.

Schwedhelm, E. and Irion, G. (1985). *Heavy Metals and Nutrients in the Sediments of the German North Sea Tidal Flats.* Cour. Forsch.-Inst. Senckenberg, 73, Senckenberg Naturforsch. Ges. Frankfurt: 119 pages (in German).

Seifert, R. (1985). *Organic Substances in European Rivers and Estuaries.* Diplomarbeit, Fachbereich Geowiss., University of Hamburg: 143 pages (unpublished manuscript, in German).

Sholkovitz, E. R. (1976). Flocculation of dissolved organic and inorganic matter during the mixing of river water and seawater. *Geochim. Cosmochim. Acta*, **40**, 831–845.

Siefert, W. (1982). Bemerkenswerte Veränderungen der Wasserstände in den deutschen Tideflüssen. *Die Küste*, **37**, 2–36 (in German).

Sündermann, J., and Zielke, W. (1983). Mathematisches Modell zur Simulation von Sturmflutereignissen in der Unterelbe. *Die Küste*, **38**, 177–200 (in German).

Szekielda, K. H. (1982). Investigations with satellites on eutrophication of coastal regions. In Degens, E. T. (Ed.) *Transport of Carbon and Minerals in Major World Rivers, Vol. 1*, pp. 13–37. SCOPE/UNEP Sonderbd., Mitt. Geol. Paläont. Inst. Univ. Hamburg, **52**.

Szekielda, K. H., McGinnis D., and Gird, R. (1983a). Investigations with satellites on eutrophication of coastal regions, Part II. In Degens, E. T., Kempe, S., and Soliman, H. (Eds.). *Transport of Carbon and Minerals in Major World Rivers, Vol. 2*, pp. 55–84. SCOPE/UNEP Sonderbd., Mitt. Geol. Paläont. Inst. Univ. Hamburg, **55**.

Szekielda, K. H., Piatti, L., and Legeckis, R. (1983b). Turbidity Zones over the Rio de la Plata region as monitored with satellites. In Degens, E. T., Kempe, S., and Soliman, H. (Eds.) *Transport of Carbon and Minerals in Major World Rivers, Vol.* 2, pp. 183–192. SCOPE/UNEP Sonderbd., Mitt. Geol. Paläont. Inst. Univ. Hamburg, **55**.

Szekielda, K. H. (1985). Investigations with satellites on eutrophication of coastal regions, Part V: Note on the Amazon salt wedge. In Degens, E. T., Kempe, S., and Herrera, R. (Eds.) *Transport of Carbon and Minerals in Major World Rivers, Vol.* 3, pp. 85–90. SCOPE/UNEP Sonderbd., Mitt. Geol. Paläont. Inst. Univ. Hamburg, **58**.

Szekielda, K. H., and McGinnis, D. (1985a). Investigations with satellites on eutrophication of coastal regions, Part III: The patch concept. In Degens, E. T., Kempe, S., and Herrera, R. (Eds.) *Transport of Carbon and Minerals in Major World Rivers, Vol.* 3. pp. 33–48. SCOPE/UNEP Sonderbd., Mitt. Geol. Paläont. Inst. Univ. Hamburg, **58**.

Szekielda, K. H., and McGinnis, D. (1985b). Investigations with satellites on eutrophication of coastal regions, Part IV: The Changjiang River and the Huanghai Sea. In Degens, E. T., Kempe, S., and Herrera, R. (Eds.) *Transport of Carbon and Minerals in Major World Rivers, Vol.* 3. pp. 49–84. SCOPE/UNEP Sonderbd., Mitt. Geol. Paläont. Inst. Univ. Hamburg, **58**.

Weichart, G. (1973). Pollution of the North Sea. *Naturwissenschaften*, **60**, 469–472 (in German).

Wellershaus, S. (1982). The turbidity cloud in the Weser Estuary. *Deut. Gewässerkundl. Mitt.*, **26**, 2–6 (in German).

Whitehouse, U. G., Jeffrey, L. M., and Debbrecht, J. D. (1960). Differential settling tendencies of clay minerals in saline waters. In Swineford, A. (Ed.) *Clays and Clay Minerals*, pp. 1–79. Natl. Conf. on Clays and Clay Minerals, 7th, Washington 1958, Proc. Pergamon Press, New York.

Wigley, T. M. L., and Plummer, L. N. (1976). Mixing of carbonate waters. *Geochim. Cosmochim. Acta*, **40**, 989–995.

Wollast, R. (1983). Interactions in estuaries and coastal waters. In Bolin, B., and Cook, R. B. (Eds.) *The Major Biogeochemical Cycles and Their Interactions*. SCOPE Rep. 21, pp. 385–407, J. Wiley & Sons, Chichester, New York.

Scales and Global Change
Edited by T. Rosswall, R. G. Woodmansee and P. G. Risser
© 1988 Scientific Committee on Problems of the Environment (SCOPE)
Published by John Wiley & Sons Ltd.

CHAPTER 14

Use of Satellite Ocean Colour Observations to Refine Understanding of Global Geochemical Cycles

JOHN J. WALSH AND DWIGHT A. DIETERLE

Department of Marine Science,
University of South Florida,
St. Petersburg, Florida 33701

ABSTRACT

Primary production in the sea accounts each year for at least 30 percent of the global plant fixation of carbon and is thus an integral part of major biogeochemical fluxes, such as the N_2 or CO_2 cycles. However, uncertainties exist in the magnitude of annual uptake of both carbon and nitrogen in the ocean, as well as in the regional or seasonal variability of these processes. The highly productive continental shelves of the world, where man's nutrient loadings are discharged, where 95 percent of the world's fishery is yielded, and where most of the organic carbon sink of atmospheric CO_2 occurs, may actually be two or three times more productive than presently estimated. Such uncertainties result from our past inabilities to accurately specify temporal and spatial inhomogeneities of phytoplankton biomass and associated productivities within shelf waters. As a result of the launch of the Nimbus-7 Coastal Zone Color Scanner (CZCS) in October 1978 and the subsequent progress with data analysis, it is now possible to determine ocean chlorophyll concentrations from space to better than ± 30 percent for values of $0-10$ μg chl l^{-1} in Case-I waters (little sediment or humic matter). On a cloudless planet, calculations of net annual shelf primary production could be made from the local change with time of phytoplankton biomass as measured by the CZCS. Time aliasing by clouds of such a satellite-derived data set can be overcome with simulation models that incorporate *in situ* data, rate parameters of biological processes, and the spatially-rich, but sparse CZCS data sets in time. Significant improvement in the estimates of shelf productivity, and in understanding the controlling mechanisms for their temporal and spatial variations, can be achieved when satellite measurements of ocean colour are combined appropriately with *in situ* observations. Such a capability is clearly needed if we are to understand biological variability of the sea in response to discrete or climatic changes, such as the impact of anthropogenic nutrient inputs on coastal fisheries, and the effects of fossil-fuel CO_2 loading to the atmosphere in terms of global habitability.

INTRODUCTION

The potential of the carbon cycle of the sea to either yield fish or store atmospheric CO_2 is a subject of continuing controversy as man's ability to modify the marine environment increases. Because the actual amount of CO_2 fixed annually during marine photosynthesis is unknown, the fate of phytoplankton, serving as a precursor either to fish carbon or to sediment carbon, is also unknown. Debates over the amount of potential fish harvest (Ryther, 1969; Alversen *et al.*, 1970) and CO_2 storage capacity (Broecker *et al.,* 1979; Walsh *et al.,* 1981) of the ocean thus hinge on the amount and fate of marine primary production. Current estimates of annual marine primary production range from 20 to 55×10^9 tons of carbon per year (Ryther, 1969; Koblentz-Mishke *et al.,* 1970; Platt and Subba Rao, 1975; DeVooys, 1979; Walsh, 1980). This range accounts for $\simeq 25$ to 50 percent of the total net global carbon fixation, assuming a terrestrial primary production of $\sim 55 \times 10^9$ tons C yr^{-1} (Woodwell *et al.,* 1978). This possible range in marine primary productivity is thus 400 to 1000 percent of present fossil fuel emissions of 5.2×10^9 tons C yr^{-1}.

Over the last century, man's ability to extract nitrogen from the atmosphere has also begun to rival that of N_2 fixation by plants. Between 1950 and 1975, world production of agricultural fertilizers increased tenfold. Anthropogenic nutrient input from agrarian runoff, deforestation, and urban sewage has already impacted local streams and ponds, some large lakes, major rivers, and perhaps even the continental shelves, with a tenfold increase of nutrient loading since 1850 (Walsh, 1984). Nitrogen is now only routinely measured in 25 percent of the world's 240 largest rivers, however, and few biological time series are available to document the coastal zone's past response to fluvial nutrient transients on even a decadal time scale.

The annual primary production of the Dutch Waddensea, for example, has apparently increased threefold between 1950 and 1970 (Postma, 1978) and presumably other shelves have responded to anthropogenic nutrient loadings as well. Rigorous analyses of past measurements of ocean colour by satellite over the last 5 years are now required to provide an adequate departure point for long-term chlorophyll time series to assess the fate of phytoplankton carbon and nitrogen biomass, as well as their productivity in the sea. With the present Nimbus-7 satellite information and that from a follow-on sensor, one can begin to address serious problems of overfishing today, as well as the future and perhaps more ominous consequences of the above linked activities —man's accelerated extraction of nitrogen from the atmosphere and addition of carbon to the atmosphere.

Large areas of the ocean, such as the central gyres, have relatively low rates of production per unit surface area (Figure 14.1), but account for a major fraction of total carbon fixation because of their large areal extent (Table 14.1). In contrast, highly productive coastal and upwelling regions account for only 10 percent of the ocean by area and probably 25 percent of the ocean primary productivity. The coastal zooplankton populations (Figure 14.2) provide the basis for more than 95 percent of the world's estimated fishery yield, however, and a large part of the proposed organic carbon sink of the atmospheric CO_2 is located in adjacent slope sediments (Figure 14.3). These various ocean provinces exhibit pronounced differences in their phytoplankton species assemblages as the evolutionary consequence of their physical habitat. They also have significant differences in spatial and temporal variability of algal biomass as a function of nutrient input and grazing losses (Walsh, 1976), with perhaps very different fates of the fixed carbon (Walsh *et al.,* 1981; Walsh, 1983): oxidation in the deep sea and burial along the continental margins.

There are two basic reasons for the uncertainty in the estimates of marine carbon fixation; both are of equal importance. First, the methodology used to estimate the rate of primary productivity (the ^{14}C method) may be in serious error (Geiskes *et al.,* 1979; Eppley, 1980). Secondly, the highly productive shelf regions exhibit a much wider range of spatial and temporal variability of biomass than the open ocean, on scales which have been very poorly sampled by classical shipboard programmes. It is in the oligotrophic (gyre) regions, where the biomass variability is *not* pronounced, that the methodology errors are greatest. This is because the oceanic phytoplankton are thought to be more sensitive to stresses related to capture and prolonged enclosure. However, the long food chains and the 90 percent recycling processes of the offshore regime provide insignificant fish harvest (Ryther, 1969) and little net biotic storage of CO_2 (Eppley and Peterson, 1980).

In the coastal regions where phytoplankton productivity and zooplankton biomass are much higher (Figs 14.1 and 14.2), the results of the ^{14}C methodology are probably more representative of the actual rate, but the spatial extent and the temporal character of the algal biomass fields are poorly known. For example, within 30 km off the Peru coast, the surface chlorophyll estimate of phytoplankton biomass ranges from 0.4 to 40.0 μg chl l^{-1} and the integrated primary production from <1 to >10 g C m^{-2} day^{-1} (Walsh *et al.,* 1980). Classical shipboard programmes consist of productivity measurements made once a day over a small spatial area, limited by the usual 10–12 knot speed of the research vessel and aliased within the above gradients (Walsh *et*

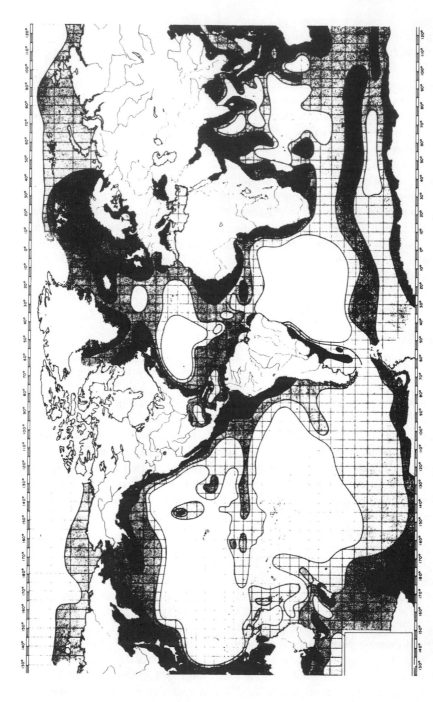

Figure 14.1 The global distribution of phytoplankton primary production (mg C m^{-2} day^{-1}) in five categories of >500, 250–500, 150–250, 100–150, and <100 (after Koblentz-Mishke *et al.*, 1970). (Reproduced with permission.)

Table 14.1 Aquatic photosynthesis and nitrogen fixation in relation to losses of sediment storage of organic carbon, of denitrification, and of methane production (after Walsh, 1984)

Region	Area (km^2)	Net primary production $(\times 10^9$ tons C yr$^{-1})$	Nitrogen fixation $(\times 10^7$ tons N yr$^{-1})$	Sediment organic carbon sink $(\times 10^9$ tons C yr$^{-1})$	Denitrification loss $(\times 10^7$ tons N yr$^{-1})$	Methane emission $(\times 10^7$ tons CH$_4$-C yr$^{-1})$
Open Ocean	3.1×10^8	18.60	0.43	0.19	0	0.36
Continental Shelf	2.7×10^7	5.40	0.27	0	2.97	0.04
Continental Slope	3.2×10^7	2.24	0.06	0.50	5.50	0.03
Freshwater Marshes	1.6×10^6	1.51	2.21	0.15	6.40	3.10
Estuaries/Deltas	1.4×10^6	0.92	0.06	0.20	1.04	0.60
Salt Marshes	3.5×10^5	0.49	0.48	0.05	1.40	0.80
Rivers/Lakes	2.0×10^6	0.40	1.88	0.13	0.26	5.10
Coral Reefs	1.1×10^5	0.30	0.28	0.01	0	0.32
Seaweed Beds	2.0×10^4	0.03	0	0	0	0.08
TOTAL AQUATIC AREA	3.75×10^8	C INPUT: 29.89	N$_2$ INPUT: 5.67	C OUTPUT: 1.23	N$_2$ OUPUT: 17.57	CH$_4$ OUPUT: 10.43

292

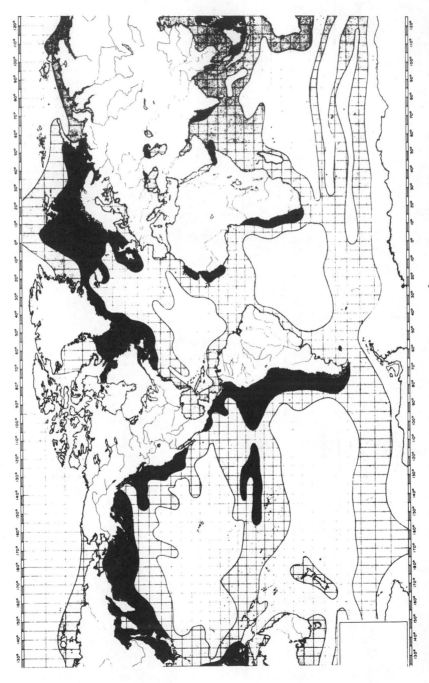

Figure 14.2 The global distribution of zooplankton abundance (mg m^{-3}) over the upper 100 m of the water column in four categories of >500, 201–500, 51–200, and <50 (after Bogorov *et al.*, 1968). (Reproduced with permission.)

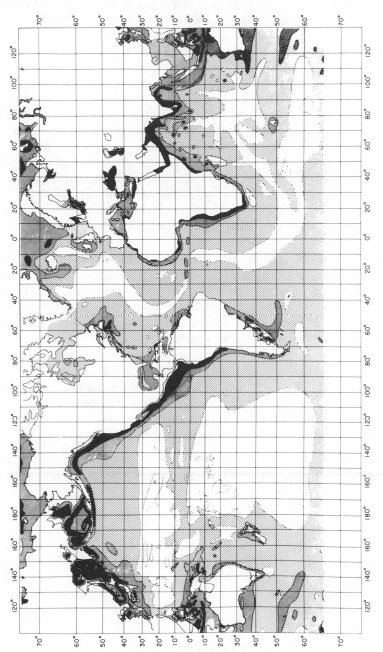

Figure 14.3 The global distribution of organic carbon (% dw) within surface sediments in five categories of >2.00, 1.01–2.00, 0.51–1.00, 0.25–0.50, and <0.25 (after Premuzic *et al.*, 1982). (Reprinted with permission from *Organ. Geochem.*, 4, 1982, Pergamon Journals Ltd.)

al., 1987a). At present, other shipboard rate measurements, such as nutrient uptake, respiration, grazing, excretion, and sinking, can only be made at a few points of the sea as well, to be later multiplied by some inadequate estimate of mean biomass in order to calculate fluxes of elements within the marine food web.

However, phytoplankton species on the continental shelves can divide every 0.5 to 2 days; without significant losses, an algal population during the spring bloom could increase at the same rate. To resolve the temporal and spatial consequences of this resultant growth process, a Nyquist sampling frequency of at least 0.25 day^{-1} is required by sampling theory (Blackman and Tukey, 1957). If one sampled every 4 hours in a typical longshore upwelling flow regime of 30 cm sec^{-1} to resolve this process, at least 5 ships would be required every 20 km^2 for the necessary biomass measurements (Kelley, 1976)! We thus arrive at a major reason for our analysis of CZCS colour data—the need to analyze distributions of biological properties at frequencies which can resolve causally the sources of their variance.

Approximately 10 percent of the annual shelf production (0.5×10^9 tons C yr^{-1}) is thought to be sequestered as organic carbon deposits (Table 14.1) on adjacent continental slopes (Walsh *et al.,* 1981). Although the anthropogenic input of nitrogen to the shelves may have increased tenfold over the last 50 to 100 years, a sufficient time series of phytoplankton data is not available to accurately specify changes in primary productivity or shelf export to continental slopes. This lack of a proper spatial and temporal perspective has hindered our understanding and, therefore, our ability to make accurate estimates of coastal productivity and subsequent carbon and nitrogen fluxes to the rest of the food web. Understanding the coastal ecosystem processes has far greater significance than their areal extent or contribution to total marine carbon fixation (Table 14.1) would suggest because:

(a) The fate of carbon and nitrogen fixed in these highly productive shelf regions is quite different from the oceanic areas of the sea, sinking to slope depocentres instead of being grazed within the water column (Walsh *et al.,* 1981; Eppley and Peterson, 1980), and

(b) Impacts of human activity are greater in the coastal region. Thus, there is a strong motivation to obtain, for the first time, an analysis of synoptic biomass information, coupled with rate process data, required to study these highly dynamic oceanographic regions over longer periods at annual and decadal time scales in addition to the much higher Nyquist frequency for resolution of the basic biological processes.

SATELLITE IMAGERY

Preliminary investigations undertaken in the 1960s by Clarke, Ewing, Lorenzen, Yentsch, and others provided evidence that the quality of light reflected from the sea surface and remotely sensed by aircraft instrumentation might be interpreted as phytoplankton biomass, i.e. chlorophyll, in the upper portion of the water column. These workers (e.g. Clarke *et al.,* 1970) were limited by their equipment to an altitude of 3 km. However, even at that altitude, the influence of the atmospheric backscatter was quite obvious as it began to dominate the colour signal reflected from the ocean surface. This raised the question of whether the rather poorly reflected ocean could be sensed through the entire atmosphere from a spacecraft, and if the contributions of the Rayleigh backscatter and aerosol backscatter could be effectively removed from the signal seen by a spacecraft. Additional NASA-supported studies in 1971 and 1972 with Lear Jet and U-2 aircraft and a rapid scan spectrometer at altitudes of 14.9 and 19.8 km, demonstrated that this concept could be used to develop spacecraft equipment for the purpose of estimating ocean water column chlorophyll from earth satellites. This became possible through the realization that problems associated with the scattering properties of the atmosphere, as well as direct reflectance of the sun from the sea surface (glint), could be either avoided or corrected (Hovis and Leung, 1977).

The first satellite-borne ocean colour sensor, the Coastal Zone Color Scanner (CZCS), was launched aboard Nimbus-7 in October 1978 with four visible and two infrared (one of which is thermal) bands, allowing a sensitivity about 60 times that of the Landsat-1 multispectral scanner. With failure of the sensor, transmission of CZCS data finally ceased in December 1986. Unlike many satellite sensors of ocean properties, the CZCS responded to more than the features of the mere surface of the sea and was sensitive to algal pigment concentrations in the upper 20 to 30 percent of the euphotic zone (Hovis *et al.,* 1980; Gordon *et al.,* 1980). The CZCS was specifically designed to detect upwelling radiance in spectral bands selected for the purpose of detecting variations in the concentrations of phytoplankton pigments. The theoretical and experimental techniques for describing the bio-optical state of ocean waters and its relationship to optical parameters that can be remotely sensed have been discussed by a number of workers (Morel and Prieur, 1977; Smith and Baker, 1978a, b; Gordon and Morel, 1983).

Simply stated, the CZCS radiance data can be utilized to estimate ocean chlorophyll concentrations by detecting shifts in sea colour, particularly in oceanic waters. Clear open ocean waters have low chlorophyll concentrations $(0.01-1.0 \, \mu g \, chl \, l^{-1})$ and the solar radiation reflected from the upper layers of these waters is blue; conversely, waters with high concentrations of chlorophyll $(> 1.0 \, \mu g \, chl \, l^{-1})$ are green (Morel and Smith, 1974). It has been demonstrated that this change in ocean colour can now provide a quantitative estimate of chlorophyll concentration (Gordon and Clark, 1980; Smith and Baker, 1982) for oceanic regions with an accuracy of 0.3 to 0.5 log C (where C is the chlorophyll concentration). The initial comparisons between CZCS imagery and surface pigments measured continuously along ship tracks carried

296

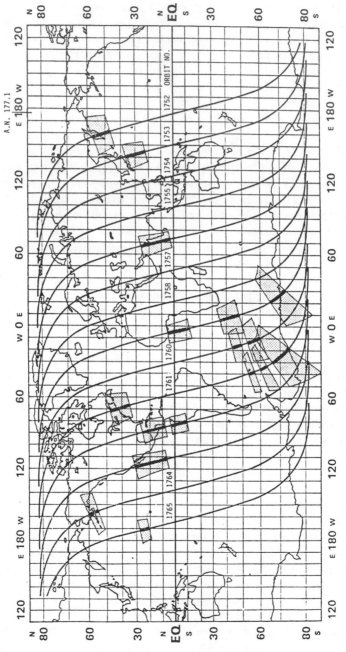

Figure 14.4 The availability of CZCS data from relatively cloud-free regions (shaded areas) during the 14 orbits of the Nimbus-7 satellite on 28 February 1979 (after OAO, 1979)

out by Gordon *et al.* (1983) and Smith and Wilson (1981) had suggested that C could be retrieved from the imagery to within about a factor of two. Smith and Baker (1982) and Gordon *et al.* (1982) have shown, however, that accuracies on the order of ±30 percent in C are possible for Morel's Case-I waters (Morel and Prieur, 1977), i.e. areas of little sediment or humic matter within the water column.

We are now in a position to systematically exploit the rich CZCS data base, obtained somewhere on the world shelves (Figure 14.4) each day over the last 5 years. Figure 14.5 of the CZCS-derived (SASC, 1984) chlorophyll distribution

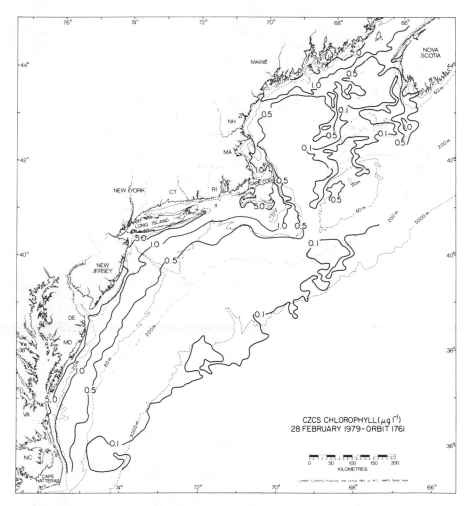

Figure 14.5   The distribution of satellite-derived chlorophyll (μg l^{-1}) over a grid of 4500 pixels, 5 nautical miles apart, within the Mid-Atlantic Bight on 28 February 1979

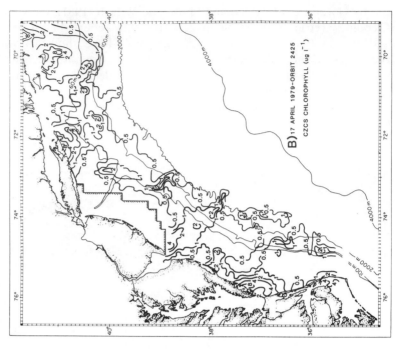

B) 17 APRIL 1979—ORBIT 2425
CZCS CHLOROPHYLL (ug l^{-1})

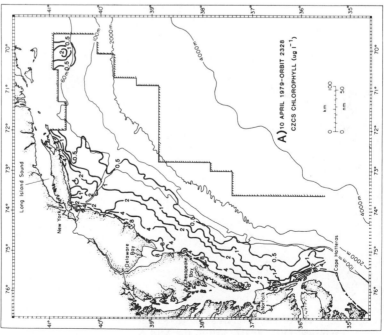

A) 10 APRIL 1979—ORBIT 2328
CZCS CHLOROPHYLL (ug l^{-1})

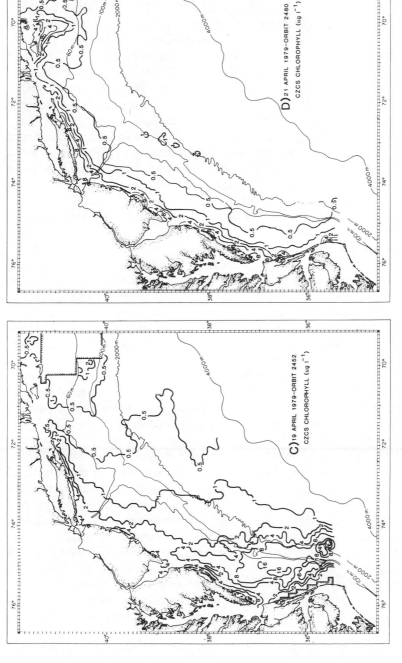

Figure 14.6 The CZCS estimate of chlorophyll distribution during a) 10 April 1979, b) 17 April 1979, c) 19 April 1979, and d) 21 April 1979 (after Walsh *et al.*, 1987a). (Reprinted with permission from *Deep-Sea Res.*, **34**, 1987, Pergamon Journals Ltd.)

during 28 February 1979 within the Mid-Atlantic Bight is an example of the information available from one of the many satellite images taken daily during the Nimbus-7 orbits around the planet (Figure 14.4). The chlorophyll isopleths of Figure 14.5 were contoured from a grid of ~4500 data points between the 10 to 3000 m isobaths. Each grid point was 5 nautical miles apart and each chlorophyll value was thus the mean of ~100 observations since the pixel resolution of the CZCS is ~800 m. The cross-isobath and parabathic gradients of this synoptic CZCS-derived chlorophyll field (Figure 14.5) are 'typical' of composites of chlorophyll data taken at sea during the same season between 1975 and 1984 (Walsh *et al.*, 1978; O'Reilly and Busch, 1984; Walsh *et al.*, 1987a). On a cloudless planet, the relevance of these aliased shipboard observations could simply be addressed by assembling a daily time series of CZCS data, computing the local rate of change of phytoplankton at 4500 to 450000 grid points, and unravelling the rate processes responsible for the amount of algal biomass left behind in the upper water column.

Unfortunately, it has not been possible to obtain CZCS measurements of the global oceans on anything close to a daily basis. On any given day (Figure 14.4), a major fraction of our watery planet is obscured by clouds. A qualitative estimate of realizable CZCS sampling characteristics was gleaned by the NASA Ocean Color Science Working Group (Walsh *et al.*, 1982) from screening a few time sequences of CZCS data for which regular sampling was attempted. This experience suggests that in a month of data collection, useful satellite data can be obtained on several days within randomly distributed clear-sky domains which are a few hundred km in extent, less frequently >1000 km in extent.

Of the nominal 2 hours of Nimbus-7 CZCS coverage taken and recorded each day, an average of approximately 30–40 percent was rejected and not processed due to total cloud cover (no significant open water areas). Furthermore, other data gaps for a particular site are derived from the inability of one satellite to provide sufficient daily overlap in swath width during the 14 orbits (Figure 14.4). As a result, only 45 useful CZCS images were available for the Mid-Atlantic Bight between 1 January and 30 June 1979, for example; the 25 percent data recovery was also not grouped in equal time increments. We present the results of 4 CZCS images taken on 10 April, 17 April, 19 April, and 21 April 1979 as a representative time series (Figure 14.6) of the types of data sets that are likely to be available from some continental shelves.

Our experience to date suggests that global CZCS coverage would yield, on average, between 10 (at the equator) and 20 (at 40 degrees N) usable images per month. The upper estimate represents a mean sampling interval of about every 1.5 days, for a given 1000 km × 1000 km boreal ocean domain, with the majority of usable data in patchy subscenes of typically a few hundred kilometres in extent, excepting an occasional clear view of most of the domain in one image. Coverage frequencies, however, fluctuate seasonally (and regionally) around these nominal estimates; coverage gaps of 2 to 3 weeks are likely to occur several times per year, with less frequent gaps of longer duration. In winter, low sun elevations will cause sampling voids of several weeks to a few months (increasing with latitude) at latitudes above 40 degrees.

Such sampling constraints mean that the CZCS data sets violate stationary assumptions, i.e. time invariant probability density functions over the time domain of interest, inherent in most statistical approaches to time series analyses of phytoplankton processes. Clearly, the present data base collected with the Nimbus-7 CZCS is thus inadequate to be directly applied to the global mapping of primary productivity, except in a qualitative sense. It is limited both in terms of sampling frequency and in terms of concurrent oceanographic experimental data necessary to bridge the interpretive gap from phytoplankton pigment distributions to net primary production by statistical methods.

Adequate data do exist in certain shelf regions, however, to develop a modelling methodology for a future global productivity assessment programme, utilizing the present and follow-on CZCS-type sensors. We present the preliminary results of simulation models of the time-dependent flow field of continental shelf waters in the Mid-Atlantic Bight to accompany the satellite data set of April 1979. Such numerical analyses will eventually allow us to interpolate between the available CZCS images with sufficient accuracy that future changes of shelf ecosystems, induced by continued overfishing or discharge of anthropogenic nutrients, can be detected on other shelves as well, allowing realistic estimates of the ocean's role in the future habitability of the planet.

1979 SATELLITE TIME SERIES

Phytoplankton standing stock as indicated by CZCS estimates of pigment concentrations at a fixed time, within a particular spatial pattern, are the result of a complicated set of biological, chemical, and physical processes with time scales ranging from seconds to seasonal, and space scales ranging from global to microscopic. The shapes and locations of chlorophyll patterns delineating the synoptic-scale shelf features are dominated, however, by physical–dynamical processes, transporting the plankton populations left in the water column as the net result of birth and death processes. Therefore, mesoscale spatial patterns ($10 < \times < 100$ km) tend to evolve over time scales ranging from several hours of phytoplankton cell division to a few days, i.e. at the time scale of a wind event (Walsh, 1976), while the synoptic-scale patterns ($100 < \times < 1000$ km) tend to evolve over time scales of the order of a few weeks to a month, i. e. above the Nyquist frequency of the probable CZCS sampling time scale. Considering these separate biological and physical scales and the CZCS sampling characteristics together, we chose (Walsh *et al.*, 1987c) to simulate water motion with respect to concurrent CZCS images over 3 weeks during 5–25 April 1979 (Figure 14.7).

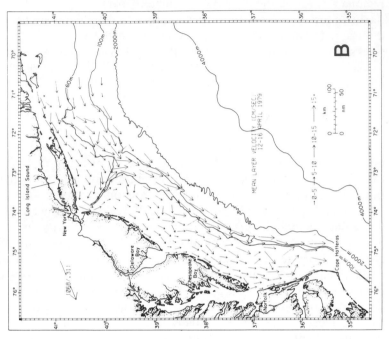

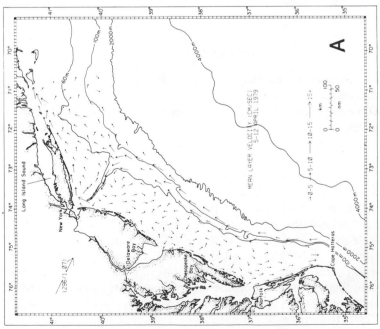

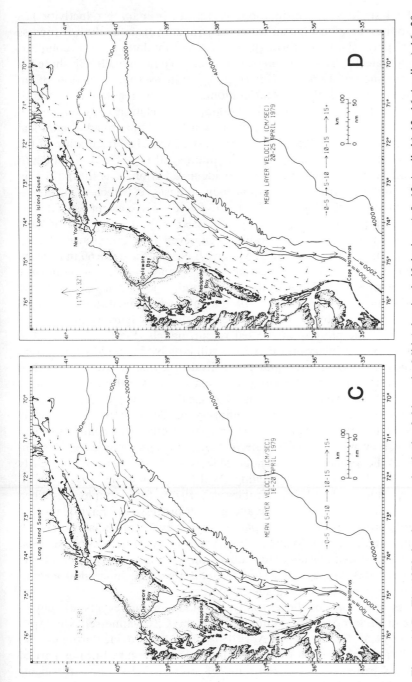

Figure 14.7 The 1979 depth-averaged velocity field on the Mid-A-lantic shelf during a) 5–12 April, b) 12–16 April, c) 16–20 April, and d) 20–25 April (after Walsh *et al.*, 1987a). (Reproduced with permission from *Deep-Sea Res.*, **34**, 1987, Pergamon Journals Ltd.)

Over the last 50 years, estimation of marine primary productivity has involved some type of mathematical model, ranging from analytical process models of photosynthesis through ecosystem simulation with complex, coupled biological–physical numerical models. These models all shared a common parameterization of primary production as some function of an initial concentration of phytoplankton biomass and the regulation of photosynthesis by light and nutrients. For example, the relation between vertical mixing, light intensity, and phytoplankton growth was quantified as a critical depth concept (Sverdrup, 1953), below which the 24 hr algal respiration of the water column exceeds the integrated daily photosynthesis. This critical depth was $h_c \simeq 0.2\ I_0 (k I_c)^{-1}$, where I_0 is the incident radiation, k is the extinction coefficient, and I_c is the compensation light intensity at which algal photosynthesis equals respiration ($\sim 0.3\ \mathrm{ly\,h}^{-1}$). Sverdrup's concept was that if $h_c < h$, the depth to which the phytoplankton are mixed as a result of wind and/or tidal stirring, no bloom would occur even in the presence of high nutrient content. At the beginning of March off New York at the 60 m isobath, the incident radiation is $< 250\ \mathrm{g\,cal\,cm}^{-2}\,\mathrm{day}^{-1}$ and h_c is 53 m within the well-mixed 60 m water column. Accordingly, high chlorophyll was not found at the 60 m isobath on 28 February 1979, but an order of magnitude more algal biomass was observed by the CZCS at depths $\leqslant 20$ m (Figure 14.5).

The utility of the more sophisticated recent models (Platt *et al.*, 1977) in predicting regional primary productivity is still largely hindered by the current meagre knowledge on the coupling of physical dynamics with biological processes on the appropriate time and space scales. Models of marine processes, whether physical, chemical, geological, or biological, are, in fact, inadequate theoretical constructs attempting to describe an incompletely known dynamic balance. Models thus lead to specific data acquisition and analysis which eventually suggest rejection of the original model, or hypothesis, in a cyclic process of oceanographic research (Walsh, 1972). Early ecological models of coastal upwelling (Walsh and Dugdale, 1971; Walsh, 1975; Walsh and Howe, 1976; Wroblewski, 1977), for example, described reasonably well the nutrient uptake and growth processes of phytoplankton, but their loss processes were poorly parameterized by unknown lateral export, grazing, and sinking terms (Walsh, 1983). With the synoptic CZCS data sets, we can begin to systematically place bounds on these loss terms by modelling the spatial changes of phytoplankton over discrete time intervals for comparison with successive CZCS images.

Rapid resuspension, offshore transport, and sinking events of phytoplankton can be inferred, for example, from the CZCS time series in April 1979 (Figure 14.6). During 17–19 April 1979, 10 ships provided ground-truth chlorophyll measurements in the Mid-Atlantic Bight (Figure 14.8) as part of the LAMPEX experiment for calibration of aircraft and satellite overflights (Thomas, 1981). The R/V *Kelez* was occupying stations in the apex of the New

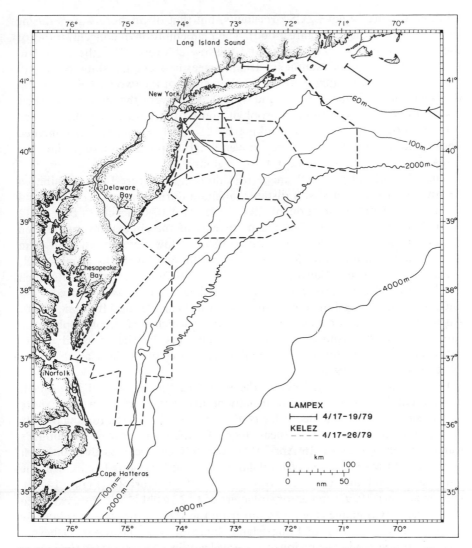

Figure 14.8 Cruise tracks of the R/V *Kelez, Advance II, Julius Nelson, Onrust, Kyma, Pathfinder, Shang Wheeler, Sub Sig II, Short Snort,* and *Lady Donna* during 17–26 April 1979 (after Walsh *et al.,* 1987a). (Reproduced with permission from *Deep-Sea Res.,* **34**, 1987, Pergamon Journals Ltd.)

York Bight during 17–19 April 1979, such that most of the chlorophyll observations on these days were taken within 25 km of the coast, south of Hudson Canyon. The rest of the *Kelez* cruise track (Figure 14.8) was performed after 19 April 1979, such that the shipboard chlorophyll composite (Figure 14.9) is that of nearshore waters during the 17 April (Figure 14.6b) and

19 April (Figure 14.6c) CZCS overflights, but that of offshore waters during the 21 April overflight (Figure 14.6d). The high chlorophyll concentrations of the coastal zone (< 20 m depth) and the low chlorophyll at the 60–100 m isobaths, southeast of Nantucket Island, measured aboard ships on 17–19 April (Figure 14.9) match quite well these regions of the two CZCS images on the same days (Figures 14.6b and 14.6c). However, the major mid-shelf resuspension of near-bottom chlorophyll, south of Delaware Bay, on 19 April (Figure 14.6c) and subsequent sinking by 21 April (Figure 14.6d) remained undetected (Figure 14.9) by the conventional shipboard surveys, i.e. the research vessels were then not in those regions.

The 10 April 1979 CZCS image of chlorophyll (Figure 14.6a) exhibits a decline of algal biomass with distance offshore during mean northwest wind forcing, i.e. from a direction of 296° True, of 1.07 dynes cm^{-2} over 5–12 April 1979 (Figure 14.7a). The mean currents of the circulation model then (Figure 14.7a) are $\leqslant 5$ cm sec^{-1}, with offshore flow between the 20–40 m isobaths and southwesterly flow between the 40–60 m isobaths at mid-shelf, south of the Hudson Canyon and north of Norfolk. In response to such northwest wind events, surface waters are pushed offshore, and the predominantly westward alongshore flow is slowed down (Beardsley and Butman, 1974). An upwelling circulation pattern is created, in which surface phytoplankton can be advected offshore and dissolved nutrients can be returned within subsurface waters to the shelf, providing the source of the next algal growth cycle (Walsh *et al.*, 1978).

During a northeast wind event, the westward flow is instead intensified, however, and weak onshore flow usually occurs at the surface, with offshore flow of subsurface water (Beardsley *et al.*, 1983). Under a wind forcing of only 0.31 dynes cm^{-2} from the northeast (068° T), the mean flow of the model's water column during 12–16 April 1979 was >10 cm sec^{-1} to the southwest over most of the shelf, except for offshore flows south of Long Island, near the Hudson Canyon, off Delaware Bay, and south of Norfolk (Figure 14.7b). During 17 April 1979, tongues or streamers of 1–2 μg chl l^{-1}, in fact, extended within the CZCS image (Figure 14.6b) from the shelf to slope waters in these areas of the shelf, south of Long Island, south of New Jersey, off Delaware Bay, and off Norfolk, as previously observed within a March CZCS time series (Walsh *et al.*, 1987a). Similar to the northwest wind forcing, it appears that northeast wind forcing could also move algal cells in some shelf areas from the coastal zone to slope waters (Figure 14.7b).

Following this northeast wind event, another mean wind forcing from the northwest (341° T) occurred during 16–20 April 1979 (Figure 14.7c), but with half the intensity (0.58 dynes cm^{-2}) of the first period (Figure 14.7a). In response to this shift in wind forcing, the mean alongshore flow is now weaker north of the Hudson Canyon and stronger south of Delaware Bay during 16–20 April, compared to the 12–16 April time period (Figure 14.7b). In the model, there was little or no offshore flow south of Long Island and New

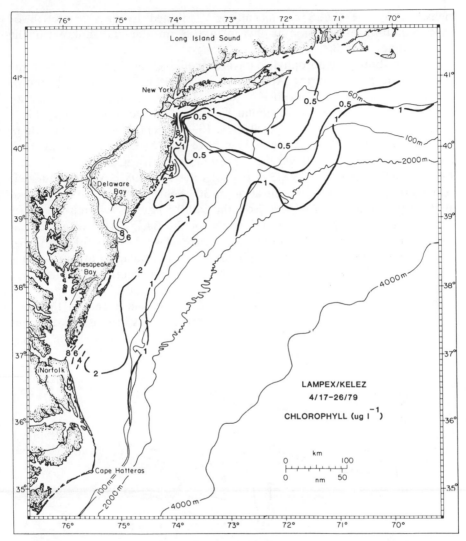

Figure 14.9 A chlorophyll (μg l^{-1}) composite of the distribution of phytoplankton biomass measured aboard ships during 17–26 April 1979 (after Walsh *et al.*, 1987a). (Reproduced with permission from *Deep-Sea Res.*, **34**, 1987, Pergamon Journals Ltd.)

Jersey, but continued offshore movement of water occurred off Delaware Bay and Norfolk (Figure 14.7c). In contrast to observations of ~ 1 μg chl l^{-1} on April 17 at mid-shelf within the latter region (Figure 14.6b), two days later, more than 16 μg chl l^{-1} was detected by the CZCS on 19 April 1979, from off Norfolk to off Cape Hatteras, 150 km south along the 40–60 m iosbaths (Figure 14.6c). At a population growth rate of only one doubling every two days (Walsh *et al.*, 1987a), the order of magnitude change in algal biomass measured by the CZCS cannot all be attributed to *in situ* growth of phytoplankton; resuspension of phytoplankton, uneaten and sunk out of the water column, is a likely source term.

Few near-bottom chlorophyll data are available for April 1979 over the Mid-Atlantic Bight. During April 1984, however, near-bottom chlorophyll concentrations of $>25\ \mu g\,chl\,l^{-1}$ were found at the 60 m isobath (Figure 14.10). These data are similar to our observations of $9-17\ \mu g\,chl\,l^{-1}$ above the 21–63 m isobaths in April 1978, $10-12\ \mu g\,chl\,l^{-1}$ above the 15–35 m isobaths in April 1980, $8-14\ \mu g\,chl\,l^{-1}$ above the 21–49 m isobaths in April 1981, $10-29\ \mu g\,chl\,l^{-1}$ above the 17–39 m isobaths in April 1982, and $9-15\ \mu g\,chl\,l^{-1}$ above the 13–43 m isobaths in April 1983. An accumulated, near-bottom chlorophyll concentration of $30\ \mu g\,chl\,l^{-1}$ within the lower 10 m of the water column at the 40 m isobath, and $0.5\ \mu g\,chl\,l^{-1}$ within the upper 30 m before a resuspension event, would yield $7.88\ \mu g\,chl\,l^{-1}$ after vertical homogenization in response to such a sequence of wind events (Figures 14.7b and 14.7c). A doubling of such a phytoplankton population after two days would then yield the estimated CZCS chlorophyll concentration of $\sim 16\ \mu g$-$chl\,l^{-1}$ seen on 19 April 1979 (Figure 14.6c).

With southerly wind forcing, surface flow is offshore and to the east, reversing the predominantly westward currents if a storm is of sufficient

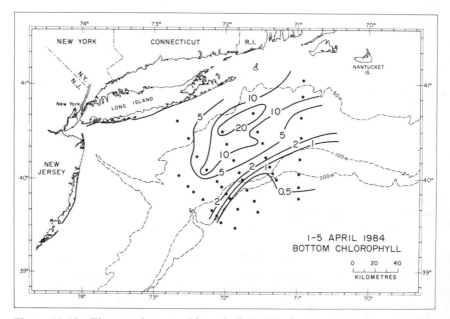

Figure 14.10   The near-bottom chlorophyll distribution during 1–5 April 1984 as measured aboard the R/V *Oceanus* (after Walsh *et al.*, 1987b). (Reproduced with permission from *Cont. Shelf Res.*, in press, Pergamon Journals Ltd.)

intensity. During 20–25 April 1979, a mean wind forcing of 0.32 dynes cm^{-2} from the south (174° T) was sufficient to drive weak mean currents (<5 cm sec^{-1}) to the northeast within the 20–50 m isobaths (Figure 14.7d), in contrast to the three previous flow fields of the model (Figures 14.7a–c). Within this simulated circulation pattern, which should lead to upwelling again, at least on the inner shelf like the case of 5–12 April 1979 (Figure 14.7a), the observed chlorophyll concentrations decline by an order of magnitude within 2 days, i.e. by 21 April 1979 (Figure 14.6d). Such a tenfold decline in near-surface chlorophyll concentrations between 19 April

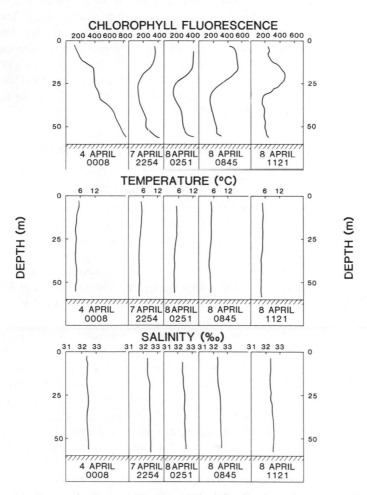

Figure 14.11 The continuous, vertical distribution of chlorophyll fluorescence off the coast of Long Island during 4–8 April 1984 (after Walsh *et al.,* 1987b). (Reproduced with permission from *Cont. Shelf Res.,* in press, Pergamon Journals Ltd.)

(Figure 14.6c) and 21 April (Figure 14.6d) implies rapid sinking and/or down-welling of the phytoplankton, since grazing stress removes less than 10 percent of the early spring bloom (Walsh *et al.,* 1978).

No continuous fluorescence data with depth are available for April 1979, but such a chlorophyll time series was obtained during 4–8 April 1984 (Figure 14.11) at the 60 m isobath, south of Long Island. It suggests a downward displacement of about 30 m day^{-1} (3×10^{-2} cm sec^{-1}) on 7–8 April 1984 (Walsh *et al.,* 1987b). Furthermore, the vertically integrated biomass of phytoplankton, in suspension during this period at the 60 m isobath, was only 54–72 percent of that previously sunk out on 4 April 1984 (Figure 14.11), i.e. loss of algal biomass has occurred during this 8 April 1984 resuspension event. After March–April wind transport events off Peru, the integrated chlorophyll biomass over the upper 40 m similarly decreased by 10–25 percent in 1976 and 50–75 percent in 1977 (Walsh *et al.,* 1980). We now attempt to quantify such a net offshore loss, or export, of phytoplankton biomass from the 60 m isobath, in terms of 1984 carbon fluxes to adjacent slope depocentres (Walsh *et al.,* 1987b).

1984 MOORED FLUOROMETER TIME SERIES

Twice as much near-surface chlorophyll was found within the satellite images during the 19 April 1979 resuspension event (Figure 14.6c) above the 40–60 m isobaths, compared to a previous 21 March 1979 event detected by the CZCS above the 20–40 m isobaths (Walsh *et al.,* 1987a). Such a temporal sequence over ~1 month implies both seasonal build up of chlorophyll within the aphotic zone and a gradual transfer of uneaten phytodetritus seaward. A 30–40 km offshore migration of the algal resuspension area, from the 20–40 m isobaths to the 40–60 m isobaths, within ~30 days from 21 March to 19 April 1979, suggests a mean net seaward movement of algal particles of 1–1.3 km day^{-1} (~1–1.3 cm sec^{-1}). Since much of the wind forcing is from the north in the Mid-Atlantic Bight during February–April, with more frequent northwest storms from off the North American continent, the average flow of the upper 30 m of the water column past 4 current meter arrays, between the 45–105 m isobaths off Martha's Vineyard (Beardsley *et al.,* 1983), was a net 7.73 cm sec^{-1} to the west and a net 1.43 cm sec^{-1} offshore during February–April 1980 and April 1979. At a rapid sinking rate of 30 m day^{-1}, the phytoplankton of the spring bloom would remain in the euphotic zone (~30 m depth), however, for one day until the next wind resuspension event.

During northeast transport events, offshore tranport also occurs, but near the bottom; during April 1979 and February–April 1980, the mean subsurface flow (>30 m) was offshore at 1.98 cm sec^{-1} past moorings at the 66–88 m isobaths, south of Martha's Vineyard (Beardsley *et al.,* 1983). Vertical decomposition of the model's flow field during the northeast wind forcing event of 12–16 April 1979 also leads to offshore flow in the bottom layer, i.e. >45 m depth (Figure 14.12). In fact, time series of the u and v components (UCMP, VCMP) of horizontal flow and chlorophyll from current meter and moored fluorometer observations tethered 3–5 m off the bottom at the 80 m isobath off Long Island 5 years later during 17 February–7 April 1984, i.e. seaward of the time series of vertical fluorescence (Figure 14.11), yield a mean offshore flow of 1.0 cm sec^{-1} and a net seaward chlorophyll transport, $\overline{up}$, of about 3.0 ng chl cm^{-2} sec^{-1} (Figure 14.13).

Using another mooring of current meters and fluorometers, 12 km farther offshore at the 120 m isobath, the average chlorophyll flux at 80 m depth near the shelf-break, during February–April 1984, can then be calculated (Walsh *et al.*, 1987b). We obtain 0.35–0.47 g C m^{-2} day^{-1}, or 35–47g C m^{-2} over at least the 100-day period of the spring bloom, assuming a C/chl ratio of 45/1 and a bottom boundary layer of 30–40 m thickness. How long might such a shelf export continue to exit the Mid-Atlantic Bight in a combination of both offshore flow events at the surface, detected by satellites, and offshore flow events at the bottom, detected by moored instruments?

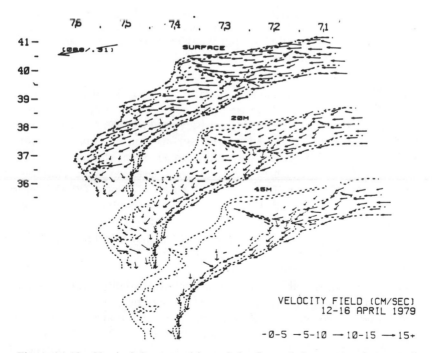

Figure 14.12 Vertical decomposition of the flow of the water column at the surface, at 20 m, and at 45 m under northeasterly wind forcing during 12–16 April 1979 (after Walsh *el al.,* 1987a). (Reproduced with permission from *Deep-Sea Res.*, **34**, 1987, Pergamon Journals Ltd.)

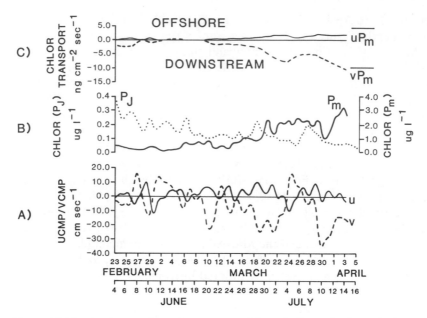

Figure 14.13 A time series of currents, chlorophyll, and phytoplankton transport past a mooring 3–5 m above the 80 m isobath, south of Long Island, during 17 February–7 April 1984 (after Walsh *et al.*, 1987a). (Reproduced with permission from *Deep-Sea Res.*, **34**, 1987, Pergamon Journals Ltd.)

During October–May, the mean diabathic component of the surface flow (0–30 m) between the 45–105 m isobaths, south of Martha's Vineyard, was 1.77 cm sec^{-1} offshore in 1979–80 (Beardsley *et al.*, 1983); weak onshore flow of 0.05–0.20 cm sec^{-1} occurred in the surface during June–August. An offshore flow of ~1.5 cm sec^{-1} over the upper 30 m of the water column across the ~1000 km length of the shelf-break between Cape Hatteras and Martha's Vineyard implies a net, seaward water transport of 0.45 Sverdrups (10^6 m^3 sec^{-1}) from the Mid-Atlantic Bight. Assuming a CZCS-derived horizontal chlorophyll gradient of 2.94 µg chl l^{-1} on the shelf and 0.44 µg chl l^{-1} within slope waters (Figures 14.6a–d), a C/chl ratio of 45/1, and a surface water flux of 0.45 Sv across the shelf break, a daily export of 0.44×10^{10} g C day^{-1} might occur.

Over a year and the area of the adjacent Mid-Atlantic continental slope (~4×10^{10} m^2), at least 41.0 g C m^{-2} yr^{-1} might be imported to this continental slope, based on this CZCS estimate. Over the area of world slopes (Table 14.1), such an annual flux would amount to 1.3×10^9 tons C yr^{-1}. Recent estimates of ^{210}Pb and ^{14}C rates of sediment mixing on the Mid-Atlantic slope, of vertical carbon gradients within the upper 10 cm of this sediment, and of anthropogenic nitrogen loading to the coastal zone suggest an accumulation rate of $9.9-16.7\,\mathrm{g\,C\,m^{-2}\,yr^{-1}}$ on the Mid-Atlantic slope (Walsh *et al.*, 1985). This is only 24–41 percent of the estimated import, suggesting a possible world slope accumulation of at least $0.3-0.5 \times 10^9$ tons C $\mathrm{yr^{-1}}$ (Table 14.1), with perhaps the difference lost to oxidation in the water column and sediments.

To examine the long-term implications of a combined two-layered offshore transport of the spring bloom during the resuspension events detected by the CZCS time series (Figure 14.6) during April 1979 and by the FTD time series (Figure 14.11) and the moored fluorometer time series (Figure 14.13) in April 1984, we analysed our winter–spring shipboard data base, taken over ten years. The chlorophyll observations at each ship station (Figure 14.14) were subdivided into those from surface (0–20 m) and subsurface (20–50 m or 20–75 m) waters on the middle shelf (20–50 m isobaths), the outer shelf (50–100 m isobaths), and the upper slope (100–1000 m isobaths). Although these station data were from different years (1973–82), each cruise was of sufficient duration (1–2 weeks) in February–May to capture the biological response to wind events at a frequency of $0.2\,\mathrm{day^{-1}}$. We have plotted the envelope of maximum chlorophyll concentration encountered by the ship each Julian day (Figure 14.14), for all the cruise data. This analysis represents the statistical ensemble of resuspension events the CZCS would be likely to detect, as maximum surface chlorophyll, over a decade of infrequent, non-stationary time series (Figure 14.6).

Maximal shipboard chlorophyll concentrations as high as 16 $\mu\mathrm{g\,chl\,l^{-1}}$ and as low as 4 $\mu\mathrm{g\,chl\,l^{-1}}$ were found, similar to the CZCS images during April 1979, with the peaks separated in time by perhaps 5 days (Figure 14.14). The slanted lines of Figure 14.14 are thus drawn 5 days apart and suggest a seaward translation of maximal chlorophyll concentrations, from mid-shelf to the slope, in response to wind forcing of the same frequency. If the algal populations of the surface maxima (0–20 m) on the mid-shelf, outer shelf, and upper slope become the subsurface chlorophyll maxima (20–46 m) 5 Julian days later in time, a sinking velocity of at least $4-5\,\mathrm{m\,day^{-1}}$ is implied. If the diatom populations of the mid-shelf spring bloom are translated 60–80 km across the the shelf to become part of the phytoplankton populations (i.e. including daughter cells produced en route) of the slope water 40 Julian days later in time, an offshore velocity of at least $1.5-2.0\,\mathrm{cm\,sec^{-1}}$ is implied.

Such inferred vertical and horizontal displacement rates of the chlorophyll ensemble over 10 years (Figure 14.14) are not inconsistent with estimates of sinking and offshore advection rates obtained from CZCS images, shipboard measurements and moored *in situ* data, during the 1979 and 1984 spring blooms. Such horizontal exchange rates also provide reasonable carbon budgets for export of particulate matter from the shelf. As additional analyses of spatially synoptic satellite data provide more time series on algal biomass changes in shelf and slope waters, we will be able to specify the interannual

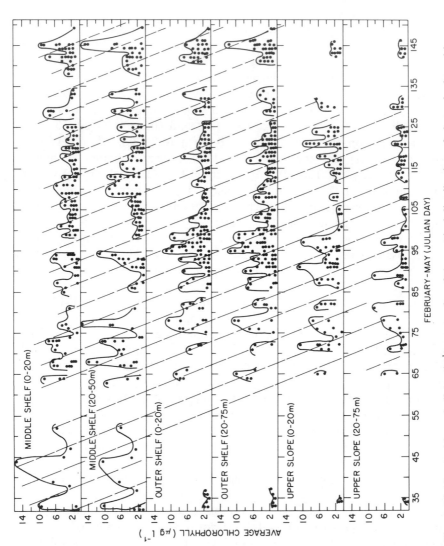

Figure 14.14 A chlorophyll (μg l^{-1}) composite of the vertical distribution of phytoplankton biomass across the shelf and slope of Mid-Atlantic Bight from over 50 cruises taken during February–May 1973–1982

and possible long-term changes of shelf export from both the Mid-Atlantic Bight and other shelves. These data will provide a basis to refine our present understanding of global biogeochemical cycles between their oceanic, terrestrial, and atmospheric reservoirs (Walsh, 1984).

ACKNOWLEDGEMENTS

This research was sponsored by NASA for reduction of the CZCS data, by DOE for acquisition of the shipboard rate data, and by NOAA for providing biomass data and the use of their research vessels.

REFERENCES

Alversen, D. L., Longhurst, A. R., and Gulland, J. A. (1970). How much food from the sea? *Science,* **168,** 503–505.

Beardsley, R. C., and Butman, B. (1974). Circulation on the New England continental shelf: Response to strong winter storms. *Geophysical Research Letters,* **1,** 181–184.

Beardsley, R. C., Mills, C. A., Vermersch, J. A., Brown, W. S., Pettigrew, N., Irish, J., Ramp, S., Schlitz, R., and Butman, B. (1983). *Nantucket Shoals Flux Experiment (NSFE 79) Part 2: Moored Array Data Report,* pp. 1–140. WHOI-83-13.

Blackman, R. B., and Tukey, J. W. (1957). *The Measurement of Power Spectra,* pp. 1 190. Dover Press.

Bogorov, V. G., Vinogradov, M. D., Varonina, N. M., Kanaeva, I. P., and Suetova, I. A. (1968). Distribution of zooplankton biomass within the surficial layer of the world ocean. *Dokl. Akad. Nauk. USSR,* **182,** 1205–1207.

Broecker, W. S., Takahashi, T., Simpson, H. J., and Peng, T. H. (1979). Fate of fossil fuel carbon dioxide and the global carbon budget. *Science,* **206,** 409–418.

Clarke, G. K., Ewing, G. C., and Lorenzen, C. J. (1970). Spectra of backscattered light from the sea obtained from aircraft as a measure of chlorophyll concentration. *Science,* **167,** 1119–1121.

DeVooys, C. G. N. (1979). Primary production in aquatic environments. In Bolin, B., Degens, E. T., Kempe, S., and Ketner, P. (Eds.) *The Global Carbon Cycle,* pp. 259–292. Wiley, New York.

Eppley, R. W. (1980). Estimating phytoplankton growth rates in the central oligotrophic oceans. In Falkowski, P. G. (Ed.) *Primary Productivity in the Sea,* pp. 231–242. Plenum, New York.

Eppley, R. W. and Peterson, B. J. (1980). Particulate organic matter flux and planktonic new production in the deep ocean. *Nature,* **282,** 677–680.

Geiskes, W. W. C., Kraay, G. W., and Baars, M. A. (1979). Current ^{14}C methods for measuring primary production: gross underestimates in oceanic waters. *Neth. J. Sea Res.,* **13,** 58–78.

Gordon, H. R. and Clark, D. K. (1980). Remote sensing optical properties of a stratified ocean: an improved interpretation. *Appl. Optics,* **19,** 3428–3430.

Gordon, H. R., Clark, D. K., Mueller, J. L., and Hovis, W. A. (1980). Phytoplankton pigments derived from the Nimbus-7 CZCS: initial comparisons with surface measurements. *Science,* **210,** 63–66.

Gordon, H. R., Clark, D. K., Brown, J. W., Brown, O. B., and Evans, R. H. (1982). Satellite measurement of the phytoplankton pigment concentration in the surface waters of a warm core Gulf Stream ring. *J. Mar. Res.,* **40,** 491–502.

Gordon, H. R., and Morel, A. G. (1983). *Remote Assessment of Ocean Color for Interpretation of Satellite Visible Imagery,* pp. 1–114. Springer-Verlag.

Gordon, H. R., Clark, D. K., Brown, J. W., Brown, O. B., Evans, R. H., and Broenkow, W. W. (1983). Phytoplankton pigment concentration in the Middle Atlantic Bight: comparison of ship determinations and CZCS estimates. *Applied Optics,* **22,** 20–36.

Hovis, W. A. and Leung, K. C. (1977). Remote sensing of ocean color. *Opt. Engineer.,* **16,** 158–164.

Hovis, W. A., Clark, D. K., Anderson, F., Austin, R. W., Wilson, W. H., Baker, E. T., Ball, B., Gordon, H. R., Meuller, J. L., El-Sayed, S. Z., Sturm, B., Wrigley, R. C., and Yentsch, C. S. (1980). Nimbus-7 Coastal Zone Color Scanner systems: description and initial imagery. *Science,* **210,** 60–63.

Kelley, J. C. (1976). Sampling the sea. In Cushing, D. H., Walsh, J. J. (Eds.) *Ecology of the Seas,* pp. 361–387. Blackwell, Oxford.

Koblentz-Mishke, O. J., Valkovsinky, V. V., and Kabanova, J. C. (1970). Plankton primary production of the world ocean. In Wooster, W. S. (Ed.) *Scientific Exploration of the South Pacific,* pp. 189–193. Nat. Acad. Sci., Washington, DC.

Morel, A., and Smith, R. C. (1974). Relation between total quanta and total energy for aquatic photosynthesis. *Limnol. Oceanogr.,* **19,** 591–600.

Morel, A., and Prieur, L. (1977). Analysis of variations in ocean color. *Limnol. Oceanogr.,* **22,** 708–722.

OAO Corporation. (1979). *The Nimbus-7 CZCS data catalog: February* 1–28, 1979, pp. 1–59. Goddard Space Flight Center Contract NASS-25937.

O'Reilly, J. E. and Busch, D. A. (1984). Phytoplankton primary production on the northwestern shelf. *Rapp. P.-v. Reun. Cons. Int. Explor. Mer,* **183,** 255–268.

Platt, T. and Subba Rao, D. V. (1975). Primary production of marine microphytes. In Cooper, J. P. (Ed.) *Photosynthesis and Productivity in Different Environments,* pp. 249–80. Cambridge Univ. Press.

Platt, T., Denman, K. L., and Jassby, A. D. (1977). Modeling the productivity of phytoplankton. In Goldberg, E. D., McCave, I. N., O'Brien, J. J., and Steele, J. H. (Eds.) *The Sea, Vol. 6,* pp. 807–856. Wiley Interscience, New York.

Postma, H. (1978). The nutrient contents of North Sea water: changes in recent years, particularly in the Southern Bight. *Rapp. P.-v. Reun. Cons. Int. Explor. Mer,* **172,** 350–352.

Premuzic, E. T., Benkovitz, C. M., Gaffney, J. S., and Walsh, J. J. (1982). The nature and distribution of organic matter in the surface sediments of world oceans and seas. *Organ. Geochem.,* **4,** 63–77.

Ryther, J. H. (1969). Photosynthesis and fish production in the sea. *Science,* **166,** 72–76.

Smith, R. C. and Baker, K. S. (1978a). The bio-optical state of ocean waters and remote sensing. *Limnol. Oceanogr.,* **23,** 247–259.

Smith, R. C., and Baker, K. S. (1978b). Optical classification of natural waters. *Limnol. Oceanogr.,* **23,** 260–267.

Smith, R. C., and Wilson, W. H. (1981). Ship and satellite bio-optical research in the California Bight. In Gower, J. R. F. (Ed.) *Oceanography from Space,* pp. 281–294. Plenum Press, New York.

Smith, R. C., and Baker, K. S. (1982). Oceanic chlorophyll concentrations as determined using Coastal Zone Color Scanner imagery. *Mar. Biol.,* **66,** 269–279.

Sverdrup, H. U. (1953). On conditions for the vernal blooming of phytoplankton. *J. Cons. Int. Explor. Mer,* **18,** 287–295.

Systems and Applied Sciences Corporation. (1984). *Users Guide for the Coastal Zone*

Color Scanner Compressed Earth Gridded Data Sets of the Northeast Coast of the United States for February 28 through May 27, 1979, pp. 1–15. SASC-T-5-5100-0002-008-84.

Thomas, J. P. (1981). Large Area Marine Productivity-Pollution Experiments (LAMPEX)—A series of studies being developed to hasten the operational use of remote sensing for living marine resources and environmental quality. In Gower, J. F. R. (Ed.) *Oceanography from Space,* pp. 403–409. Plenum Press, New York.

Walsh, J. J. (1972). Implications of a systems approach to oceanography. *Science,* **196,** 969–975.

Walsh, J. J. (1975). A spatial simulation model of the Peru upwelling ecosystem. *Deep-Sea Res.,* **22,** 201–236.

Walsh, J. J. (1976). Herbivory as a factor in patterns of nutrient utilization in the sea. *Limnol. Oceanogr.,* **21,** 1–13.

Walsh, J. J. (1980). Concluding remarks: marine photosynthesis and the global cycle. In Falkowski, P. G. (Ed.) *Primary Productivity in the Sea,* pp. 497–506. Plenum Press, New York.

Walsh, J. J. (1983). Death in the sea: enigmatic phytoplankton losses. *Prog. Oceanogr.,* **12,** 1–86.

Walsh, J. J. (1984). The role of the ocean biota in accelerated global ecological cycles: a temporal view. *Bioscience,* **34,** 499–507.

Walsh, J. J., and Dugdale, R. C. (1971). A simulation model for the nitrogen flow in the Peruvian upwelling system. *Inv. Pesq.,* **35,** 309–330.

Walsh, J. J., and Howe, S. O. (1976). Protein from the sea: a comparison of the simulated nitrogen and carbon productivity in the Peru upwelling system. In Patten, B. C. (Ed.) *Systems Analysis and Simulation in Ecology, Vol.* IV, pp. 47–61. Academic Press, New York.

Walsh, J. J., Whitledge, T. E., Barvenik, F. W., Wirick, C. D., Howe, S. O., Esaias, W. E., and Scott, J. T. (1978). Wind events and food chain dynamics within the New York Bight. *Limnol. Oceanogr.,* **23,** 659–683.

Walsh, J. J., Whitledge, T. E., Esaias, W. E., Smith, R. L., Huntsman, S. A., Santander, H., and DeMendiola, B. R. (1980). The spawning habitat of the Peruvian anchovy, *Engraulis ringens. Deep Sea Res.,* **27,** 1–27.

Walsh, J. J., Rowe, G. T., Iverson, R. L., and McRoy, C. P. (1981). Biological export of shelf carbon is a neglected sink of the global CO_2 cycle. *Nature,* **291,** 196–201.

Walsh, J. J., Barnes, W. B., Brown, O. B., Carder, K. L., Clark, D. K., Esaias, W. E., Gordon, H. R., Holyer, R. C., Hovis, W. A., Kirk, R. J., Lasker, R., McCarthy, J. J., McElroy, M. A., Mueller, J. L., Perry, M. J., and Smith, R. C. (1982). *The MAREX (Marine Resources Experiment) Program. A report of the Ocean Color Science Working Group,* pp. 1–102. Goddard Space Flight Center, Greenbelt, Maryland.

Walsh, J. J., Premuzic, E. T., Gaffney, J. S., Rowe, G. T., Harbottle, G., Stoenner, R. W., Balsam, W. L., Betzer, P. R., and Macko, S. A. (1985). Organic storage of CO_2 on the continental slope off the Mid-Atlantic Bight, the Southeastern Bering Sea, and the Peru coast. *Deep-Sea Res.,* **32**(7), 853–883.

Walsh, J. J., Dieterle, D. A., and Esaias, W. E. (1987a). Satellite detection of phytoplankton export from the Mid-Atlantic Bight during the 1979 spring bloom. *Deep-Sea Res.,* **34,** 675–703.

Walsh, J. J., Wirick, C. D., Pietrafesa, L. J., Whitledge, T. E., Hoge, F. E., and Swift, R. N. (1987b). High frequency sampling of the 1984 spring bloom within the Mid-Atlantic Bight: synoptic shipboard aircraft, and *in situ* perspectives of the SEEP-I experiment. *Cont. Shelf Res.* (in press).

Walsh, J. J., Dieterle, D. A., and Meyers, M. B. (1987c). A simulation analysis of the fate of phytoplankton within the Mid-Atlantic Bight. *Cont. Shelf Res.* (in press).

Woodwell, G. M., Whittaker, R. H., Reiners, W. A., Likens, G. E., Delwiche, C. C., and Botkin, D. B. (1978). The biota and the world carbon budget. *Science,* **199**, 144–146.

Wroblewski, J. S. (1977). A model of phytoplankton plume formation during variable Oregon upwelling. *J. Mar. Res.,* **35**, 357–394.

Scales and Global Change
Edited by T. Rosswall, R. G. Woodmansee and P. G. Risser
© 1988 Scientific Committee on Problems of the Environment (SCOPE)
Published by John Wiley & Sons Ltd.

CHAPTER 15

The Ocean System—Ocean/Climate and Ocean/CO₂ Interactions

HANS OESCHGER

Physics Institute, University of Bern,
Sidlerstrasse 5, CH-3012 Bern,
Switzerland

ABSTRACT

This chapter serves the purpose to demonstrate the achievements and the potential of studies of information on earth and planetary system processes recorded in natural archives, such as ocean and lake sediments, peat bogs, tree-rings, and polar ice. Especially valuable information carriers are the radioactive and stable isotopes of elements like C, H, and O. As an example ^{14}C analyses, beside dating, make the reconstruction of solar modulation of cosmic radiation possible, enable estimates of the CO_2 fluxes between atmosphere and ocean and of the turnover characteristics of ocean water. CO_2 measurements on air occluded in polar ice enabled reconstruction of the anthropogenic CO_2 increase and showed that the atmospheric CO_2/air ratio varied also in the preindustrial era. Detailed studies of a series of environmental parameters indicate the existence of a climatic flip–flop mechanism in the North Atlantic area during the Wisconsin.

An essential point is the observation that the earth system mechanisms are so complex that for real progress in their understanding, studies making use of modern analytical methods which reveal the experiments the system performed in the past play an important role.

INTRODUCTION

The central focus of the proposed International Geosphere–Biosphere Programme (IGBP) is 'to describe and understand the interactive physical, chemical and biological processes that regulate the total earth system, the unique environment for life, the changes that are occurring in this system, and the manner in which they are influenced by human actions'.

This holistic view of the environment emerged from studies of the global geophysical and biogeochemical experiments which man has been performing since about 150 years ago, but also from the reconstruction of experiments which nature itself has performed in the past. These experiments show the complex interactions between the individual parts of the environmental system on which hitherto research has tended to concentrate.

In the holistic studies, tracers as the radioactive and stable isotopes play a major role. Much of the discussion in this paper is therefore devoted to information on environmental systems and especially ocean system processes revealed from isotope studies. Lesser emphasis is given to the more classical theoretical and experimental studies which will certainly be fundamental to the understanding of the ocean system and its interactions in the future. During the last few years, due to the rising atmospheric CO_2 content and its probable impact on climate, the ocean/CO_2 cycle interaction and the ocean/climate interaction have received considerable attention by scientists and those interested in man's impact on the environment. In this chapter, therefore, main emphasis is given to these topics and recommendations are made regarding the improvement of knowledge on the interaction of the ocean with the other parts of the environmental system.

THE OCEAN'S ROLE IN ENVIRONMENTAL PROCESSES

Among the great number of ways in which the ocean influences the environment, we emphasize the following:

—Ocean circulation helps to distribute energy over the globe.

—The oceans are the main source of water vapour which carries latent heat, condenses and falls as precipitation, partly over the continents.

—The atmospheric CO_2 concentration is determined by the partial pressure of dissolved CO_2 in the ocean surface. This partial pressure in turn is determined by the chemical equilibria of the carbonate system.

– Atmospheric and oceanic circulation interact in a very complex manner. This nonlinear system at present seems to oscillate between different modes of operation. An example is the quasi-periodical suppression of the upwelling of subsurface water in the Equatorial Pacific Ocean combined with a changing wind system. This so-called El Niño phenomenon leads to strong increases in the temperature of Equatorial Pacific surface water and has a significant influence on weather and climate of a great part of the earth's surface.

—During the last glaciation the ocean showed a different circulation pattern and seems to have oscillated between two quasi-stable modes of operation.

This points to the sensitivity of the ocean circulation to changing boundary conditions, as sea-level, salinity, sea-ice cover, solar irradiation, etc.

—Related to man's impact on the environment, the ocean plays an important role since it is the main sink for CO_2 emitted by human activities, and since it dampens (due to its heat capacity) changes of the energy balance at the earth's surface induced by the increase of 'greenhouse' gases. Today, the uptake of CO_2 and excess heat can be estimated with some confidence. However, the open question remains if the ocean in this period of global change will operate as it did during the last decades and centuries.

TRACER STUDIES OF ENVIRONMENTAL SYSTEM PROCESSES AND THEIR HISTORY

As a basis for the discussion of the physical, chemical, and biological role of the ocean in a broad spectrum of processes on the earth's surface, in the following we introduce a concept of an environmental system (E.S.) research which evolved from isotope and tracer studies.

Figure 15.1 illustrates this research concept. The E.S. includes the entirety of physical, chemical, and biological processes, acting upon the earth's surface and in the atmosphere. The various parts of the system interact in various dynamic sequences and are in contact with the planetary and galactic systems. The E.S., as defined here, agrees largely with the climate system as it is generally defined, but stronger emphasis is given to chemical and biological processes. Special attention is drawn to the parameters which can be studied in natural archives and therefore enable the reconstruction of ancient system states.

The main components of the E.S. are the atmosphere, the hydrosphere, including the oceans and continental waters, the cryosphere, consisting of the polar ice sheets, sea ice, and mountain glaciers, the biosphere, consisting of the marine and continental living organisms, and the lithosphere, with bedrock and sediments which interface with the hydrosphere.

The energy of the sun is driving the dynamic processes in the E.S. It causes atmospheric circulation and oceanic mixing, and due to evaporation and precipitation, the cycling of water. The energy balance determines the climatic conditions at individual locations on the earth's surface. It is affected by scattering and reflection of short wavelength solar radiation in the atmosphere and on the earth's surface, and by the infrared radiation emission, absorption and re-emission by the surface and by water vapour and gases ($CO_2, O_3, \ldots$) in the atmosphere.

The dynamic cycles of some elements, such as C, N, and O, involve interactions between several of the E.S. components.

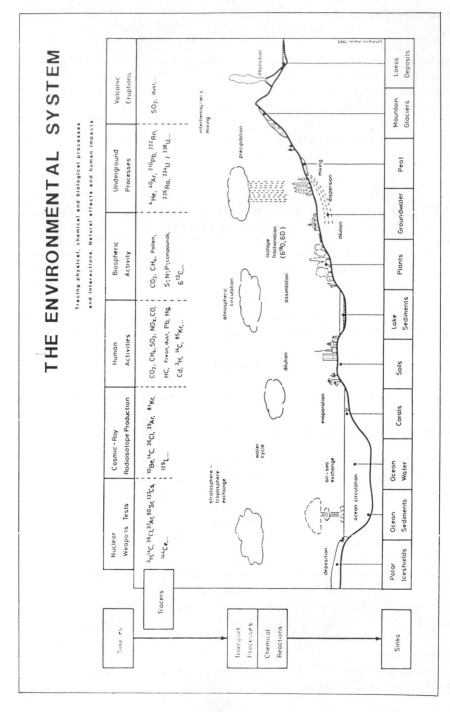

Figure 15.1   Diagram of the environmental system research concept based on isotopes and tracers

Dust particles and aerosols are injected into the atmosphere by wind action, volcanic eruptions, biospheric processes, and human activities. They reappear on the earth's surface as dry fallout or wet deposition.

Of special interest for our approach to the understanding of the E.S. processes are the radioactive and stable isotopes which constitute ideal tracers for a variety of processes and their dynamics. Radioactive isotopes have different origins:

(a) They are formed by interaction of cosmic radiation with atoms in the upper atmosphere (e.g. ^{10}Be, ^{14}C, ^{36}Cl, ^{39}Ar, ^{81}Kr)

(b) They are also introduced to the E.S. as a result of man's use of nuclear fusion and fission (e.g. ^{3}H, ^{37}Ar, ^{85}Kr)

(c) They are released from the earth's crust as products of the natural decay series of U and Th (e.g. ^{222}Rn, ^{210}Pb).

Since halflives vary from days (^{37}Ar) to hundreds of thousands of years (^{10}Be, ^{36}Cl, ^{81}Kr), information about time constants of natural processes over a very wide range is attainable. Each isotope has its characteristic field of applications which may reach far beyond that of merely dating.

Stable isotopes (e.g. ^{2}H, ^{13}C, ^{15}N, ^{18}O) are other important sources of information. Phase transitions, chemical reactions and diffusion processes produce small changes in the natural isotope ratios. They reflect the conditions at which the processes occurred. Elements originating from different natural reservoirs can often be distinguished based on their different isotopic composition. Samples of air, water, and ice, organic materials and sediments, taken from any part of the E.S., thus contain information on its static characteristics, like the partitioning of water between atmosphere, cryosphere, and ocean, but also its dynamic characteristics, like mixing and circulation, and exchange processes in and between the different system components. This and other information can also be derived from the concentration levels of chemical elements and molecules, pollen and dust. A complete set of all these parameters defines the state of the E.S. As 'fingerprint parameters' they are continuously recorded in the natural archives, as in polar ice sheets, mountain glaciers, ocean and lake sediments, and organic materials like tree-rings or peat and coral deposits. Analyses of sequential samples allow the reconstruction of the historical evolution of the E.S.

Mathematical simulation models play an important role in this research. The models are based on the fundamental physical, chemical, and biological knowledge. Since the various processes are very complex, it is often necessary to simplify the equations and to use empirical procedures to arrive at conclusions. Of course, the models should produce system responses to various perturbations which closely agree with observations. Man's impact upon the environment is an example of a present perturbation. Past system perturbations and responses are revealed from the natural archives.

THE CO$_2$/OCEAN INTERACTION

The global carbon cycle

Carbon is essentially exchanged between the four reservoirs shown in Figure 15.2. Due to the burning of fossil fuels and the human impact on the biomass (deforestation and changing land use), the carbon concentrations in the atmosphere and ocean are rising. The atmosphere contains at present some 700 Gt (1 Gt = 1 Gigaton = 10^{12} kg) carbon in form of CO$_2$. Precise, continuous atmospheric CO$_2$ measurements have only been performed since 1958. Since then the CO$_2$ concentration has increased from 315 ppm to 345 ppm in 1985. The atmospheric CO$_2$ significantly influences the terrestrial radiation balance. It is expected that its increase will lead to a global warming.

The ocean contains roughly 40 000 Gt, about 60 times the atmospheric amount. The carbon is present in inorganic form as dissolved CO$_2$ [CO$_2$, aq] ($\sim 1\%$), as bicarbonate [HCO$_3^-$] ($\sim 90\%$) and carbonate ions [CO$_3^{2-}$] ($\sim 9\%$), but also as dissolved organic carbon compounds, ca. 3% of the inorganic C. The sum [CO$_2$, aq] + [HCO$_3^-$] + [CO$_3^{2-}$] is often referred to as ΣCO$_2$ or total CO$_2$.

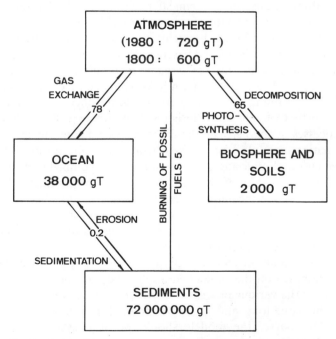

Figure 15.2   Global carbon system main reservoirs and fluxes. Units are gigatons (10^9 T) C for amounts, and gigatons C per year for fluxes. Steady state assumed

The size of the carbon pool on the terrestrial biosphere is difficult to estimate because of large regional differences in its distribution. The values range between 2000 and 3000 Gt or 3 to 4 times the amount of atmospheric CO_2.

In sediments, carbon is stored in huge amounts, estimated at 60 000 000 Gt.

Of course, each of these main reservoirs represents itself as a very complicated system and may be subdivided further.

Carbon is exchanged between the different reservoirs. The exchange tries to bring the system to an equilibrium state. In the steady state in each reservoir the influxes compensate the losses.

In quasi-equilibrium states the atmospheric CO_2 content is determined by pCO_2, the CO_2 partial pressure averaged over the ocean surface. As shown in Figure 15.3, pCO_2 is significantly varying, mainly with latitude. It is determined by the chemical equilibria between the dissolved carbonate species H_2CO_3, HCO_3^-, CO_3^{2-} and is essentially a function of ΣCO_2, alkalinity, and temperature. ΣCO_2 and alkalinity, however, are depleted in the surface ocean because biologically produced organic and carbonate particles sink to greater depth, where most of them get oxidized (organics) or redissolved (carbonates). The surface depletion of ΣCO_2 is between 10 and 20% compared to the deep water value.

The marine biological activity depends mainly on the availability of the nutrients phosphate and nitrate. In most of the ocean surface, these nutrients

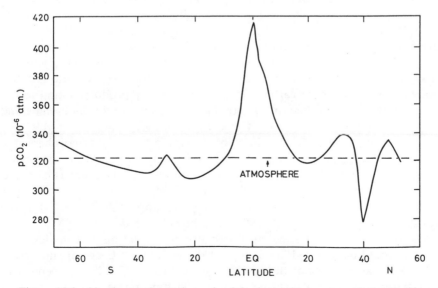

Figure 15.3   North–south section of pCO_2 in surface water of the Pacific Ocean. The equatorial maximum is due to upwelling of CO_2-rich subsurface water. The dashed line shows the atmospheric CO_2 concentration. After Broecker *et al.* (1980)

are essentially consumed and at very low concentrations. This is not the case in the Pacific Equatorial Ocean due to upwelling of nutrient and CO_2-rich subsurface water. This leads to a maximum in pCO_2 in the water in this region (see Figure 15.3). Also, in the Southern Ocean around Antarctica the nutrients are abundant in surface waters, probably since vertical circulation or lack of light are limiting the photosynthetic activity.

The essential role, which the marine biological activity plays in controlling the atmospheric CO_2 concentration, is demonstrated by the estimate that an ocean without biological activity would have an average ocean surface pCO_2 of between two and three times the preindustrial atmospheric CO_2 concentration of 280 ppm, and that in a stagnant ocean, in which all the nutrients were consumed, pCO_2 in the surface water would correspond to a an atmospheric CO_2 concentration of half the atmospheric value. Thus the complex interaction of physical, chemical and biological processes in the ocean determines such an important E.S. parameter as the atmospheric CO_2 content. It is evident that changes in the oceanic behaviour lead to changes in the atmospheric CO_2 concentration.

The carbon isotopic ratios in the CO_2 cycle/ocean System

Natural carbon consists of the two stable isotopes ^{12}C ($\sim 98.9\%$) and ^{13}C ($\sim 1.1\%$). The interaction of secondary cosmic ray neutrons with ^{14}N nuclei in the atmosphere leads to the production of radioactive ^{14}C which gets oxidized to $^{14}CO_2$. In the atmosphere the $^{14}C/^{12}C$ ratio is of the order of 10^{-12}; the halflife of ^{14}C is 5730 yr and the specific activity of living organic matter *ca.* 13.5 dpm g^{-1} C. Figure 15.4 shows the $^{14}C/^{12}C$ ratios in the atmosphere/surface ocean/subsurface ocean system. The atmospheric $^{14}C/^{12}C$ ratio (1.00) is defined as 100% modern carbon. It is determined by production and radioactive decay of ^{14}C in the atmosphere and its net fluxes into the oceanic and biomass carbon reservoirs. The $^{14}C/^{12}C$ ratios in the surface ocean (0.95) are determined by the net ^{14}C flux into this reservoir from the atmosphere, the radioactive decay, and the net flux into the subsurface ocean. In the subsurface ocean ($^{14}C/^{12}C = 0.84$), the net ^{14}C in-flux from the surface ocean is balanced by the radioactive decay. Based on the ratio of 0.84/0.95 for the $^{14}C/^{12}C$ ratios in the subsurface and the surface ocean, an apparent ^{14}C age for the subsurface ocean water of *ca.* 1000 yr can be calculated.

A slowing down or a speeding up of ocean circulation and mixing over millenia would be reflected in changes of the apparent ^{14}C age of subsurface water. Since about 93% of the cosmic ray produced ^{14}C is in the deep ocean and only about 4% in the surface ocean and the atmosphere, a change in ocean dynamics would only slightly affect the $^{14}C/^{12}C$ ratio of the carbonate system in the subsurface ocean. The change would be mainly visible in shifting surface ocean and atmospheric $^{14}C/^{12}C$ ratios.

INFORMATION ON CO_2 AND CLIMATE SYSTEM

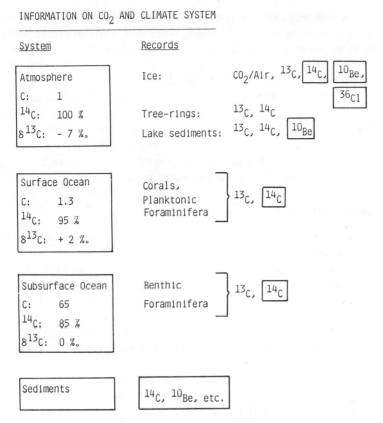

Figure 15.4   Information on CO_2 and climate system. CO₂ content, carbon isotopes, and their natural records

In Figure 15.4 also the $^{13}C/^{12}C$ ratios in atmospheric CO_2 and in the carbonate system of the ocean reservoirs are given. They are expressed as permil deviations ($\delta^{13}C$) from a standard. The mean $\delta^{13}C$ value of the ΣCO_2 in the surface ocean is by *ca*. 1.5‰ higher than that in the subsurface ocean. This difference is due to fractionation during formation of organic matter. In aquatic plants $\delta^{13}C$ is by about 20‰ lower than in the ΣCO_2 in the water in which the plants grow. Animals consuming the plants incorporate in their tissues carbon of nearly the same isotopic composition as the plants. Due to the sinking down of organic particles, carbon with somewhat lower $\delta^{13}C$ is withdrawn from the surface ocean, leaving behind ΣCO_2 slightly enriched in ^{13}C compared to the subsurface ocean. The difference between the mean surface and the mean subsurface $\delta^{13}C$ values in ΣCO_2 thus is a measure of the average biological activity in the surface ocean which in turn determines pCO_2 and correspondingly the atmospheric CO_2 content.

The carbon isotopic ratios in the atmosphere and in the ocean are recorded in various natural archives, like polar ice, tree-rings, and sediments. In natural ice, air bubbles constitute samples of the air at the time of ice formation. From the extracted ancient air, not only can $^{13}C/^{12}C$ and $^{14}C/^{12}C$ ratios of the CO_2 be measured but so too can the CO_2/air ratio.

Surface dwelling planktonic foraminifera reflect the $^{14}C/^{12}C$ and $^{13}C/^{12}C$ ratios of ΣCO_2 in surface water, and, likewise, bottom dwelling benthic foraminifera those of the bottom water. As both kinds of foraminifera are deposited in ocean sediments, parameters relevant to the carbon cycle and its dynamics cannot only be monitored in atmosphere and ocean from present into the future but can also be reconstructed from the past.

Disturbances of the carbon cycle

The release of CO_2 due to human activities has led to disturbances of the atmospheric CO_2 content and the isotopic ratios $^{13}C/^{12}C$ and $^{14}C/^{12}C$ in the exchanging carbon reservoirs. In addition, the $^{14}C/^{12}C$ ratio in the atmosphere, biosphere, and ocean has been considerably increased as a consequence of nuclear weapon tests. On the other hand, there are also natural changes in both the carbon contents of the reservoirs and their isotopic ratios.

The analysis of these perturbations yields clues regarding ocean circulation and mixing and their time history as well as changes in the ocean surface chemistry.

Table 15.1 shows the expected changes in one or several of the atmospheric parameters (CO_2/air, $^{13}C/^{12}C$, $^{14}C/^{12}C$) related to the CO_2 cycle due to human or natural perturbations of the carbon cycle.

The emission of fossil CO_2 into the atmosphere leads to an increase of CO_2/air and a decrease of $^{14}C/^{12}C$ and of $^{13}C/^{12}C$, since in the fossil CO_2 all the ^{14}C atoms have decayed, and since the $^{13}C/^{12}C$ ratio in fossil fuels is *ca.* 18‰ lower than in the atmospheric CO_2. The emission of CO_2 from biomass burning also leads to a CO_2 increase and a decrease of the $^{13}C/^{12}C$ ratio. The

Table 15.1 Carbon cycle disturbances. Expected changes of atmospheric parameters

	pCO_2	^{14}C	^{13}C
EXTERNAL FORCING			
Cosmic radiation		×	
Biomass destruction	×		×
Fossil fuels	×	×	×
Nuclear tests		×	
INTERNAL SYSTEM CHANGES			
CO_2 partitioning	×	×	×
Cycle dynamics		×	

atmospheric $^{14}C/^{12}C$ ratio is not affected, since in biomass this ratio is close to that of the atmosphere.

During atmospheric nuclear weapon tests, huge amounts of ^{14}C were produced. After the test series in 1960 and 1961 the $^{14}C/^{12}C$ ratio in tropospheric CO_2 had about doubled by 1963 (Nydal and Løvseth, 1983, and Figure 15.5). Also in the upper ocean layers, the $^{14}C/^{12}C$ ratio started to rise, and by detailed measurements on ΣCO_2 in ocean water samples, the penetration of the nuclear weapon ^{14}C pulse into the ocean could be followed (Broecker *et al.*, 1980).

To estimate the uptake of a disturbance at the atmosphere–ocean interface often simple one-dimensional (1D) diffusive models have been used. One example is the box-diffusion (BD) model (Figure 15.6), which consists of a surface mixed layer and a diffusive subsurface ocean with constant eddy diffusivity K. The eddy diffusion in these 1D geochemical models describes all kinds of transport processes, projected on a vertical axis. In particular it also represents mixing along inclined isopycnal surfaces.

The box-diffusion model is calibrated using the natural ^{14}C distribution in the ocean and/or the penetration of bomb-produced ^{14}C. It can be validated against two observed carbon cycle perturbations:

(a) the decrease of atmospheric $^{14}C/^{12}C$ due to fossil CO_2 (Suess effect)

(b) the natural ^{14}C variations as observed in tree-rings.

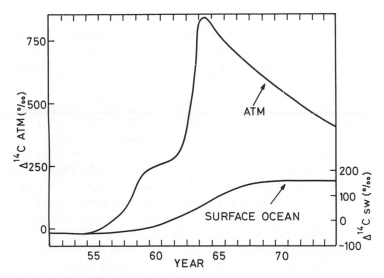

Figure 15.5　$\Delta^{14}C$ of atmospheric CO_2 (left-hand scale) and in ocean surface water (right-hand scale) in response to the input of bomb-produced ^{14}C. After Broecker and Peng (1982)

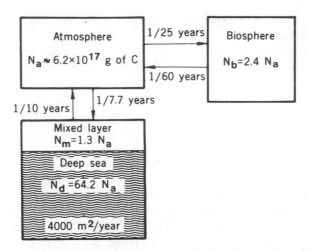

Figure 15.6 Four-reservoir model of CO_2 exchange consisting of atmosphere and mixed layer (well mixed boxes), and the long-term biosphere and an eddy diffusive deep sea. Oeschger *et al.* (1975)

The atmospheric $^{14}C/^{12}C$ ratio decreases due to the emission of ^{14}C-free fossil CO_2; the natural ^{14}C variations are most probably caused by fluctuations of the ^{14}C production which can be derived from the ^{10}Be deposition history in polar ice cores.

(a) Suess effect: The BD model calibrated with bomb-produced ^{14}C predicts a $^{14}C/^{12}C$ decrease of 1.8% (Siegenthaler, 1983), while best observational results, taking into account ^{14}C production variations due to changing solar modulation, lead to a value of 1.7% (Stuiver and Quay, 1981).

(b) Recent AMS measurements [Beer *et al.*, 1983] of the cosmic ray produced isotope ^{10}Be in an ice core from Milcent, Greenland, have provided an approximate record of the radioisotope production variations in the atmosphere due to solar modulation. ^{10}Be, unlike ^{14}C, is attached after production to aerosols and in a few months to about 1 year deposited with precipitation on the surface. Polar ice cores therefore provide a continuous record of the ^{10}Be content of snowfalls, reflecting production of this radioisotope by cosmic radiation. Figure 15.7 (top) shows the observed ^{10}Be concentrations for the period 1200 to 1800 AD. A strong increase is observed for the period of the quiet sun (1640 to 1710), indicating reduced shielding due to solar magnetism. The ^{14}C variations during the corresponding period (Figure 15.7, bottom) show quasi-periodic variations with a period of *ca.* 200 years. Comparison of the ^{14}C variations observed in tree-rings (dashed curve in Figure 15.7, bottom)

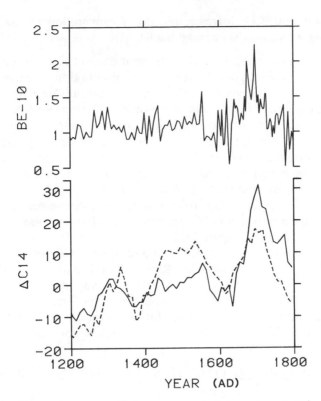

Figure 15.7 Top: [10]Be concentration in units of 10^4 atoms/g ice as a function of age in the Milcent ice core (Greenland). The age was determined based on annual variations of δ^{18}O. Bottom: [14]C in the atmosphere. Dashed curve: tree-ring measurements; solid curve: [14]C model-calculated based on production rates determined from the [10]Be data. Beer *et al.* (1983)

with those calculated by the BD-model (Figure 15.7, bottom, solid line), assuming that the ^{14}C production variations are proportional to those of [10]Be, shows a satisfactory agreement. This demonstrates that the [10]Be and the ^{14}C variations are caused largely by the same phenomenon, i.e. solar modulation of cosmic radiation and that the BD-model is capable of providing correct attenuations and phase shifts for perturbations of this type. However, it needs to be emphasized that the BD-model is a simple simulation model and that the development of physical ocean-circulation models incorporating the carbon cycle is highly desirable. An interesting approach is the model by Maier-Reimer and Hasselmann, 1987.

The anthropogenic CO₂ increase and the deconvolution of the CO₂ input history using an ocean CO₂ uptake model

Next we discuss the anthropogenic CO_2 input due to deforestation and during the last decades mainly due to fossil fuel consumption. Precise continuous measurements of the atmospheric CO_2 content have only been performed since 1958 on Mauna Loa, Hawaii, and at South Pole Station (Keeling *et al.*, 1982). Data prior to 1958 are afflicted with uncertainties due to a variety of reasons. As mentioned before, the bubbles in polar ice constitute physically occluded samples of ancient air. During the last decade the analysis of the air bubbles in ice cores has been perfected, and information on the air occlusion mechanism has been gained. At Siple Station, Antarctica, in a joint US–Swiss operation, an ice core has been obtained which enabled an astonishingly precise reconstruction of the atmospheric CO_2 increase (Neftel *et al.*, 1985) since the first half of the last century (Figure 15.8).

These results indicate a preindustrial atmospheric CO_2 concentration near 280 ppm before about 1800 AD. Raynaud and Barnola, 1985, quote a preindustrial value of 260 ppm. Reasons for the discrepancy are being investigated. Measurements on ice cores from the South Pole support our preindustrial value and indicate that during the past preindustrial millenium the atmospheric CO_2 concentration showed only minor changes (< 10 ppm).

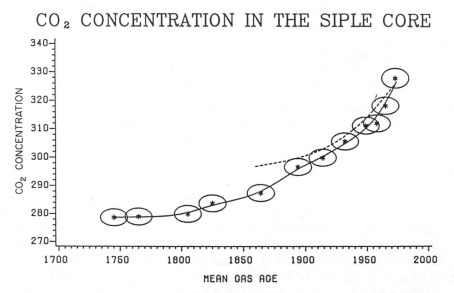

Figure 15.8   Measured mean CO_2 concentration plotted against the estimated mean gas age. The horizontal axis of the ellipses indicates the close-off time interval of 22 y. The uncertainties of the concentration measurements are twice the standard deviation of the mean value. Neftel et al. (1985)

The increase, as shown in Figure 15.8, can therefore be unambiguously attributed to human activities. It started in the first half of the last century, probably mainly due to deforestation. In the middle of the 20th century the input due to fossil fuel consumption became dominant.

In the recent past it has also been possible to obtain $\delta^{13}C$ data on CO_2 extracted from the Siple ice core. These data will add to the establishment of the anthropogenic CO_2 input history and the disentangling of the different sources.

Taking all these results together, one gets the impression that a more and more consistent picture of the human impact on the CO_2 cycle is evolving: The atmospheric CO_2 concentration has been increasing monotonously since the first half of the 19th century. The major sink of the emitted CO_2 is the ocean, and we seem to know the response of the ocean system to a perturbation at the atmosphere/ocean interface sufficiently well to give a reasonable estimate of the anthropogenic CO_2 production function.

Of course, all the elements needed to assess in detail the human impact on the carbon cycle, i.e. the CO_2/air, ^{13}C, and ^{14}C records, the CO_2 uptake by the ocean and possibly by the biosphere and so forth, are still afflicted with uncertainties, and further strong research efforts are required.

The long-term ^{14}C variations: trend in ocean dynamics or in ^{14}C production?

The ^{14}C record in tree-rings not only shows short-term fluctuations as discussed earlier, but also a long-term trend. In Figure 15.9 we see that 9000 to 10 000 years ago the atmospheric $^{14}C/^{12}C$ ratio was 8 to 10% higher than before nuclear weapon testing started in the early 1950s (Suess, 1970). The question is often posed, is the decrease from higher $^{14}C/^{12}C$ ratios at the end of the last glaciation to lower ones during the last 2000 years a result of a change in production, e.g. due to an increasing geomagnetic shielding of cosmic radiation, or does it reflect a speeding up of ocean circulation and mixing which brings the atmospheric and mixed layer $^{14}C/^{12}C$ ratios closer to that of the subsurface ocean?

In our natural archives we can find in principle the answers to this question.

The first test is to follow the ^{10}Be profile in ice cores back into the last glacial period (Figure 15.10). An interesting result is the absence of an increase until about 5000 years before present. Further back, when approaching the glacial–postglacial transition, higher ^{10}Be concentrations can indeed be observed. However, in this time range a different climatic regime is approached. Lower precipitation rates lead to a reduced dilution of atmospheric constituents and therefore higher ^{10}Be concentrations. Thus it is difficult to reconstruct the ^{10}Be fluxes at the glacial–postglacial transition and to normalize them to present conditions.

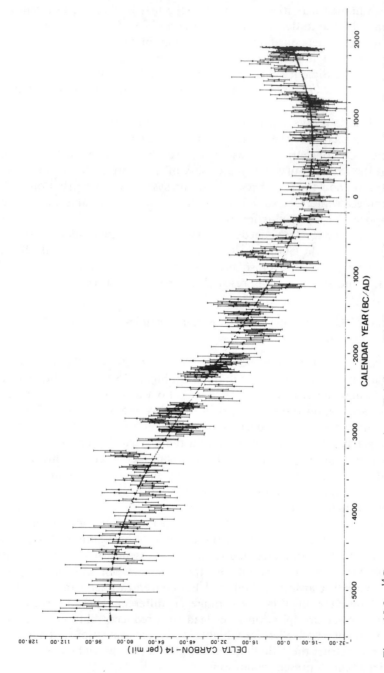

Figure 15.9 ^{14}C measurements on tree-rings (Bristlecone Pine), corrected for the radioactive decay. Deviations from a standard are given in permil. Neftel *et al.* (1981)

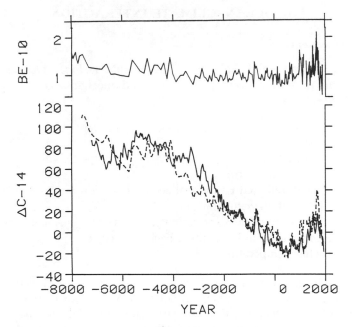

Figure 15.10   Upper part: [10]Be concentration measured on the Camp Century ice core plotted versus time. Lower part: comparison between ^{14}C data (dotted line) calculated with the carbon cycle model using the [10]Be concentrations and the ^{14}C data (solid line) measured on tree-rings. Beer *et al.* (1988)

[14]C measurements on foraminifera in ocean sediments offer another possibility to answer the question raised above. The ^{14}C/^{12}C ratio in surface water is reflected in shells of planktonic (ocean surface dwelling) foraminifera, and that of bottom water in shells of benthonic (ocean bottom dwelling) foraminifera. In high accumulation deep sea cores, where bioturbation only slightly affects stratigraphy, samples of coexisting foraminifera of the two types can be collected for ^{14}C analysis.

This experiment has been successfully performed on a sediment core from the South China Sea, and interestingly enough the apparent age difference between the two foraminifera species remained, within the error limits, essentially constant (Andrée *et al.*, 1986). At least, based on the information from this part of the ocean, no major change in ocean circulation and mixing is indicated in the Holocene. If this finding would hold for the entire ocean, the long-term trend in atmospheric ^{14}C could not be explained by a shift in the ocean dynamics but rather in higher ^{14}C production at the end of the last glaciation and the glacial–postglacial transition.

THE OCEAN–CLIMATE INTERACTION

The climate system

The average global energy balance at the earth's surface gives indications regarding the most important parameters and mechanisms governing the earth's climate. It can be expressed as follows:

$$S\pi R^2 (1 - A) = 4\pi R^2 \sigma T_S^4 (1 - B)$$

with S = solar constant
          R = radius of the earth
          A = albedo (reflected fraction of solar irradiation)
          σ = Stefan–Boltzmann constant
          B = fraction of infrared radiation (emitted from surface), absorbed in
               atmosphere, and re-emitted back to surface
          T_S = surface temperature

The ocean also plays an important role regarding storage and distribution of heat in interaction with atmospheric circulation. Variations of the parameters S, A, and B lead to changes in the passive parameter T_S, the earth's surface temperature. Information on all these four parameters are recorded in natural archives.

The varying emission of solar plasma leads to changes in the magnetic shielding of galactic cosmic radiation in the inner part of the solar system. This produces variations in the cosmic ray flux reaching the earth. They are reflected as variations in the production of radioactive nuclei in the earth's atmosphere, as discussed above. The changes in the production rate are recorded in tree-rings as $^{14}C/^{12}C$ ratio changes and in precipitation as changes in the concentrations of the cosmogenic radio-isotopes ^{10}Be and ^{36}Cl. Solar properties like luminosity and ultraviolet emission might also influence climate. If they are related to the solar plasma emission—which seems to be plausible—changes in the ^{10}Be content in ancient deposited snowfalls therefore reflect changes in solar parameters (like S) which influence climate.

Atmospheric turbidity alters the earth surface's albedo (A). It is influenced by volcanic eruptions which lead to stratospheric dust layers. Solid electrical conductivity measurements on ice cores enable the identification of volcanic dust layers and therefore contribute to the reconstruction of the history of atmospheric turbidity.

During periods of glaciation, due to the enhanced continental and sea-ice cover, the albedo (A) is significantly increased. The $^{18}O/^{16}O$ ratio in the continental ice is considerably depleted compared to the ocean water. This leads to small enrichments in the oceanic $^{18}O/^{16}O$ ratio during glaciations which are reflected in the carbonate shells of foraminifera deposited in ocean sediments and enable, therefore, to estimate the history of the ice-cover of the continents.

As mentioned before, in natural ice essentially undisturbed air samples from the ancient atmosphere are stored. Measurements of the gas composition reveal variations of the contents of infrared active gases like CO_2 and CH_4, which influence the parameter B.

Variations in the $^{18}O/^{16}O$ ratio in precipitations reflect changes in temperature (T_S). Relative to sea-water, in cold periods precipitations are strongly depleted in ^{18}O, while in warm periods the depletion is less. Ancient precipitation is stored in cold glacier ice. Under favourable conditions, the high resolution of isotopic information in ice cores enables even the reconstruction of seasonal ^{18}O variations as far back as 10 000 years.

Climatic change of the last 100 to 1000 years

Unlike the case of CO_2 increase and its interpretation, the global climatic trend of the last 100 years, as estimated in Figure 15.11, cannot yet be unambiguously attributed to changes of the parameters S, A, and B. There exist different models which attempt to reconstruct a possible change in solar luminosity. Precise measurements of S from satellites might help to identify

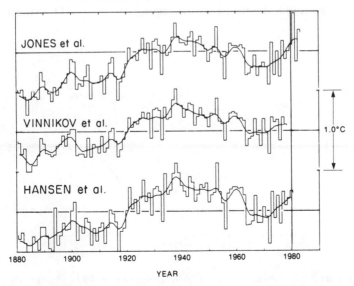

Figure 15.11 Surface air temperature changes representative of the Northerm Hemisphere land masses from Vinnikov *et al.* (1980). Hansen *et al.* (1981), and Jones *et al.* (1982). Smooth curves were obtained by using a 10-year Gaussian filter. Figure and caption from Wigley *et al.* (1985)

trends, perhaps correlated with solar activity and with radioisotope production. Attempts to reconstruct the history of the albedo *A* often make use of information on the atmospheric turbidity, based on the sulphate content recorded in Greenland ice cores which may or may not be a valuable measure of global albedo variations.

It would be most valuable if, in addition to these parameters determining the globally averaged energy balance of the earth, information regarding changes and anomalies in atmospheric and oceanic circulation could also be obtained. As mentioned before, the El Niño phenomenon is connected with considerable anomalies of the Pacific Ocean/atmosphere coupling. As an example, in December 1982, in parts of the Pacific Equatorial Ocean, the sea surface temperature had been *ca.* 3 to 5 °C above the average. It should be possible to reconstruct such anomalies from the $^{18}O/^{16}O$ ratios in low latitude glaciers and possibly in other natural archives like tree-rings. Alpine type and polar glaciers should also enable the reconstruction of atmosphere/ocean coupling anomalies, based on the deposition history of continental dust.

As Figure 15.12 indicates, the 1982–83 El Niño event also showed an effect on the global atmospheric CO_2 content. If such a change of the order of 1 ppm yr^{-1} had lasted over a longer period of time, it would have led to a change in the atmospheric CO_2 content of such a size that it should be detectable by very precise CO_2 measurements on ice cores. Parallel $^{13}C/^{12}C$ measurements would enable us to distinguish between CO_2 changes due to

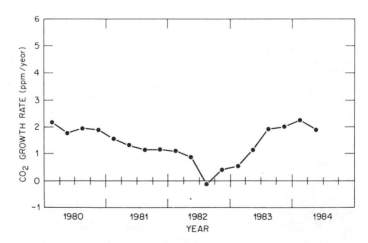

Figure 15.12   Year-to-year global atmospheric CO_2 change by season, plotted in the forward year. Data are from the 24-station global flask network of NOAA/GMCC [Komhyr *et al.*, 1985; R. H. Gammon, NOAA, personal communication, 1984]. Seasonal global means are weighted. Figure and caption from Gammon *et al.*, 1985

varying ocean surface properties and those due to varying terrestrial biomass. The precision of CO_2 concentration measurements on ice cores has by now been improved to *ca.* ± 1.5 ppm, and CO_2 concentration shifts of a few ppm during periods of a somewhat different climate, like the Medieval Warm period or the Little Ice Age, would indicate a possible oceanic climatic change component.

Long-term climatic change

Variation of the orbital parameters of the earth and the glacial cycles

Even for constant solar luminosity the intensity of solar radiation reaching the earth's surface and its latitudinal distribution undergo changes due to variations in the geometry of the earth's orbit around the sun. These changes, a result of the gravitational interaction in the planetary system, involve the eccentricity of the orbit and the obliquity and the precession of the earth's axis. They produce significant changes in the pattern of the incoming radiation on time-scales from several thousand to several hundred thousand years. These changes can be extrapolated many hundred thousand years into the past and can also be calculated into the future. Milankovitch (1930) suggested that these variations have caused the glacial periods. The averaged irradiation over the globe is not much changed by these orbital variations. The important effect is the modulation of the seasonal cycle. Summer insolation varies by as much as $\pm 5\%$ at latitudes at which continental ice existed. A comparison of the northern summer insolation history with the ice mass record, reflected in the foraminifera of deep-sea cores, shows that periods of rapid ice disappearance coincide with peaks of higher summer insolation. Furthermore, frequency analyses of the sedimentary oxygen isotope record reveal the characteristic frequencies of the orbital changes. It is widely accepted today that the orbital element changes play a major role in controlling long-term climate.

The modulation of the earth's climate by changes in the orbital parameters can also be seen in the $\delta^{18}O$ record of an Antarctic ice core from Vostok Station. (Lorius *et al.*, 1985). In contrast to this Antarctic $\delta^{18}O$ profile, spanning the last 150 000 yr, in the Greenland ice cores the orbital effect is much less pronounced, and shortterm climatic variations there seem to play an important role.

Short-term variations: Information from Greenland ice cores and lake sediments

Superimposed on the long-term glacial cycles, there are rather drastic short-term climatic changes. An example is the so-called Little Ice Age in the period from the middle of the 16th to the middle of the 19th century, followed by a

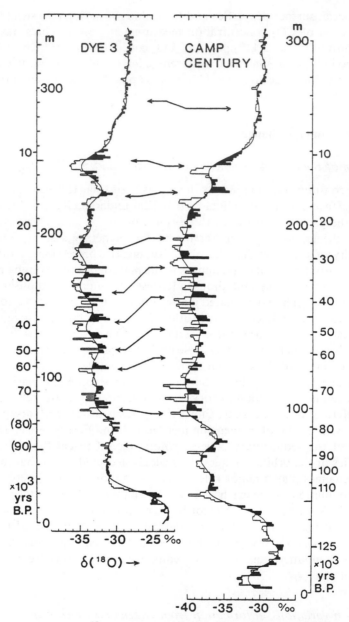

Figure 15.13 $\delta^{18}O$ profiles measured in Copenhagen along the
Dye 3 (0 to 1982 m depth) and the Camp Century (0 to 1370 m
depth) ice cores plotted on a common linear time-scale based on
considerations discussed by Dansgaard *et al.* (1982)

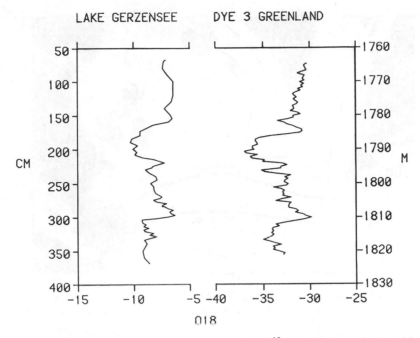

Figure 15.14   Comparison of a section of the δ^{18}O profile from the Dye 3 ice core (right) with the δ^{18}O record in lake Gerzensee (left). The strong similarities suggest that both records represent the same sequence of climatic events and thus the same time period. Oeschger *et al.* (1984)

significant temperature increase, leading, for example, in Central Europe to a strong retreat of the Alpine glaciers. Information on short-term climatic changes is not available from deep sea sediment cores: bioturbation leads to a continuous mixing of the uppermost sediment layers, and the low accumulation rates do not allow a sufficient time resolution. More detailed climatic information is stored in ice cores and lake sediments. In Figure 15.13 the δ^{18}O recores of the recently recovered Dye 3 core are compared with that of the Camp Century core (Dansgaard *et al.*, 1982). The good correlation between the two δ^{18}O records is clearly visible. This not only holds for the glacial –postglacial transition 10 000 years ago but also for most of the pronounced δ^{18}O oscillations in the ice from the Wisconsin stage. The shifts of δ^{18}O at the transition to the Holocene 13 000 and 10 000 years BP are also recorded in carbonate deposits of Central European lakes. Figure 15.14 demonstrates the excellent correlation. The rapid δ^{18}O oscillations reflect relatively warm periods in the generally cold Wisconsin stage. Some of them may have their counterparts in pollen profiles of peat bogs or lake sediments from Central Europe. Such pronounced correlations are not found in climatic records from

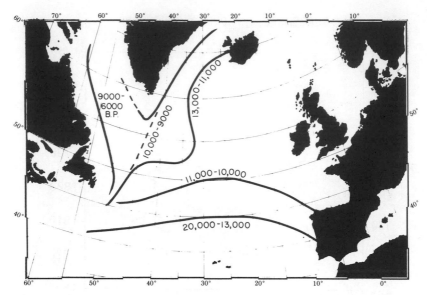

Figure 15.15   Location of the North Atlantic polar front for different time
periods [Ruddiman and McIntyre, 1981]. Deglacial retreat was interrupted by
the readvance from 11–10 kyr

the North American continent. Ruddiman and McIntyre (1981) found evidence in North Atlantic sediment cores for changes of the North Atlantic polar
front (Figure 15.15), leading to deviations of the Gulf Stream which then
caused the climatic variations in Central Europe.

Ice core parameters during the Wisconsin and the Wisconsin/Holcene transition

In the discussion of experimental evidence on the carbon cycle, we mentioned
that during the last thousand years the atmospheric CO_2 concentration had
been essentially constant in the range of 280 ± 5 ppm. But does this hold for
periods of major climatic change? It is one of the most surprising observations
on ice cores that at the end of the Wisconsin the CO_2 concentration ranged
between 180–200 ppm and then increased parallel to the $\delta^{18}O$ shift to values of
260 to 300 ppm (Stauffer *et al.*, 1984). The CO_2 measurements on the recently
drilled Dye 3 ice core confirm this observation (Figure 15.16). A rapid
transition in both $\delta^{18}O$ and CO_2 is observed around 13 000 BP. The CO_2
transition seems to lead the $\delta^{18}O$ transition in the ice core by a few metres. This
depth difference might be explained by the fact that the gases get occluded in
cold ice at a depth of *ca.* 100 ± 30 m. The phase shift might therefore reflect
the original trapping depth of air, and it cannot be excluded that the

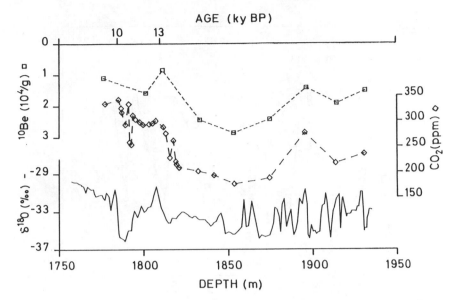

Figure 15.16   From top to bottom: ^{10}Be concentration, CO_2 concentration (ppm) and $\delta^{18}O$ data obtained for the Dye 3 ice core. The tentative time marks are suggested by the comparison with ^{14}C dated European lake sediments (see Figure 15.14)

parameters CO_2/air and $\delta^{18}O$ changed simultaneously. A high CO_2 concentration is also found in the Dye 3 ice at a depth of 1890 m, corresponding to an age of approximately 40 000 BP. This value coincides with one of the high $\delta^{18}O$ periods during the Wisconsin. Figure 15.17 shows the result of a detailed study of the $\delta^{18}O$/CO_2 relationship in the 30 000 to 40 000 year old section of the Dye 3 core. All the rapid $\delta^{18}O$ oscillations are accompanied by simultaneous, perfectly correlated CO_2 oscillations. Though the observation of rapid atmospheric CO_2 concentration changes during the Wisconsin needs further confirmation by measurements on other ice cores to exclude artefacts due to melt-layers or interaction with the impurities in the ice lattice, these experimental results have inspired the discussion of mechanisms which might produce rapid atmospheric CO_2 changes of the observed extent.

Other parameters vary in a similar way over this time interval. Microparticle concentrations show strong $\delta^{18}O$ correlated variations, concentrations being more than a factor of six higher during cold periods than during warm ones. The cold/warm ratios are about 1.5 for chlorine and nitrate and about 4 for sulphate (Finkel and Langway, 1985). Because of the larger sample requirements for the ^{10}Be analyses, measurements could not be made with the same resolution as for the other species, but in core sections with high $\delta^{18}O$, the ^{10}Be values are also 2 to 2.5 times higher than compared to low $\delta^{18}O$ periods.

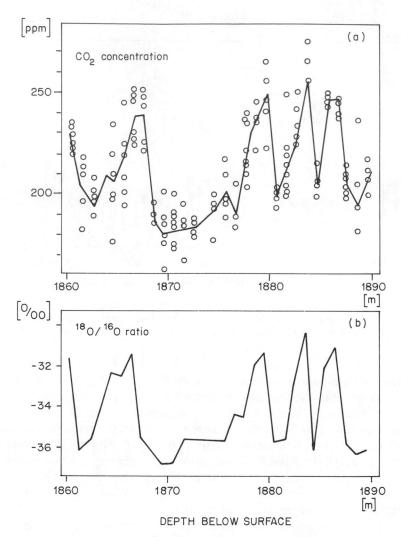

Figure 15.17 CO_2 and $\delta^{18}O$ values measured on ice samples from Dye 3 (the 30 m increment corresponds to about 10 000 years). a) Circles indicate the results of single measurements of the CO_2 concentration of air extracted from ice samples. The solid line connects the mean values for each depth. b) The solid line connects the $\delta^{18}O$ measurements done on 0.1 m core increments. Stauffer *et al.* (1984)

It is difficult to interpret these phenomena in detail. The relative concentration shifts are not the same for all species. Therefore, they cannot only be the result of a simple dilution modulation of a constant impurity flux by a variable water flux system.

The bistable climatic system

Not much is known for sure about the mechanisms determining the state and the stability of the climate system. The most remarkable observation in the data set from the Dye 3 core is the apparent bistable nature of the observed variations. Cold and warm period values of all parameters tend to occur in bands the width of which is significantly smaller than the difference between them. The correlated variation of the parameter sets suggests the existence of a bistable climate system during the Wisconsin, one set describing the warm and the other the cold climate state.

The rapid climatic oscillations during the Wisconsin are probably internally forced changes of the climate system. During the transition times, the essential external parameters and boundary conditions like insolation pattern, sea-level, and continental ice-cover probably changed in a very limited range. It seems therefore possible that for a given set of boundary conditions the atmosphere–ocean system has more than one mode of operation. After a period of one to two thousand years, the stabilizing mechanisms of one mode of operation seem to become ineffective, and the system organizes itself in a new mode. Based on the Greenland ice core records and European climatic records, one gets the impression that during the ice age essentially two separated climatic modes existed. About 10 000 BP boundary conditions as high latitude summer insolation seem to have prevented the system from switching once more back into the cold state. Based on European continental climatic information, we know that around 13 000 BP almost Holocene climatic conditions existed. Around 11 000 BP the system switched back for the last time, until present, to the cold mode, leading to almost glacial conditions in Europe. After the transition to the warm state around 10 000 BP, the stability was already so high that a return to the cold state was impossible. Since then the system has remained essentially in the warm state, and it is important to note that during the Holocene, the climate, as reconstructed by $\delta^{18}O$, has been much more quiet than during the Wisconsin. It seems that the large amounts of continental ice played an important role in the pronounced variability of the climate system during the last glaciation. Broecker *et al.* (1985) suggested that the rapid Wisconsin climatic variations reflect switches of deep water formation between the North Atlantic and the North Pacific Ocean.

The ocean and atmospheric CO_2 changes

At present, the low atmospheric CO_2 concentrations at the end of the last glaciation have been observed in three ice cores from Antarctica and two ice cores from Greenland. It is therefore highly probable that the atmospheric CO_2 content did indeed show these fluctuations. Measurements on the Vostok ice core, Antarctica, even show a similar CO_2 shift from low 'Riss' glaciation

CO_2 concentration to Eemian interglacial CO_2 concentration in the range of those observed for the Holocene. The CO_2 changes therefore seem to be an inherent mechanism of the glacial cycle.

Regarding the rapid CO_2 changes, observed in the Greenland cores during the Wisconsin, however, the final proof that they reflect atmospheric changes could not yet be given.

As was indicated earlier, the atmospheric CO_2 concentration depends on the chemical composition of the ocean surface water which in turn is strongly influenced by marine biological activity. Broecker (1982) suggested that the observed CO_2 content changes must be the result of alterations in the nutrient element chemistry of sea-water. A higher phosphate content in the glacial ocean would have led to a stronger depletion of total CO_2, and thus pCO_2 in the surface ocean. In one of his models he suggested that the source of this additional phosphate was erosion of nutrient-rich organic sediments on the continental shelves exposed due to the lower sea-level in glacial times. This hypothesis, though extremely important for the initiation of attempts to explain the observed CO_2 changes, now seems to disagree with the observations which ask for rather rapid CO_2 changes. Today, therefore, the reason for the CO_2 changes is searched for in changes of the highly interactive system of physical, chemical, and biological processes in the ocean system. In the Southern Ocean around Antarctica, the nutrients phosphate and nitrate are abundant in surface waters in contrast to most of the other part of the ocean. The biological activity seems to be limited around Antarctica due to rapid vertical mixing or lack of solar light. Several authors showed that a decrease of the vertical mixing in the Southern Ocean, assuming the biological productivity to remain constant, would lead to a lower surface concentration of nutrients, total CO_2 and consequently pCO_2. Changes in atmospheric CO_2 by 50 ppm within a few 100 years could be produced by halfing the surface/deep water exchange rate.

It was previously mentioned that the average difference of $\delta^{13}C$ in surface and deep ocean water is a measure for the ocean's average biological activity. The time history of $\delta^{13}C$ in surface and deep ocean water is recorded in coexisting planktonic and benthic foraminifera in ocean sediment samples. Shackleton and Pisias (1985) provided for a Pacific sediment core a record of the ice volume based on $\delta^{18}O$ measurements and of the difference of the $\delta^{13}C$ ratios in surface and deep ocean dwelling foraminifera. They assumed the $\delta^{13}C$ difference to be representative for the average oceanic biological activity and derived a record of the atmospheric CO_2 content of the last 350 000 years (Figure 15.18). Their data confirm the low CO_2 concentrations at the end of the last glaciation as measured in ice cores. A spectral analysis of the CO_2 variation showed significant amplitudes for the Milankovitch frequencies of the changing orbital parameters. These studies show that both CO_2 and ice volume lag the orbital changes. And there is most exciting evidence—which

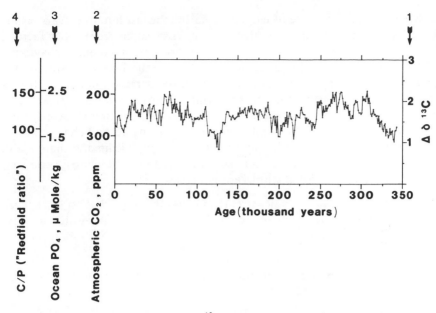

Figure 15.18   Surface-to-deep water ^{13}C difference. Scales are shown (1) in permil (by subtraction); (2) in parts per million by volume atmospheric CO₂ concentration implied by scale 1; (3) in mean phosphorus content of ocean deep water that would be required to produce scale 1; (4) in mean Redfield ratio that would be required to explain scale 1(a) (note that scales 3 and 4 are presented as alternatives) Shackleton and Pisias (1985)

should be confirmed by additional measurements—that the changes of CO₂ seem to lead those in ice volume by a few thousand years.

CONCLUSIONS

The ocean's role in the anticipated global change

The short-term reaction of the atmosphere/ocean CO₂ system to the anthropogenic CO₂ immission is rather well known. The preindustrial atmospheric CO₂ concentration (as well as that of other trace gases) can be reconstructed within narrow uncertainty limits. Using the reconstructed and measured atmospheric CO₂ increase and models of the CO₂ uptake by the oceans, the CO₂ immission originating from deforestation and from fossil fuel consumption can be calculated, and a satisfactory agreement with independent production estimates is obtained.

The rising concentrations of CO₂ and other trace gases are estimated to have already led to a significant global temperature increase. However, at present it is not possible to deduce in a scientifically rigorous manner to what degree the global temperature increase of *ca.* 0.5 °C during the last hundred years results from an increasing greenhouse effect and to what extent from other factors, like changing solar parameters, atmospheric turbidity or oceanic behaviour.

There exists considerable uncertainty regarding the long-time role of the ocean in global change. The expected temperature increase will lead to density changes of surface ocean water due to thermal expansion, influx of fresh water from melting continental ice but also due to changes in the evaporation –precipitation balance. These potential changes in the ocean behaviour seem to be the greatest uncertainty in the prediction of regional climatic change. But a changing ocean might also affect the future trend of the atmospheric CO_2 concenration and the global temperature increase:

—A slowing down of oceanic circulation and mixing, due to a reduced latitudinal temperature gradient and reduced ocean surface water density, might reduce the ocean's uptake of excess CO_2 and its transport to greater depths.

—A changing oceanic circulation pattern might also influence the baseline CO_2 concentration as discussed earlier

—The dampening of the anticipated temperature rise due to the ocean's large heat capacity would be reduced if vertical ocean mixing slowed down.

The ocean/CO_2 and ocean/climate coupling at present and in the past

The interaction between ocean, CO_2, and climate is clearly demonstrated by the El Niño events which repeat themselves within 3 to 7 years and lead to significant climatic anomalies as well as small, but clearly identifiable, irregularities in the atmospheric CO_2 increase. In 1982 these phenomena were especially pronounced, and sometimes the question is posed, will the anticipated global change manifest itself in more prominent anomalies of ocean circulation.

An example of the strong influence of ocean circulation on climate (mainly on the European continent) are the events at the glacial–postglacial transition reconstructed from palaeoclimatic information. A strong link between changes in the North Atlantic Ocean circulation and rapid changes of the European climate have been observed. The rapid changes of $\delta^{18}O$ and other parameters in the Greenland ice cores further suggest the existence of a bistable climatic system North Atlantic—Europe.

The observation that climatic events are accompanied by changes in the atmospheric CO_2 concentration is of great interest for two reasons:

(a) The major climatic shifts over the last 150 000 years are strongly correlated with shifts in the atmospheric CO_2 concentrations. During the coldest periods of the two last glaciations (18 000 BP and 140 000 BP) the CO_2 concentrations were of the order of 180–200 ppmv. The CO$_2$ concentrations rose almost simultaneously with the climate shifts to values of 270–290 ppmv in the subsequent interglacial. Thus CO$_2$ shifts by a factor of about 1.4 seem to be inherent to major climatic variations. The CO$_2$ shifts between glacial and interglacial stages are probably mainly forced by changes in oceanic circulation and/or marine biological activity.

(b) The link between ocean circulation and CO$_2$ anomalies perhaps offers the possibility to trace back changes in oceans by very precise high resolution measurements of CO$_2$ concentrations of the air occluded in polar ice cores. With the precision achieved at present, it should be possible to detect atmospheric CO$_2$ concentration changes of $\geqslant 2$ to 3 ppm over periods of $\geqslant 20$ years. This might provide otherwise not obtainable information on oceanic circulation changes which could then be linked to climatic deviations like the Mediaeval Warm Period and the Little Ice Age in Europe. Such information would significantly help us to understand major climatic mechanisms involving the ocean.

RECOMMENDATIONS

The importance of improving the knowledge of ocean processes relevant to climate and CO$_2$ has been widely recognized by the scientific community. This is exemplified by a series of international scientific projects related to questions raised in this chapter which are coordinated by the *Committee on Climatic Changes and the Ocean (CCCO)*:

The World Ocean Circulation Experiment (WOCE) is concerned with improving our understanding of ocean circulation mechanisms and the ability to model global circulation.

The tropical Ocean-Global Atmosphere Project (TOGA) is focused on the attempt to understand and describe the tropical oceans and the global atmosphere as a coupled system for the purpose of predicting its variations on time-scales of months to years.

A new programme is being planned by CCCO to *Improve the Understanding of the Carbon Cycle in the Ocean*. The main aspects are:

(a) Measurements over time of the main constituents of the CO$_2$ system in surface and subsurface ocean waters and the rates and locations of CO$_2$ exchange between air and sea

(b) studies of the interaction between biological activity in the sea and atmospheric and oceanic CO$_2$, including determinations of organic particle fluxes from the euphotic zone into deeper waters and the chemical transformations of the particles. The physical and chemical measurements planned in WOCE and TOGA should be extended to improve predictions of the response of the carbon cycle to changing climate during the next century.

Another activity in the frame of CCCO is *palaeoclimatology*. Considerable progress has been made in this field, as discussed in this article, due to the availability of new analytical methods (e.g. accelerator mass spectrometry) and the application of fundamental principles of geochemistry to the reconstruction of past climate and oceanic properties. Studies of primary importance comprise reconstructions of variations of the climate and the CO_2 systems, process studies and model experiments involving the biochemistry and the dynamics of the ocean, with special emphasis on the past abrupt climatic changes.

Efforts should particularly concentrate on an observational programme to obtain time series of environmental parameters suitable to identify important mechanisms and to validate models. The needs for new deep ice cores, new high resolution deep sea sediment records, and lake and other continental records are stressed.

In the following we want to expand on the importance of palaeoclimatological studies and propose a systematic coordinated observational programme together with modelling attempts of the ocean/climate/CO_2 interaction.

In the next few decades a considerable part of the progress in our understanding of the ocean system and its role in environmental processes will be due to information on past variations in ocean properties in relation with other changes in the entire environmental system. Therefore a new project should be initiated which coordinates and reinforces palaeoclimatic research, making use of state of the art analytical techniques to study the information contained in both oceanic and continental records. This project should be aimed at providing:

—links between oceanic and continental events,

—a basis for the development and validation of a wide hierarchy of models involving the coupled atmosphere—ocean—kryosphere system including biological processes.

Such a project could be called *Palaeoenvironmental Traverse*. Time periods of major interest are:

—*1950 until present:* Period of detailed instrument data on ocean and atmosphere with anomalies in ocean circulation.

—*1850–1950:* Period of rising human impact on the environment (CO_2 emission, consequences of population increase, changes in atmospheric composition).

—*Last 1000 years:* Period of significant climatic deviations like the Mediaeval Warm and the Little Ice Age in Europe.

—*Altithermal (6000 to 8000 BP):* Period of warmer earth similar to that anticipated for the first half of the next century due to human impacts.

—*Glacial–Postglacial Transition (15 000 to 9000 BP):* Period of pronounced climatic events involving ocean and kryosphere.

—*Wisconsin:* Period of rapid climatic oscillation between a cold and a warm climatic system state in the North Atlantic region.

REFERENCES

Andrée, M., Oeschger, H., Broecker, W., Beavan, N., Klas, M., Mix, A., Bonani, G., Hofmann, H. J., Suter, M., Wölfli, W., and Peng, T.-H. (1986). Limits on the ventilation rate for the deep ocean over the last 12 000 years. *Climate Dynamics*, **1**, 53–62.

Beer, J., Siegenthaler, U., Oeschger, H., Andrée M., Bonani, G., Suter, M., Wölfli, W., Finkel, R., and Langway, C. C. (1983). Temporal ¹⁰Be variations. *Proc. 18th Int. Cosmic Ray Conf., Bangalore*, **9**, 317.

Beer, J., Siegenthaler, V., Bonani, G., Finkel, R. C., Oeschger, H., Sutter, H. and Wölfli, W. (1988). Information on past solar activity and geomagnetism from ¹⁰Be in the Camp Century ice core. *Nature* (in press).

Broecker, W. S. (1982). Ocean chemistry during glacial time. *Geochimica et Cosmochimica Acta*, **46**, 1689–1705.

Broecker, W. S., and Peng, T.-H. (1982). *Tracers in the Sea*. Eldigio Press, Lamont-Doherty Geological Observatory, Palisades, NY 10964: 690 pages.

Broecker, W. S., Peng, T.-H., and Engh, R. (1980). Modelling the carbon system. *Radiocarbon*, **22**, 565–598.

Broecker, W. S., Peteet, D. M., and Rind, D. (1985). Does the ocean–atmosphere system have more than one stable mode of operation? *Nature*, **315**, 21–26.

Dansgaard, W., Clausen, H. B., Gundestrup, G., Hammer, C. U., Johnsen, S. F., Kristinsdottir, P. M., and Reeh, N. (1982). A new Greenland deep ice core. *Science*, **218**, 1273–1277.

Finkel, R. C., and Langway, C. C., Jr. (1985). Global and local influences on the chemical composition of snowfall at Dye 3, Greenland: the record between 10 ka BP and 40 ka BP. *Earth and Planetary Science Letters*, **73**, 196–206.

Gammon, R. H., Sundquist, E. T., and Fraser, P. J. (1985). History of carbon dioxide in the atmosphere. In Trabalka, J. R. (Ed.) *Atmospheric Carbon Dioxide and the Global Carbon Cycle*, pp. 25–62. U.S. Department of Energy, Washington, DC.

Hansen, J., Johnson, D., Lacis, A., Lebedeff, S., Lee, P., Rind, D., and Russell, G. (1981). Climatic impact of increasing atmospheric carbon dioxide. *Science*, **213**, 957–966.

Jones, P. D., Wigley, T. M. C., and Kelly, P. M. (1982)). Variations in surface air temperatures: Part I. Northern Hemisphere 1881–1980. *Monthly Weather Review*, **110**, 59–70.

Keeling, C. D., Bacastow, R. B., and Whorf, T. P. (1982). Measurements of the concentration of carbon dioxide at Mauna Loa Observatory, Hawaii. In Clark, W. C. (Ed.) *Carbon Dioxide Review*, **1982**, 377–385. Oxford University Press, New York.

Komhyr, W. D., Gammon, R. H., Harris, T. B., Waterman, L. S., Conway, T. J., Taylor, W. R., and Thoring, K. W. (1985). Global atmospheric CO₂ distribution and variations from 1968–1982. NOAA/GMCC Flask Sample Data. *J. Geophys. Res.*, **90**, 5567–5596.

Lorius, C., Jouzel, J., Ritz, C., Merlivat, L., Barkov, N. J., Korotkevitch, Y. S., and Kotlyakov, V. M. (1985). A 150 000-year climatic record from Antarctic ice. *Nature*, **316**, 591–596.

Maier-Reimer, E., and Hasselmann, K. (1987). Transport and storage of CO_2 in the ocean—an inorganic ocean-circulation carbon cycle model. *Climate Dynamics*, **2**.

Milankovitch, M., (1930). Mathematische Klimalehre und astronomische Theorie der Klimaschwankungen. In *Handbuch der Klimatologie, I (A),* Koppen, W., and Geiger, R. (Eds.) Gebrüder Bornträger, Berlin.

Neftel, A., Oeschger, H., and Suess, H. E. (1981). Secular non-random variations of cosmogenic carbon-14 in the terrestrial atmosphere. *Earth Planet. Sci. Lett.*, **56**, 127–147.

Neftel, A., Moor, E., Oeschger, H., and Stauffer, B. (1985). Evidence from polar ice cores for the increase in atmospheric CO_2 in the past two centuries, *Nature*, **315**, No. 6014, 45–47.

Nydal, R., and Løvseth, K. (1983). Tracing bomb ^{14}C in the atmosphere 1962–1980. *J. Geophys. Res.*, **88**, 3621–3642.

Oeschger, H., Beer, J., Siegenthaler, U., Stauffer, B., Dansgaard, W., and Langway, C. C. (1984). Late-Glacial climate history from ice cores. In Hansen, J., and Takahashi, T. (Eds.) *Climate Processes and Climate Sensitivity*, pp. 299–306. Am. Geophys. Union.

Oeschger, H., Siegenthaler, U., Schotterer, U., and Gugelmann, A. (1975). A box diffusion model to study the carbon dioxide exchange in nature. *Tellus*, **27**, 168–192.

Raynaund, D., and Barnola, J. M. (1985). An Antarctic ice core reveals atmospheric CO_2 variations over the past few centuries. *Nature*, **315**, 309–311.

Ruddiman, W. F., and McIntyre, A. (1981). The North Atlantic Ocean during the last glaciation. *Palaeogeogr., Palaeoclimatol., Palaeoecol.*, **35**, 145–214.

Shackleton, N. J., and Pisias, N. G. (1985). Atmospheric carbon dioxide, orbital forcing, and climate. In Sundquist, E. T., and Broecker, W. S. (Eds.) *The Carbon Cycle and Atmospheric CO_2: Natural Variations Archean to Present*, pp. 303–317. Geophysical Monograph, 32, AGU Washington, DC.

Siegenthaler, U. (1983). Uptake of excess CO_2 by an outcrop-diffusion model of the ocean. *J. Geophys. Res.*, **88**, 3599–3608.

Stauffer, B., Hofer, H., Oeschger, H., Schwander, J., and Siegenthaler, U. (1984). Atmospheric CO_2 concentration during the last glaciation. *Annals of Glaciology*, **5**, 160–164.

Stuiver, M., and Quay, P. D. (1981). Atmospheric ^{14}C changes resulting from fossil fuel CO_2 release and cosmic ray flux variability. *Earth and Planetary Science Letters*, **53**, 349–362.

Suess, H. E. (1970). Bristlecone-pine calibration of the radiocarbon time-scale 5200 BC to the present. In Olsson, I. U. (Ed.) *Radiocarbon Variations and Absolute Chronology*, pp. 303–311. John Wiley & Sons.

Vinnikov, K. Ya., Gruza, G. V., Zakharov, V. F., Krillov, A. A., Kovyneva, N. P., and Rarikova, E. Ya, (1980). Current climatic changes in the Northern Hemisphere. *Meteorologiya i Gidrologya*, **6**, 5–17.

Wigley, T. M. L., Angell, J. K., and Jones, P. D. (1985). Analysis of the temperature record. In McCracken, M. C., and Luther, F. M. (Eds.) *Detecting the Climatic Effects of Increasing Carbon Dioxide*, pp. 55–90. U.S. Department of Energy, Washington, DC.

Index